High-Throughput Imaging Methods

High-Throughput Imaging Methods

Cells and Tissues

edited by

Manas Ranjan Gartia

Ram V. Devireddy

JENNY STANFORD
PUBLISHING

Published by

Jenny Stanford Publishing Pte. Ltd.
101 Thomson Road
#06-01, United Square
Singapore 307591

Email: editorial@jennystanford.com
Web: www.jennystanford.com

British Library Cataloguing-in-Publication Data
A catalogue record for this book is available from the British Library.

High-Throughput Imaging Methods: Cells and Tissues

ISBN 978-981-5352-08-5 (Hardcover)
ISBN 978-1-003-77417-4 (eBook)
DOI: 10.1201/9781003774174

Contents

Preface xiii

1. High-Throughput and Label-Free Analysis **1**

Mahima Gupta, Sreyashi Das, and Ram V. Devireddy

1.1 Introduction 2
1.2 Fundamentals of High-Throughput Analysis 5
1.2.1 Microarrays 15
1.2.2 High-Throughput Sequencing 17
1.3 Fundamentals of Label-Free Analysis 18
1.3.1 Mass Spectrometry 19
1.3.2 Surface Plasmon Resonance 20
1.3.3 Atomic Force Microscopy 23
1.3.4 Acoustic Wave Sensors 24
1.3.5 Electrical Detection Techniques 26
1.3.6 Hyperspectral Imaging 29
1.4 Integrating High-Throughput and Label-Free Techniques 31
1.5 Applications in Life Sciences 34
1.5.1 Drug Discovery and Development 35
1.5.2 Screening of Potential Drug Candidates 36
1.5.3 Pharmacokinetics and Pharmacodynamics 36
1.5.4 Proteomics and Genomics 36
1.5.4.1 Protein identification and quantification 37
1.5.4.2 DNA/RNA analysis 37
1.5.5 Clinical Diagnosis 38
1.5.5.1 Biomarker discovery 38
1.5.5.2 Disease diagnosis and monitoring 39
1.5.6 Role of HSI and Electrical Detection in Life Sciences 39
1.6 Applications in Organoids and Tissue Engineering 40

1.6.1 Organoids 40
1.6.2 Tissue Engineering 43
1.7 Applications in Materials Science 44
1.7.1 Characterization of Nanomaterials 45
1.7.2 Surface and Interface Analysis 46
1.7.3 Quality Control in Manufacturing Processes 46
1.8 Technological Advancements 46
1.9 Challenges and Solutions 49
1.9.1 Challenges in HTS 49
1.9.2 Proposed Solutions and Best Practices 51
1.10 Conclusion 52

2. High-Throughput Analysis of Stem Cells 69

Mohan K. Dey, Sreyashi Das, Kirti Agrawal, Ram V. Devireddy, and Manas R. Gartia

2.1 Introduction 69
2.1.1 Definition of High-Throughput Analysis 69
2.1.2 Stem Cells Morphology and Gene Expression 73
2.1.3 Overview of High-Throughput Analysis of Stem Cells 74
2.1.4 Purpose of This Chapter 76
2.2 High-Throughput Analysis Techniques 76
2.2.1 Microarray Analysis 76
2.2.2 Flow Cytometry 80
2.2.3 Mass Spectrometry 84
2.2.4 Next-Generation Sequencing by Machine Learning Algorithm 89
2.3 Applications of High-Throughput Analysis in Stem Cell Research 94
2.3.1 Identification of Stem Cell Markers 94
2.3.2 Differentiation Tracking and Analysis of Stem Cells Signaling Pathways 98
2.3.3 Disease Modeling 100
2.3.4 Drug Development 105
2.3.5 Tissue Engineering 109
2.3.6 Understanding the Role of Epigenetic Modifications 109

2.4 Challenges of High-Throughput Analysis of Stem Cells 115
2.4.1 Data Interpretation by Multi-Omics Using High-Throughput Analysis 115
2.4.2 Navigation of Biological Variation Complexity 116
2.4.3 Cost and Availability of High-Throughput Equipment 117
2.4.4 How Can Machine Learning Help to Alleviate Some of the Issues with Cell Heterogeneity and Associated Errors: Discussion? 118
2.5 Conclusion 119

3. High-Throughput Detection Approaches Based on Surface Plasmon Resonance 131

Fenni Zhang, Jinbiao Ma, Xiaoyin Liu, Jiahao Xu, Yanli Lu, and Qingjun Liu

3.1 Introduction 132
3.2 Typical SPR-Sensing Approaches in High-Throughput Detection 134
3.2.1 Detection Principle of SPR 134
3.2.2 High-Throughput Instrumentation of SPR 137
3.3 Application of SPR Sensing Approaches in Molecular Interaction Detection 140
3.3.1 Molecular Interaction Detection with High-Throughput CSPR 140
3.3.2 Molecular Interaction Detection with SPR 143
3.3.3 Molecular Interaction Detection with SPRM 146
3.4 Application of SPR-Sensing Approaches in Cell Biology 148
3.4.1 Cell Biology Study with SPRi 148
3.4.2 Cell Biology Study with SPRM 151
3.5 Conclusion and Outlook 153

4. Hyperspectral Imaging of Optically Resonant Surfaces for Multiplexed Sensing of Biomaterials **159**
Qinming Zhang, Shirin Parvin, Liang Dong, and Meng Lu
4.1 Introduction 160
4.2 Hyperspectral Imaging System 161
4.2.1 A Label-Free Microarray Using Photonic Crystal Biosensor 164
4.3 Label-Free Exosome Microarray Assay 168
4.4 Detection of Macrophage-Specific Exosomes 173
4.5 Conclusion 176

5. Single-Molecule Junctions for Electrical Detection of Biomolecules **181**
Ian Bunker and Kun Wang
5.1 Introduction 181
5.2 Electrical Conductance of Single DNA Molecules 186
5.2.1 Impact of Environmental Condition and Molecular Conformation 187
5.2.2 Sequence and Length Dependance 190
5.3 Electronic Diode and Switch Made of Single DNA 194
5.4 SMJ for Biosensing and Sequencing 197
5.5 Summary and Outlook 200

6. Tissue Engineering Approach: 3D Printing **207**
Mahima Gupta and Ram V. Devireddy
6.1 Introduction 208
6.2 Adipose Tissue-Derived Adult Stem Cells 210
6.3 Fabrication of Adult Stem Cell Sheets 211
6.3.1 Nunc™ Dishes with UpCell™ Surface 212
6.3.2 Cell Sheets Using MC Hydrogel-Coated Dishes 213
6.4 In Vitro Degradation, Swelling, and Osmotic Stability of MC Gels 214
6.5 Thermal Reversibility of Attachment 215
6.6 Analysis of Cell Sheets Using Histology 216

6.7 Cell Viability in Cell Sheets 217
6.8 Confocal Microscopy 219
6.9 Osteogenic Potential of ASCs Embedded in Cell Sheets 219
6.10 Polyethylene Glycol Hydrogels 222
6.11 Cell Viability and Proliferation in ECM-PEG Hydrogels 224
6.12 In Vivo Proliferation 226
6.13 3D Printing in Tissue Engineering 227
6.14 Hyperspectral Imaging in Tissue Engineering and 3D Bioprinting 229
6.15 Clinical Problem 231
6.16 Closing Remarks 233

7. Organoids: In Vitro Alternatives to Study In Vivo Events and the Possibility of Using 3D Bioprinting and Hyperspectral Imaging for Better Outcomes 245

Saumya Rai, Mohankumar B. S., Qudsiya Mohiuddin, Bhawana, Sanya Sharma, and Santosh K. Misra

7.1 Introduction 246
7.2 Types of Organoids and Optimized Applications 248
7.2.1 Cardiac Organoid 249
7.2.2 Characterization of Heart Organoids 251
7.2.3 Application 252
7.3 Lung Organoid 254
7.3.1 Characterization of Lung Organoids 256
7.3.2 Application 257
7.4 Brain Organoid 259
7.4.1 Characterization of Brain Organoids 261
7.4.2 Application 262
7.5 Bone Organoid 264
7.5.1 Composition of Bone Organoid 264
7.5.2 Characterization of Bone Organoid 264
7.5.3 Application 265
7.6 Liver Organoid 267
7.6.1 Characterization of Liver Organoid 269
7.6.2 Application 270
7.7 Kidney Organoid 271
7.7.1 Characterization of Kidney Organoid 273

7.7.2 Application 274
7.8 Retinal Organoid 276
7.8.1 Characterization of Retinal Organoid 278
7.8.2 Application 278
7.9 Types of Bioink and Its Role in Generating Organoids for Different Applications 280
7.9.1 Cell-Scaffold-Free Bioinks 281
7.10 Cell-Scaffold-Based Bioink 281
7.10.1 Collagen-Based Bioink 284
7.10.2 Alginate-Based Bioink 284
7.10.3 HA-Based Bioink 285
7.10.4 Fibrin-Based Bioink 285
7.10.5 Multicomponent Bioink 286
7.11 Types of 3D-Bioprinting 286
7.11.1 3D-Bioprinting Techniques 290
7.11.2 Inkjet 3D Bioprinting 291
7.11.3 Laser-Assisted 3D Bioprinting 291
7.11.4 Micro-Extrusion 3D Bioprinting 292
7.11.5 Stereolithography-Based Bioprinting 292
7.12 Current Challenge in 3D Bioprinting Technology 293
7.13 Use of Hyperspectral Imaging in Improving 3D Bioprinted Products 293
7.14 Characterization of Organoids 297
7.15 Scanning Electron Microscopy 297
7.16 Transmission Electron Microscopy 298
7.17 Laser Scanning Confocal Microscopy 298
7.18 Super-Resolution Fluorescence Microscopy 299
7.18.1 Stimulated Emission Depletion (STED) Microscopy 299
7.18.2 Single-Molecule Localization Microscopy 299
7.19 Flow Cytometry 300
7.20 Biochemical Characterization 300
7.20.1 Real-Time PCR (RT-PCR) 300
7.20.2 Gene and Transcriptome Profiling 301
7.20.3 Metabolic Activity 301
7.20.3.1 Characterization of scaffold material 302

7.21 Challenges in Bioprinting Organoids and Limitations of HSI in Improving the Outcomes 302
7.22 Future Prospects 303
7.23 Conclusion 304

8. Integration of Electronics with Biological Systems 329

Anish Maharjan, Julio Zuazola, and Robert Herbert

8.1 Introduction 330
8.2 Wearable Electronics 332
8.2.1 Integration with Skin 332
8.2.2 Wearable Electrochemical Sensors 337
8.2.2.1 Sensor parameters 338
8.2.2.2 Sweat sensing 339
8.2.2.3 Wound sensing 343
8.2.2.4 Saliva monitoring 345
8.2.3 Biopotential Monitoring 347
8.2.3.1 ECG 348
8.2.3.2 EMG 351
8.2.3.3 EOG 353
8.2.3.4 EEG 355
8.3 Biohybrid Electronics 357
8.4 Conclusion and Future Outlook 359

Index 373

Preface

Recent developments in high-throughput and label-free analysis have led to significant advancements in biochemical, biophysical, and cellular assays, including cell imaging, metabolomics, drug delivery, cancer research, and tissue engineering. In this text, we explore how these techniques and methods are now being applied to assess biological events at or below the scale of individual cells and even intracellular components.

High-throughput analysis (HTS) gained widespread acceptance and universal application in the pharmaceutical and biotechnology industries in the early 80s. The initial use of HTS allowed for the screening of a large number of synthetic compounds in microscopic-sized in vitro assays (e.g., the 96-well microtiter plates to 384-well plates). This allowed from assessing 20 to 50 compounds per week to screen several hundreds or thousands in a week. Spurred by the industry's commitment to accelerate the process of drug discovery and development, HTS has significantly advanced over the years. It is important to note that most of these initial screening assays have traditionally utilized reporter labels to quantify biological responses relevant to the disease state of interest. However, there are limitations associated with the use of labels that may be overcome with temporal measurements that are possible with label-free methods. However, potential users of the label-free high throughput screening technologies remain skeptical. This reluctance can be attributed to the as-yet unclear physical basis of the signals generated and their contextual relevance to cell biology and signal transduction. The hope is that the collection of articles in this text will provide an overview of these new and emerging technology platforms that complement and significantly advance traditional screening assays.

Chapter 1 presents an overview of high throughput and label-free analysis in various molecular biology and material science domains. Starting with a definition of HTS, this chapter presents an overview of label-free assessment technologies, including high-throughput imaging. The chapter also presents several important technologies that form the foundation for high-throughput and label-free analyses and delineates their unique capabilities. Some technologies described include mass spectroscopy, surface plasmon resonance (SPR), atomic force microscopy, acoustic wave sensors, electrical detection approaches, and hyperspectral imaging. The chapter concludes with case studies and applications in life sciences (drug discovery, proteomics, genomics, clinical diagnosis of disease markers, organoids/tissue engineering) and in materials science (nanomaterial characterization) for the synergistic approach of high throughput analysis with label-free methods.

Chapter 2 presents an overview of high throughput and label-free analysis specifically targeted toward stem cell applications. As is well known, stem cells are unique in that they can distinguish into different types of cells and divide indefinitely. Thus, they are a workhorse in the field of regenerative medicine and tissue engineering. Specifically, a thorough overview and discussion of the various high-throughput technologies and platforms used to study stem cells, such as microarray-based gene expression analysis, flow cytometry, high-content imaging, and machine learning algorithms, is followed by a discussion of the advantages and limitations of these technologies. This chapter concludes by exploring the recent developments in the field of high-throughput analysis of stem cells, including the integration of machine learning algorithms and the use of high-throughput techniques for drug discovery and development.

Surface plasmon resonance (SPR) is a powerful analytical technique for label-free detection of molecular interactions and is presented in Chapter 3. A comprehensive review of the fundamental principles and instrumentation of various high throughput SPR methods, including the conventional SPR, SPR imaging, SPR microscopy, localized SPR, microfluidics, and arrays to achieve high throughput detection is presented. The latest advancements in biomolecular sensing and cell analysis using high throughput SPR methods are then discussed, allowing for the study of complex biological processes and cellular heterogeneity, functional characterization of individual cells, and the identification of potential therapeutic targets. The chapter concludes by addressing the current challenges associated with high throughput SPR and presents an author's perspective on its potential for single-molecule and single-cell analysis in biomedical applications.

Chapter 4 introduces the implementation of hyperspectral imaging (HSI) technology for the detection of biomarkers using optical biosensors. Specifically, the biosensors described are uniquely capable of assessing optical resonances that are sensitive to the absorption of target biomaterials, such as cells, exosomes, proteins, and nucleic acids. With concurrent use of HSI, these biosensors allow for the capture of the distribution of optical resonance across both spatial and spectral dimensions. Given the high spatial and spectral resolutions offered by HSI microscopy, this facilitates the characterization of optical transducers, including metasurfaces, photonic crystal, and plasmonic devices, and subsequently, for label-free analysis of microarray-based assays, cell-based assays, and even single nanoparticles. The potential of this integrated technology is then demonstrated by characterizing a novel exosome microarray built on a photonic crystal biosensor.

Chapter 5 shifts the focus onto what is known as single-molecule electronics (SMEs) or single molecules acting as functional electronic components in nanodevices. Optimizing SMEs, by definition, requires probing and understanding the electrical properties of single-molecule structures. Consequently, several experimental platforms and methods allowing for the precise measurement of charge transport through individual molecules have been developed over the years and are presented in this chapter. Intriguingly, these single-molecule experimental techniques have also made it possible to electrically monitor crucial chemical and biological processes with single-molecule precision, representing the ultimate analytical limit of molecular or biological science, thus allowing for the study and manipulation of biological systems at the nanoscale of nanobiotechnology. This chapter presents the inherent complexities in measuring phenomena at the molecular level and how nanobiotechnology and its transformative tools have opened previously unattainable avenues of research, reshaping our understanding of fundamental biomolecules.

Chapter 6 delves into the significant progress that has been made in using cellularized extracellular scaffolds to regenerate various tissues, including bone, cartilage, heart, blood vessels, nerves, liver, and many other tissues. Although this chapter is seemingly unrelated to the earlier chapters, it does represent the future avenues of applications for high-throughput, label-free analysis and should be read in conjunction with the earlier chapter on stem cells.

Similarly, Chapter 7 expands on the tissue engineering fabrication methods presented in Chapter 6 to delineate recent advances in three-dimensional tissue structures, or organoids. Specifically, a description of assessing and pre-screening in vivo events in a variety of organoids (cardiac or heart, lung, brain, bone, liver, kidney, retina) is presented. This is followed by a description of the types of bio-inks used (collagen-based, alginate-based, hyaluronic acid-based, fibrin-based, multicomponent) and their role in the variety of three-dimensional printing techniques (inkjet, laser-assisted, micro-extrusion, stereolithography based) of organoids. The chapter then discusses the use of HSI methods (described earlier in Chapter 4) in characterizing and creating the next generation of organoids.

Chapter 8, the final chapter, presents the evolution and integration of the techniques/methods presented in Chapters 6 and 7 in cell analysis and organ fabrication by discussing the nascent field of soft electronics, that is, integrated electronics for monitoring and measurement in an *in vivo* setting. Specifically, with the development of stretchable and flexible designs and materials, electronics can now be integrated into skin, tissues, and tissue models for unobtrusive monitoring. Additionally, wearable bioelectronic devices enable continuous and unobtrusive human health monitoring. The chapter describes several tissue models embedded with soft electronics as well as the key requirements for wearable electronics. The chapter concludes with a description of biohybrid electronics, where tissue and electronics are integrated for stimulation or monitoring of tissue.

Acknowledgments: We are grateful to all the authors for their advice and suggestions on how best to structure this book. We are especially grateful to all of our long-time collaborators and colleagues, as well as to all of our graduate and undergraduate mentees and students, without whose unparalleled dedication and perseverance none of this would have been possible. Both editors also acknowledge the National Science Foundation and the National Institutes of Health for their financial support over the years.

Manas Ranjan Gartia and Ram V. Devireddy
Department of Mechanical and Industrial Engineering
Louisiana State University
Baton Rouge, LA, 70803 USA
March 2026

Chapter 1

High-Throughput and Label-Free Analysis

Mahima Gupta, Sreyashi Das, and Ram V. Devireddy
Department of Mechanical Engineering,
Louisiana State University,
Baton Rouge, LA 70803, USA
rdevir1@lsu.edu

In the last decade, high-throughput analysis has grown to be a critical tool in molecular biology, genome analysis, proteomics, drug development, and material sciences, facilitating effective strategies for assessing biological and chemical specimens while saving time and resources. Recent studies have shown that high-throughput label-free approaches allow for real-time monitoring and study of molecular interactions and cellular processes in the absence of external tags or markers, which is critical in areas such as genomics, proteomics, drug discovery, and clinical diagnostics. Additionally, these procedures have significance because of their accuracy, rapid evaluation, and noninvasive aspects. According to the current research, advancements in sensing technology, especially

High-Throughput Imaging Methods: Cells and Tissues
Edited by Manas Ranjan Gartia and Ram V. Devireddy

ISBN 978-981-5352-08-5 (Hardcover), 978-1-003-77417-4 (eBook)
www.jennystanford.com

DOI: 10.1201/9781003774174-1

hyperspectral imaging (HSI), have increased the effectiveness associated with these procedures. Therefore, this chapter explains the principles and working mechanisms of distinct types of high-throughput and label-free analytical concepts, reviews technological breakthroughs, and examines their applications in clinical settings.

1.1 Introduction

High-throughput and label-free analysis is an established technology used in various areas of molecular biology and material science [1]. It has transformed conventional techniques in analytical and bioanalytical chemistry by introducing efficient and noninvasive methods for the analysis of biological and chemical samples [2]. As there is constant progress of science and technology, efficient and cost-effective techniques are required that can screen and analyze large amounts of data [3–6]. High-throughput analysis techniques allow the simultaneous testing of various biochemical samples, thereby saving testing time and improving resource utilization [7]. In addition to that, label-free methods do not require external tags or markers which allow more immediate and direct real-time observations and analysis of molecular interactions and cellular processes. Collectively, these two techniques are playing an essential role in molecular biology, genomics, proteomics, drug discovery, and materials science [8]. High-throughput analysis employs automated and parallel approaches to effectively analyze massive datasets, making it essential in domains such as genomics, proteomics, drug development, and clinical diagnostics. It employs high-performance computing, machine learning, and deep learning algorithms [7, 9]. High-throughput analysis enhances the discovery process while minimizing human error, which upgrades numerous research areas and plays an important role in developing new treatments and technologies, enabling researchers to advance their work more effectively [7–10]. Molecular targeting techniques, such as genomes, epigenomics, transcriptomics, and proteomics, are employed to find possible diagnostic and medical targets for illnesses such as cancer, thus adding to tumor biology research [10, 11].

Label-free techniques are critical in current research because they allow for accurate and quick evaluations of biological and chemical interactions while maintaining the integrity of biological networks, thus rendering them suitable for critical high-content examinations [12–15]. Their noninvasive nature allows real-time monitoring of dynamic processes, which is important in applications such as live-cell imaging and kinetic research [16]. The main challenge in advanced biological research is to acquire detailed measurements of cells at a single-cell resolution in a high-throughput manner [17]. Conventional approaches such as flow cytometry employ fluorescent stains for analysis, but they might have unintended consequences such as DNA damage. Label-free analysis removes extraneous stains, offering a direct view of how cells function [18, 19]. In this field, a remarkable upgrade occurred in flow cytometry imaging when the high-throughput capabilities of conventional flow cytometry were combined with the detailed image acquisition of fluorescence microscopy [20, 21]. Researchers analyzed cell shape and activity employing a mix of bright field and dark field photography, as well as robust software and machine-learning techniques, which are inexpensive, noninvasive, and eliminate the intricate structure and adverse reactions of fluorescent stains, facilitating deeper biological investigation [22]. High-throughput and label-free approaches are used in bioanalytical applications, such as drug discovery. High-throughput screening (HTS) enables rapid detection and characterization of medicinal compounds, while label-free methods such as MS and surface plasmon resonance (SPR) investigate how molecules interact without color or marker interference [23]. These strategies cut costs, enhance accuracy, and enable more complicated data collection, which is critical for comprehending complex biological systems and materials [24–26]. High-throughput imaging (HTI) is an effective approach for studying the complicated cellular behavior and fundamental functioning mechanisms of molecules in cells. Additionally, advanced microscopy, along with automated data analysis, allows scientists to better understand cellular aberrant behavior under disease or experimental situations. It also allows instantaneous collection of extensive data from thousands of images when scientists investigate the biological response to genetic modification or chemical treatments.

The last couple of years have seen some notable advances in the development of sensing technologies, particularly those related to real-time, nondestructive, evaluation systems. Important among these has been HSI, which has become a popular technology [27]. This approach combines the strengths of traditional imaging and spectroscopy to facilitate the capture and processing of different light wavelengths across the electromagnetic spectrum. While conventional imaging systems capture information in a limited number of bands, HSI acquires data over an extensive range of wavelengths, allowing an integrated analysis of surface and subsurface features of materials [28]. It has been very useful in many fields that require high accuracy in the measurement of quality, such as agriculture and food safety [29]. HSI allows noninvasive measurement of important factors such as chemical composition and internal structure based on the reflectance, transmission, or absorption of light [30]. This feature makes it even more functional and accurate compared to the general imaging systems, as these usually limit their study to the surface features and often cannot highlight internal features. Another important characteristic of the novel sensing techniques is electrical detection. Biological components can interact with electrical platforms to accomplish sensitive detection of a wide range of biomolecules quickly by electrical biosensors. This can detect changes in electrical properties, such as conductivity or impedance, due to biological interactions, hence playing a vital role in biosensing and diagnostic applications. Combining HSI with electrical detection makes a strong tool for modern analysis. They have complementary strengths as HSI provides detailed spectral and spatial data across different materials, while electrical detection provides high precision in detecting molecular interactions. A combination of these two techniques is an innovative approach to scientific research [31].

This chapter aims to explain the principles of high-throughput and label-free analysis. The section intends to increase awareness and understanding of the technologies and principles that form the basis of key methodologies, such as microarrays, mass spectrometry, and electrical sensing techniques. Furthermore, high-throughput and label-free methodology are highlighted across various scientific and technical domains. Also, modern technological developments and industrial challenges are reviewed. It also intends to analyze case studies in life sciences, material science, and tissue engineering to show practical applicability.

1.2 Fundamentals of High-Throughput Analysis

The healthcare sector revolves around a thorough evaluation of chemical libraries and biomolecular targets. High-throughput innovations, digitization, and commercial constituent collections have broadened the number and types of compound collections. Analyzing both pathogen and human genomes has resulted in the identification of potentially new biological targets, but there is no evident starting point for target optimization. Over the last two decades, theranostic and drug development techniques have grown dramatically due to technological breakthroughs, such as miniaturization, automation, and robots, facilitating high-throughput and ultra-high-throughput analysis [32] High-throughput analysis, often called high-throughput screening (HTS), is a high-level multiparametric approach to analyze and treat a large number of compounds or samples, simultaneously generating a large amount of data in short periods with the goal of improving the experimental methods by preserving resources and time in processing [33, 34]. For instance, in the 1990s, biopharmaceutical and biotech enterprises leaned predominantly on 96-well MTP microtiter plates for chemical handling and analysis. During the past 10 years, plate variants with larger well densities, such as 384-well MTP, have gained prominence due to automation and miniaturization [32]. Along with automation and downsizing, this strategy employs powerful computational methods that generate and assess enormous amounts of data for research advancement, drug moiety efficacy, and 3D tissue culture models [35]. For example, the multicellular tumor spheroids (MCTS) model has demonstrated considerable promise in cancer pharmacological research and has been combined with a variety of HTS approaches, including both hardware and software developments of 3D phenotypic HTI systems (Table 1.1). It highlights the recent progress that has been made in the application of HTI and HCS for drug discovery with organoids and MCTS employing microfluidics, micropatterning, and automated robotic systems.

Table 1.1 Available 3D high-throughput imaging and high-content imaging systems

3D HTI Systems	Description	Company	Key Technology	Application	Reference
Opera Phenix	The confocal spinning disk high-resolution imaging microscope gathers images in a variety of modes, acquiring slides at remarkable speeds in both 2D and 3D settings.	PerkinElmer	Confocal (laser-based)	Examination of mitochondrial dysfunction and associated oxidative stress, such as superoxide levels, membrane potential, and mitochondrial area; analyze both live and preserved neural cell lines to evaluate and characterize lysosomal attributes.	[36]
Operetta	It enables the display of cell models in 3D and XYZ viewers, evaluates volumetric and 3D-related phenotypes data, records fast cellular changes at up to 105 fps, and also offers excellent noise elimination. Harmon® high-content imaging and assessment software provides precise target targeting, faster collection with shorter processing times, and user-friendly algorithms.	PerkinElmer	Confocal (laser based)	Detects chemical clusters which associate with target proteins, as well as interactions among drug aggregators and the targeted protein of interest, but it additionally has the ability to identify the critical aggregation concentration (CAC) of a specific versatile small molecules.	[37, 38]

3D HTI Systems	Description	Company	Key Technology	Application	Reference
IN Cell Analyzer 6000	It can take images of up to 1536-well plates in a single scan. It has several different emissions and excitation filters, which enable fluorophores to be observed concurrently. The platform automates data collection and estimation, resulting in reliable outcomes across different samples and test situations. It includes user-friendly software for experiment execution and evaluation.	GE Healthcare	Confocal (laser based)	An excellent choice for laboratories doing high-content testing, discovering drugs, and cell imaging investigation, since it provides high-throughput scanning features and automated analysis options. The researchers explored the cytoplasmic extension of renal tubular cells (RTCs), the quantity of RTCs and CaOx monohydrate (COM) crystals within the cytoplasm, and the proportion of COM crystal adhesion per cell.	[39, 40]
ImageXpress Ultra	Confocal automated microscopy system with up to four lasers, adjustable scan size, software configurable pinhole, and format independence, capable of scanning 6- to 1536-well microplates or microscopic slides.	Molecular Devices	Confocal (laser based)	Gain of function chromatin remodeling, drug delivery.	[41, 42]

(*Continued*)

Table 1.1 (*Continued*)

3D HTI Systems	Description	Company	Key Technology	Application	Reference
ImageXpress Micro Confocal	It offers brightfield label-free imaging, widefield, and confocal fluorescent imaging and offers greater image resolution and sensitivity with up to 4× increase in signal, which leads to lower exposure times.	Molecular Devices	Confocal (laser based)	Functional screening of miRNA, apoptosis, autophagy, various molecular mechanisms.	[43, 44]
Arrayscan XTI	It offers label-free imaging and can add white light illumination to high content screening (HCS) experiments, allowing imaging of morphological features and interleaving the resulting data with fluorescence data from the same sample, highly sensitive camera with 4.5 μm/pixel resolution, 14-bit dynamic range, and 2208 × 2208 pixel array.	ThermoFisher	Confocal (LED based)	Cell counting, vesicle detection, and morphometric measurements.	[45, 46]

3D HTI Systems	Description	Company	Key Technology	Application	Reference
CellInsight CX7	A high-efficiency sCMOS camera with 95% peak quantum efficiency and low background noise, enhancing high-throughput performance and providing 65,536 gray detection shades. Laser illumination provides flatter and brighter images with low background noise.	ThermoFisher	Confocal (LED based)	Live-cell experiments, developmental disorders.	[47]
Cell Voyager 7000S	The wide-view multi-beam confocal scanner unit offers high-speed image acquisition, with a maximum acquisition rate of 38 fps. It can capture images of whole wells in 384 well plates in just one shot. The unit also offers an optional stage incubator for long-term imaging over three days, allowing for the kinetics of live cells to be observed and analyzed.	Yokogawa	Confocal (laser-based)	Live cell observation, molecular mechanisms, differentiation of ES cells.	[48, 49]

(*Continued*)

Table 1.1 (*Continued*)

3D HTI Systems	Description	Company	Key Technology	Application	Reference
CQ1	Compact, lightweight, and easy-to-use benchtop device that offers rich feature extractions, high-speed scanning, and Nipkow Spinning Disk Confocal technology, reducing phototoxicity and photobleaching and allowing integration for laboratory automation.	Yokogawa	Confocal (laser-based)	Measurement of spheroids, colonies, and tissue sections, enables analysis of time-lapse and live-cell.	[50, 51]
Acumen Cellista	It offers high-throughput screening (HTS), allowing for image and analysis of 96- to 1536-well plates in 8 minutes. Its robust data, multiplexing capabilities, and intuitive operation allow for easy training for multiple staff members, resulting in increased throughput and reduced costs.	TTP LabTech	Laser-scanning fluorescence	Drug discovery, targeting, and analysis.	[52]

3D HTI Systems	Description	Company	Key Technology	Application	Reference
ImageXpress Velos	Laser scanning imaging system offers high-content cell-based assays, colony morphology studies, and multiplexed spot and bead arrays. This versatile benchtop instrument provides speed and flexibility not available with microscopes or confocal scanners.	Molecular Devices	Laser-scanning fluorescence	Cell-based assays, homogeneous antibody detection, multiplexed phospho-protein detection, proliferation.	[53]
XLS	It uses large field-of-view optics to map macro-structures and accelerates cell population querying. It enables the characterization of heterogeneous samples and the identification of rare sub-populations. The system offers unlimited configurations, screens for over 50,000 wells per day, and reliable assays with a 3-log dynamic range.	Molecular Devices	Phase contrast/ widefield	Cellular pathways.	[54]

(*Continued*)

Table 1.1 (*Continued*)

3D HTI Systems	Description	Company	Key Technology	Application	Reference
Cell3 iMager	Multiple images are captured while varying the focus in the Z direction and combined into a single image. Supports various SBS plate formats such as F-bottom, V-bottom, U-bottom, etc.	InSphero	Brightfield	Multi-color fluorescence imaging.	[55]
3D Imaging Systems		**Company**	**Key Technology**		
A1 MP+/ A1R MP+	Multiphoton confocal microscopes offer faster, sharper imaging from deep within living organisms.	Nikon	Multiphoton/ confocal	Suitable for brain research, neuroscience, and in vivo imaging in both upright and inverted microscopes.	[56]
Lightsheet Z.1		ZEISS	Light-sheet fluorescence		
IncuCyte ZOOM	Advanced live-cell imaging platform that automates cell behavior quantification supports multiple plates, flasks, and dishes and allows remote data analysis.	Essen Bioscience	High-definition phase contrast/ fluorescence	Live-cell imaging.	[57]

The principles of high-throughput analysis include parallelization, scalability, and data integration. The feature of parallelization enables the processing of many samples simultaneously, thus greatly enhancing throughput. Scalability ensures that the same methods can be applied for both small- and large-scale studies with no performance compromise. In these high-throughput systems, microplates are used, each containing a large number of wells, each with a different reaction mixture, which can then be analyzed using robotic systems combined with advanced detection methodologies [35]. The results are often generated in real-time, thus ensuring speed and efficiency. Data integration is based on the capability of assembling and analyzing complex datasets, sometimes using specific software in addition to statistical methods to generate meaningful information [58].

Additionally, over the past few years, automation and robotics technologies have become the key to high-throughput analysis that generates consistent and reproducible results. Integration of robotic systems into HTS workflows allows laboratories to substantially increase throughput and facilitate high-speed evaluation of large arrays of biological and chemical entities. The key focus of robotic systems in HTS is to lower the level of human involvement in different tasks, often extremely time-consuming tasks, such as liquid handling [59]. Pipetting robots, for example, would engage in the dispensing activity of liquids into microplates while ensuring measurements with precision and minimizing human error [60]. Such robots, along with other automated equipment like barcode printers, plate readers, and incubators, are integrated through a process control system (PCS), which functions in enabling continuous processes and parallel observation.

Automation increases throughput while maintaining the consistency of experimental conditions, a very important aspect in generating reliable data. Robotic systems can analyze thousands of samples in a few days. Thus, automation considerably reduces the time and number of resources needed when compared to manual screening [61]. Moreover, robotics also enhances laboratory safety by reducing human contact with hazardous substances. On the other side, their programming and maintenance require complex systems; therefore, skilled operators are needed to monitor and resolve issues arising during performance [62]. Despite the advances in automation, several errors still arise; these include robotic misalignment and hardware failures. The elimination of such challenges requires urgent intervention by operators responsible for system performance and changes where needed. Moreover, the study of optimization of robotic workflows is still largely developmental; it focuses on improving reliability and minimizing downtime along the lines of screening.

The application of high-throughput technologies has resulted in key advancements in research while also introducing inherent limitations. The main advantages include handling larger numbers of samples at any given time, which is a major benefit of high-throughput technologies. This enables large-scale studies in a short period that normally would have required using conventional methods. Automation and robotics improve efficiency by minimizing human error and experimental variability, resulting in improved reproducibility, which is important in both scientific research and industrial applications [63]. Such systems, besides being capable of high throughput, are very versatile as well. Their applications range from the fields of drug discovery and genomics to environmental testing and food safety analysis. They show the ability to generate data at very high speeds and finally allows researchers to investigate complex biological and chemical systems. In particular, HTS technologies give insight into molecular interaction and cellular functions, contributing to the identification of new therapeutic targets and biomarkers [64].

Despite enormous advantages, several limitations also exist for high-throughput technologies. Due to vast volumes of data, data management and interpretation are complex. High-throughput experiments may generate several million data points; hence, proper bioinformatic tools and expertise are needed for their analysis. Poor data processing or statistical interpretation may result in incorrect conclusions, limiting any effective benefits achieved by these technologies [65]. Other limitations include high-cost investment at the initial stages. A single, complete high-throughput system requires huge financial investments in its initial stage for equipment and software alone. Though long-term benefits generally outweigh the costs, smaller research laboratories or startups find it rather challenging to afford such technologies without external funding or partnerships. Furthermore, some of the high-throughput methodologies, in particular those relying on automation, could offer less flexibility when handling non-standard or very complex samples [66]. The rigid construction of robotic workflows cannot allow frequent adjustments or an individually tailored approach.

Several important technologies form the foundation for high-throughput analyses, each providing unique capabilities for specific types of studies.

1.2.1 Microarrays

Microarrays are essential for rapid screening since they can precisely evaluate tiny samples in addition to examining many cellular and molecular processes in drug development and biological investigation [67]. These techniques are especially used in testing the effects of pharmaceuticals, genetic manipulations, and other types of molecular signatures on cellular behavior.

Microarrays are solid substrates that hold biomolecules, cells, or agents, permitting concurrent evaluation of many samples while enhancing experimental efficiency. Advanced robotic identifying systems can detect nanoliter amounts of materials, thus rendering them vital in large-scale screens [68, 69] Furthermore, microarrays are excellent instruments for investigating cell–substrate interactions, influencing cellular operations, aiding three-dimensional culture of cells, thus being crucial to tissue engineering, regeneration therapy, pharmaceutical testing, and the investigation of proteins and genes [70]. For example, Zhou et al. created a microarray chip-based HTS approach for a self-assembling peptide that targets the neurofilament light chain, a biomarker for Alzheimer's disorder. The peptide targeting the neurofilament light chain nanosheet showed strong binding affinity (KD = 1.39×10^{-9} mol/L), and specificity, thus being feasible to detect Alzheimer's disease noninvasively and to evaluate plasma concentrations of neurofilament light chain, shown in Fig. 1.1A [71] Additionally, Schmitz et al. invented a polymer microarray for evaluating antimicrobial substances for cell adherence. They found that antimicrobial peptide sequence for normal human dermal fibroblast (NHDF) boosted adherence, implying the potential of the high-throughput technique shown in Fig. 1.1B [72].

Additionally, a wide range of applications with examples make microarrays a core component of modern biological investigations, offering a compelling way to study complex biological systems with high accuracy and efficiency, as discussed in Table 1.2 [73].

Table 1.2 High-throughput cell-based screening methods in drug discovery and stem cell biology

Method for array fabrication	Application	References
Robotic spotting	Cellular microarrays for high-throughput toxicology assays	[67, 74]
	Cellular microarrays for high-throughput compound screening	[75–77]
	Transfection microarrays for gene function screening	[78, 79]
	Peptide and small-molecule microarrays for high-throughput cell-adhesion assays	[80]
	Cellular microarrays for the identification and characterization of multiple antigen-specific populations of T cells	[81]
	Biomaterial microarrays for microscale testing of biomaterial interactions with stem cells	[82, 83]
	Extracellular matrix microarrays for probing embryonic stem cell fate	[84, 85]
	Cellular microarrays of signaling microenvironments for adult stem cell differentiation	[86, 87]
Photo-assisted and soft-lithography approaches	Microfluidic cell culture arrays for cytotoxicity determination and high-throughput cell-based assays	[88, 89]
	Miniaturized, multiwell culture of human liver cells for drug development	[90]
	Protein surface arrays for modulation of T cell responses and stem cell growth and differentiation	[91–96]
	Microwell arrays for single-cell analysis and stem cell culture	[97–102]
	Microfluidic cell culture arrays for stem cell research	[103–106]

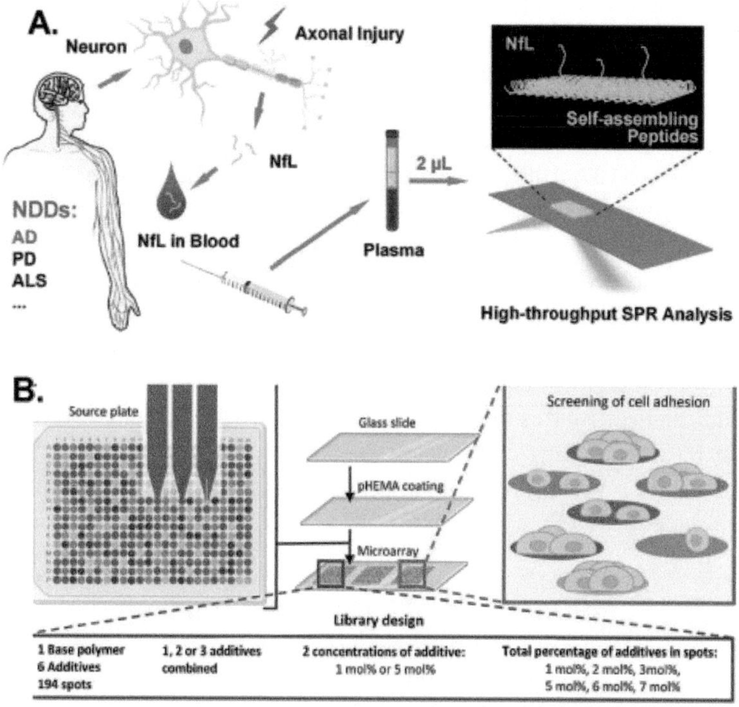

Figure 1.1 Applications of microarray in HTS; (A) An illustration of the processes for microarray chip-based HTS of NfL self-assembling peptide molecules. (B) A representation of the supramolecular microarray architecture for NHDF adhesion assessment.

1.2.2 High-Throughput Sequencing

High-throughput sequencing (HTS) has transformed genomics, epigenomics, and transcriptomics by facilitating the simultaneous sequencing of millions of strands of DNA. Unlike standard Sanger sequencing, HTS produces gigantic datasets in just one run, which enables better knowledge of genetic attributes, illness, stages of development, and cellular operation. RNA-seq enables scientists to evaluate transcriptome alterations, correlate gene expression throughout developmental stages, and examine epigenetic variations by implementing methods, such as ChIP-seq and Methyl-seq [33]. Additionally, HTS has facilitated the identification of epigenetic markers, including DNA methylation and histone alteration trends throughout the whole genome. DNA methylation, a fundamental epigenetic marker, is associated with cancer thus, comprehending it is vital to treatments. MethylC-seq and RRBS provide a comprehensive mapping of methylation patterns and histone alteration locations, which might assist in targeted therapy attempts [107, 108]. Ribosome profiling now joins that exclusive club, revealing precisely which mRNA transcripts are actively translated into proteins. Several commercial sequencing platforms have emerged to expand the potential of HTS [109]. In all these techniques, Illumina leads much of the market. Illumina technology relies on a process called bridge amplification, which makes clusters of identical DNA sequences for analysis. Different HTS techniques and their attributes are summarized in Table 1.3. Recent advances in HTS technology continue to expand the boundaries of genetic research, revealing new insights into disease causes and potential therapeutics [59].

1.3 Fundamentals of Label-Free Analysis

Label-free analysis is an approach for detecting and quantifying biomolecules or chemical interactions in the absence of external labeling. It utilizes inherent characteristics, including mass, refractive index, and electrical impulses, to study molecular or cellular processes. This approach is critical for the continuous surveillance of biological molecules in their native state, delivering precise and dynamic data on complicated biological reactions as shown in Fig. 1.1A [18]. Label-free approaches allow for ongoing surveillance of events without affecting the subject matter framework, therefore being critical for studying dynamic biological systems wherein extrinsic labels may interfere with or modify the normal activity of intended biomolecules [4]. These technologies are used in various fields, including drug discovery, diagnostics, and environment monitoring. There are distinct kinds of label-free techniques as described in the following section.

These label-free analysis methods have several advantages, including the investigation of biomolecular interactions that can be performed in real time without disabling labels or tags, thus maintaining the accuracy of results, especially in monitoring dynamic processes [110]. Besides, sensitivity is generally very high in such techniques, with the possibility to detect even slight variations in mass, refractive index, or electrical signals. However, there are limitations, too. Many label-free methods require advanced and very expensive equipment. This might be limiting for a smaller laboratory or one with limited resources. Secondly, in some of the label-free techniques, complex mixtures or systems where multiple interactions occur at the same time may be difficult to interpret [111, 112]. Nevertheless, even with these disadvantages, label-free analysis is an invaluable approach to investigating biological and chemical systems in a more natural, noninvasive way.

1.3.1 Mass Spectrometry

Mass spectroscopy (MS) is an advanced analytical approach that identifies and quantifies compounds according to their ratio of mass to charge. It offers a label-free strategy for molecular identification and characterization that aids in identifying proteins, peptides, and biomolecules [113]. This method ionizes materials and examines their behavior, permitting it to identify minute molecular weight variations that are vital to identifying post-translational alterations and other structural alterations. Its capacity to investigate complicated mixtures makes it valuable for investigating distinct biomolecules [114]. Recent advances, such as Tandem mass spectroscopy MS/MS, have increased its power by allowing ion fragmentation, which obtains higher-order information on biomolecule structure and composition. This technique analyzes compounds by breaking them down into fragments and measuring their mass. This technique occurs by first ionizing materials, resulting in a combination of ions, subsequently choosing precursor ions with an appropriate mass-to-charge ratio (MS1) and splitting them up (MS2) to form product ions that can be identified. Additionally, fragmenting to create a different set of product ions, the procedure may be expanded to first-generation product ions (MS3), as shown in Fig. 1.2A [115]. Its application in high-throughput proteomic studies has generated large datasets aimed at understanding biomolecular interactions and pathways [116]. For instance, Hendriks et al. employed rapid evaporative ionization mass spectrometry (REIMS) and matrix-assisted laser desorption ionization mass spectrometry imaging (MALDI-MSI) to identify the absolute survival duration of Glioblastoma multiforme, an aggressive kind of brain cancer. A model was constructed employing 45 samples with a 97.7% accurate prediction accuracy for 28 cases. Significant discriminators were discovered for short-term and long-term survival, with 66.7 and 69.4% precision, significantly shown in Fig. 1.2B [117].

Mass spectrometry (MS) has been used in advanced methods, including liquid chromatography-mass spectroscopy (LC-MS), which promotes the detection of compounds at minimal concentrations and facilitates complete profiling of biological specimens, biomarker identification, drug discovery, and clinical evaluation, each of which provides critical information on biological pathways and pharmacological targets.

1.3.2 Surface Plasmon Resonance

Surface plasmon resonance (SPR) is a comprehensive optical procedure that enables the real-time tracking of biomolecular interactions without labeling, especially in protein-ligand, antibody-antigen, and DNA–protein interaction investigations, therefore improving knowledge of such chemical interactions. The working principle is based on the measurement of changes in the angle or wavelength of light, which is directly related to the events of molecular binding at the sensor interface. The SPR phenomenon happens when incoming beam energy is coupled to surface plasmons at the metal-dielectric interface, minimizing the total internal reflection of light. SPR biosensors are made up of an optical component, a transducing medium, and an electrical equipment, which generates a label-free real-time response shown in Fig. 1.2C [118, 119] The real-time measurements allow quantification of kinetic parameters, such as the rates of association and dissociation, and affinity constants, which are important to understand the strength and nature of molecular interactions.

SPR has indeed proven to be quite versatile in the process of drug discovery and biomolecular research, specifically in cancer biomarker detection [120]. For instance, prostate-specific antigen (PSA), a significant prostate cancer marker, was observed in serum employing an integration of colorimetric and SPR sensors, with a limit of detection of 0.009 ng/mL [121]. SPR enables the analysis of small molecules that interact with larger protein targets, hence facilitating potential drug candidate screening [122, 123]. Quantitative information about binding kinetics, thus facilitates the assessment of the efficiency and specificity of newly developed drugs. Loo et al. developed an aptamer-based biobarcode (ABC) assay to detect cytochrome-c, a cell death signal produced by malignant cells. In order to capture Cyto-c, magnetic particles (MMPs) modified with antibodies (Ab) and a Cyto-c-specific aptamer (MMP-Ab-Cyto-c-aptamer) were employed. The DNA biobarcode was fused with probes designed particularly for RNase H identification by SPR. This ABC test identified phenylarsine oxide as a possible treatment chemical capable of killing multidrug-resistant liver malignancies at nanomolar concentrations shown in Fig. 1.2D [124]. More recently, improvements in SPR technology, such as the development of SPR imaging SPRi, extend its application up to high-throughput platforms where thousands of interactions can be monitored simultaneously [125]. Due to the advancement of SPR, it has now become a powerful technique for not only traditional biomolecular interactions but even the latest complicated systems involving protein–protein networks and multi-component reactions. Corn et al. devised an SPRi detector for sequence-specific and quick miRNA identification. The technology combines sequence-specific complexation, poly(A) polymerase extension, and oxidation reactions, thereby rendering it rapid and inexpensive [126].

SPR can detect the interactions in a label-free format with no requirement for either fluorescent or radioactive labeling. It eliminates the limitations of traditional assays, including the simplification of experimental design and reduction of potential artifacts, leading to more accurate and reproducible data. Additionally, SPR sensors can assist in diagnosis and therapy by evaluating blood serum and plasma-based markers. Current investigations should emphasize generating a multi-purpose aptamer-based instrument for cancer cell detection and binding, with sensitivity, repeatability, and miniaturization in mind. Due to its versatility, combined with real-time data acquisition and a label-free nature, has covered the role of SPR as a base technology in basic and applied biomedical research, especially in drug development and diagnostics.

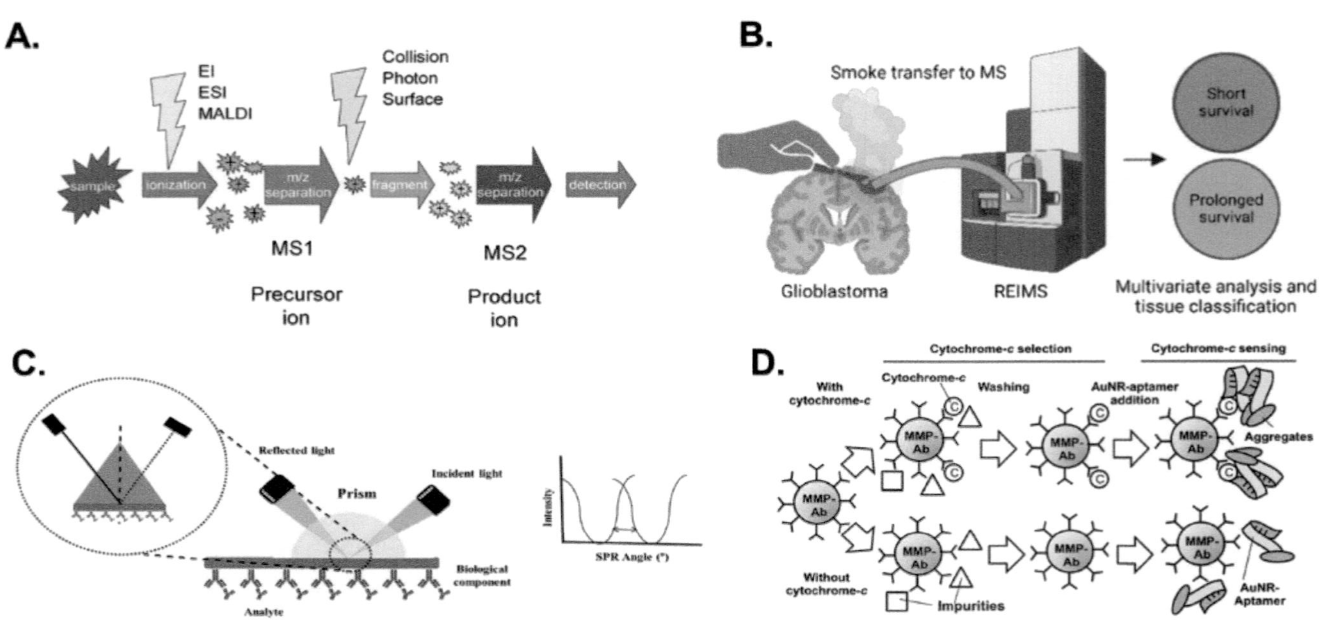

Figure 1.2 Principle and application of different label-free analyses. (A) Schematic representation of the ionization procedure of tandem mass spectroscopy (MS/MS). (B) An overall illustration of the integration of two distinct mass spectroscopy techniques, MALDI-MSI and REIMS, to predict the survival period of Glioblastoma multiforme by checking the lipidomic difference of the Glioblastoma multiform tumor samples [117]. (C) An illustration of the SPR principle, which comprises a monochromatic laser beaming on a surface, producing plasmons at an appropriate angle, and identifying the reflected light as impacted by the analyte associated with the biological component. (D) An illustration of gold nanorod localized SPR to detect cytochrome c particles in serum. Cyto-c molecules are trapped on micro-magnetic particle surfaces, and aptamers are used to form sandwich interactions that record the SPR phenomenon [124].

1.3.3 Atomic Force Microscopy

Atomic force microscopy (AFM) is a high-resolution technique that uses three-dimensional imaging of surfaces. It uses a sharp tip on a flexible cantilever that is scanned across the sample, detecting changes in surface topography due to variations in the forces between the tip and the surface. There are several modes of operation, depending on the type of sample, including contact, tapping, and non-contact modes, which represent one of the advantages of AFM as a very versatile tool for visualizing not only soft biological materials but even harder surfaces down to atomic-scale resolution. AFM can also analyze samples in their native states; for instance, samples that are in liquid environments can also be analyzed, which allows researchers to observe the behavior and interactions of live cells without perturbing their natural conditions. AFM experiments present force-distance curves that may be leveraged for measuring cell stiffness and viscoelastic attributes. For instance, Hartmann et al. used AFM and its data analysis software to establish a protocol to measure the elastic modulus of the basement membrane in human and murine lung tissue, which possesses diagnostic potential and offers new perspectives on developmental, physiological, and pathological events within living organisms [127]. Yet, manual involvement, cantilever probes, and nanomechanical evaluations, as well as live cell biomechanical investigations on AFM are laborious and low-throughput. So, people have utilized AI to measure the live cell attributes along with AFM [128]. A team developed a deep learning framework for automated cell sample selection based on shape for AFM probe tracking during biomechanical mapping, enabling a 60× speed-up and reduced time for detecting specific cell shapes in a group of cell specimens [129].

Besides its capability for imaging, AFM is excellent in force spectroscopy, probing interactions of biomolecules, including the binding forces of receptor-ligand pairs. In this process, AFM can quantify the strength of molecular interactions at the single-molecule level and provide insights into the mechanical properties of cells and proteins that can be difficult to derive with other methods. Using functionalized tips coated with specific biomolecules, AFM can probe interactions such as antibody–antigen binding or DNA–protein associations [130]. For instance, Thus, the possibility of measuring and visualizing interactions in real time under physiological conditions has turned AFM into an indispensable tool in investigations of such intricate biological processes as cell adhesion, protein folding, and biomaterial characterization.

1.3.4 Acoustic Wave Sensors

Acoustic wave sensors measure changes in the propagation of acoustic waves along a surface, which are predominantly due to changes in mass. A very powerful example of this technology is the quartz crystal microbalance (QCM), where small mass changes result in a shift of the resonant frequency of a vibrating quartz crystal. This mass sensitivity is one of the aspects that makes QCM particularly valuable in studying molecular interactions, such as in biomolecular binding and adsorption processes. QCM sensors are versatile and intended for analyzing thin film deposition, detection of biological agents, and monitoring chemical interactions at a molecular scale. Apart from QCM, Surface Acoustic Wave (SAW) is a piezoelectric sensor composed of interdigitated transducers (IDTs) and a layer of material such as quartz, lithium niobate, lithium tantalite, zinc oxide, or aluminum nitride; it was originally established in 1885 [131–133]. These sensors are also in wide application because they can detect changes in the velocity of surface waves [134]. These sensors have high efficiency in label-free detection by quantifying the changes in wave behavior due to mass adsorption or changes in viscosity. They can function in various environments like liquid phases, making them suitable for applications in real-time biosensing. Recent investigations focus on reducing sample volume and increasing sensitivity, further extending their use in fields related to environmental monitoring, medical diagnostics, and materials science. Biological detection requires chemical and biological changes. SAWs may be formed by supplying a field of electricity to IDTs, resulting in mechanical power as acoustic waves in accordance with fluctuations in the electrical input information caused by electromechanical interaction, as shown in Fig. 1.3A [135].

Although this type of sensor is rapidly being employed as real-time, label-free screening, challenges, including instrument downsizing, microfluidics manufacturing, and significant attenuation in liquid settings, limit their extensive implementation. Ntimtsas et al. developed a SAW microarray chip-based portable apparatus for real-time monitoring of biomolecules that comprises a paper-based capillary flow channel with a microcontroller, RF components, a micropump for flow control, and sensing channels. The capillary flow conduit enables continuous specimen flow while dampening the acoustic sound to the lowest possible level. The technology exhibited continuous tracking and linear relationships across amplitude and phase fluctuations, as shown in Fig. 1.3B [136].

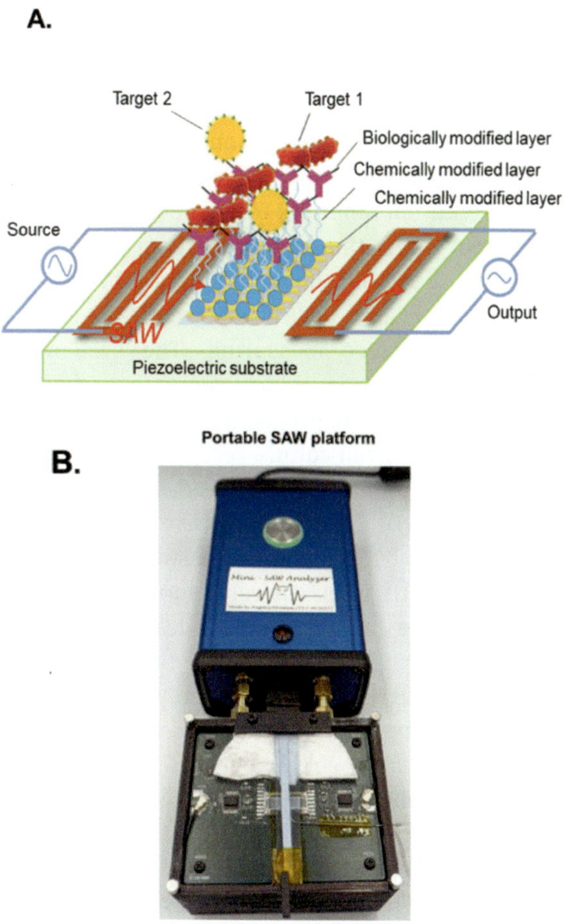

Figure 1.3 Working mechanism and application of acoustic wave sensors. (A) Schematic representation of the fundamental framework of a SAW biosensor. (B) An illustration of a portable SAW microarray chip for biomolecular detection.

1.3.5 Electrical Detection Techniques

The electrical detection approach measures the changes in electrical properties—such as conductance, capacitance, or impedance—that occur due to various biomolecular interactions (e.g., between small molecules, DNA, or proteins) on the sensor surface. There are distinct types of electrical detection techniques. Changes in conductance and impedance can indicate binding events, conformational changes, or ions. These techniques can sometimes give false-positive results, allowing real-time observation of biomolecular interactions without labels. Among these, impedance-based biological measurements entail delivering an intermittent electric field to a single as well as to a group of cells, resulting in real-time tracking of cell events, biomarker concentrations, and monolayer development and migration. These technologies provide precise spatial resolution and can be coupled with microfluidics to provide high-throughput flow cytometry without labels, thus rendering them simpler to build but more difficult to understand shown in Fig. 1.4A [137–143]. Chen et al. developed an impedance device with over 9000 gold electrodes for concurrent impedance measuring. The system's structure facilitates spatial correlation, which generates a heat map. The impedance modeling is fairly precise, resulting in an advantageous instrument for potential cell culture research such as cell counting shown in Fig. 1.4B [140].

Dielectrophoresis (DEP) is a noninvasive technology that alters cell populations by applying an inhomogeneous electric field reliant on their polarization. It is extensively employed across different sensors that influence individual cells, despite isolation can be problematic considering intrinsic differences and unexpected consequences [144–151]. The latest study has emphasized on protein and DNA segregation, with DEP's accomplishment resulting in commercially available equipment for sorting and isolating populations of cells, therefore boosting cell separation capacities [152].

Another type of electrical detection technique include field effect transistors (FETs), which are utilized in high-sensitivity biosensors irrespective of the electronics sector. Devices work on semiconductors, altering impedance utilizing an electric field. When a chemical attaches to a surface receptor, it alters the surface potential, influencing the width of the channel and electrical current as shown in Fig. 1.4C [153]. Bergveld pioneered the ion-sensitive FET (ISFET) in 1972, primarily for pH monitoring, but eventually extended to heavy metals. Industrial interest in transistor technologies facilitates downsizing, thereby rendering them appropriate for low-cost DNA sensors [154–159]. The development of enzyme FETs, which were first employed to detect penicillin, led to in the creation of a diverse spectrum of enzymes and analytes. The modification of FET substrates has increased the identification of analyte and specificity, with new technologies concentrating on clinical use. FET surfaces can be modified with nanowires, lowering the detection range to low concentrations [160–164]. A research project employed polypyrole nanotubes that had been functionalized with immunoglobulin G. Figure 1.4D depicts how silicon nanowire FETs are increasingly being employed in biosensors to boost the current [164].

In modern times, advances in nanomaterials—mostly CNTs and graphene—have considerably increased the sensitivity of electrical detection methods. These nanomaterials provide a greater surface area through which interactions can occur, improving the signal-to-noise ratio and enabling sensor miniaturization [165]. Electrical biosensors have become one of the highly used tools in diagnosis, environmental monitoring, and point-of-care devices because they provide rapid, precise, and portable solutions. The straightforward integration with electronic devices makes sensors particularly suitable for the development of lab-on-chip systems that can perform multiplex analyses, enabling real-time monitoring.

Electrical detection methods are very important in developing sensitive biosensors for the real-time and label-free detection of a wide range of biomolecules. Such sensors detect changes in electrical properties, such as conductance, capacitance, or impedance, due to binding events between target molecules, including proteins, DNA, or small ions, interacting with the sensor surface. They, hence, hold great potential for diagnostic applications, especially point-of-care devices requiring fast and accurate results. Recent advancements using nanomaterials such as carbon nanotubes and graphene have greatly increased sensitivity to the point where even low-abundance biomarkers can be detected [165]. Apart from health diagnostics, these sensors also have applications in environmental monitoring through the detection of pollutants such as pathogens and heavy metals, hence providing a portable and economically viable alternative for on-site testing.

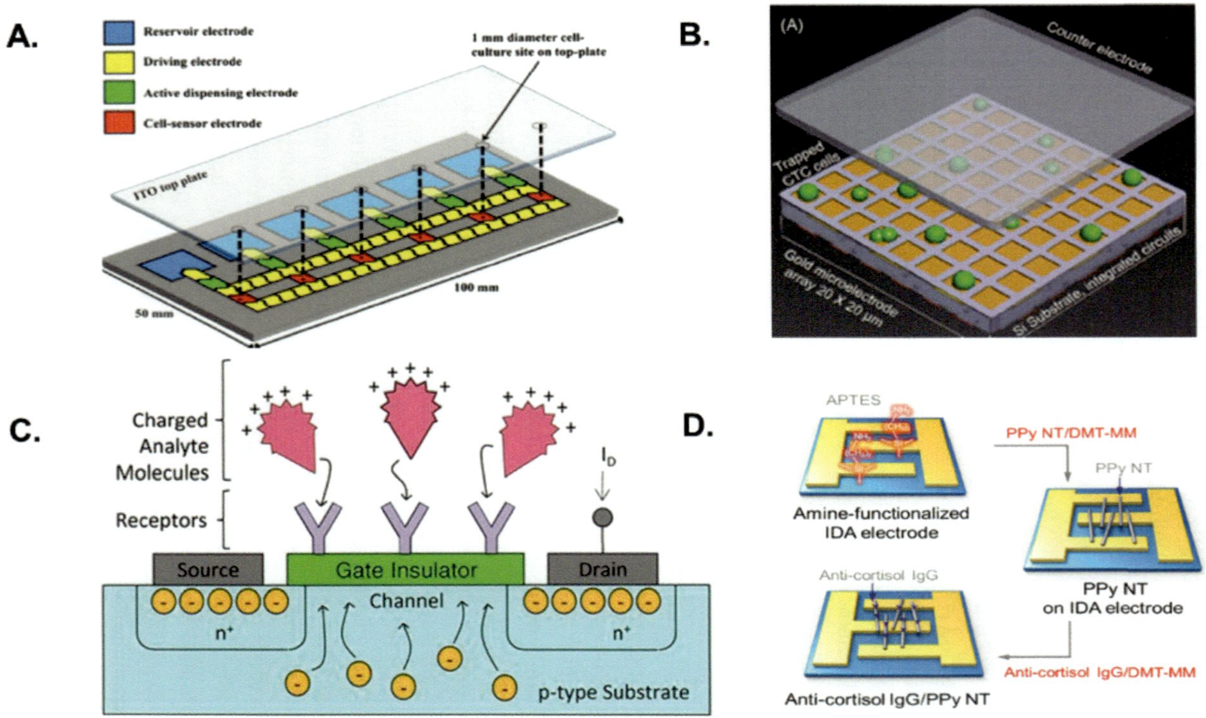

Figure 1.4 Working mechanism and application of impedance-based electrical sensor. (A) Schematic illustration of an electronic microfluidic apparatus for cell culture research and impedance measurement. (B) A representation of an impedance microelectrode matrix for the counting of cells. (C) The working principle of an FET biosensor. (D) Illustration of the development technique for an immunoglobulin G (IgG)/polypyrrole nanocomposite FET-type sensor.

1.3.6 Hyperspectral Imaging

HTI is an effective approach for studying complicated cellular behavior and the fundamental functioning mechanisms of molecules in cells. HTI approaches can be functionally grouped into three classes: screening, profiling, and deep imaging. HTI is a procedure in which cells are treated with altering chemicals to study their effects on cellular characteristics. Automated microscopy collects images of cell treatments and employs automated image processing to generate multiparametric cellular attribute databases, which are employed in screening tasks to assess the impact of significant perturbations on molecular processes. In contrast, profiling tests generate multiparametric morphological profiles for statistical learning. Automated microscopy and HCA are used in deep imaging tests to collect datasets of specified cellular properties for a significant number of cells per experimental condition. This facilitates the exact quantification of infrequent but biologically essential occurrences as shown in Fig. 1.5A.

Among various types of HTI techniques, HSI integrates conventional imaging into the spectroscopic methods of data acquisition for obtaining complex spatial and spectral information over a wide spectrum of wavelengths [166]. Despite typical imaging methods relying on a limited number of color bands, HSI acquires continuous spectral information in each pixel, generating a "data cube" that can provide detailed spectral profiles of materials being analyzed. It helps in capturing spatial and spectral information with highly accurate identification and analysis of various materials with unique spectral fingerprints [166, 167]. The major components of a typical HSI system generally include a light source, a device for wavelength dispersion, and detectors capable of capturing a wide range of wavelengths from ultraviolet to infrared. This approach combines the strengths of traditional imaging and spectroscopy to facilitate the capture and processing of different light wavelengths across the electromagnetic spectrum. While conventional imaging systems capture information in a limited number of bands, HSI acquires data over an extensive range of wavelengths. This allows an integrated analysis of surface and subsurface features of materials as shown in Fig. 1.5B. This approach enables non-destructive, label-free analysis, making HSI an invaluable tool for various applications where precise identification and differentiation are required.

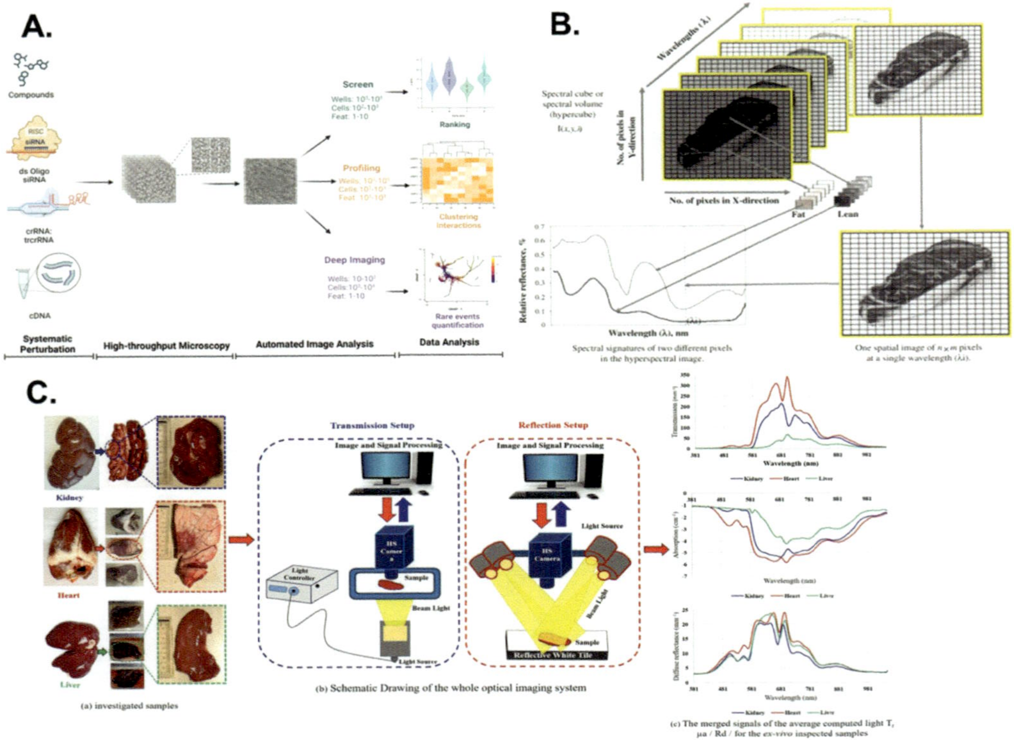

Figure 1.5 Working mechanism and application of HSI. (A) Principle and workflow of HTI. (B) An illustration of a hyperspectral image shows an interaction across spectral and spatial aspects, with each pixel reflecting a spectrum carrying chemical composition data. (C) A representation of HSI-based organ identification as it assists in distinctive spectral patterns of heart, kidney, and liver tissues.

HSI has been widely used in various applications in several fields, such as in life science, it is mainly used for biosensing, detecting abnormal growths, biomedical imaging, environmental monitoring, and agriculture. Biosensing allows the identification of different types of tissues or molecules based on their particular spectrum properties, which is important for disease diagnosis, tissue type identification, and material chemical composition analysis [167]. In material science, it helps to find defects in materials, control quality, and monitor chemical processes in real-time.

Aref et al. investigated the utilization of HSI in cancer to detect spectral alterations in biological tissues of the kidney, heart, and liver during endoscopy or open surgeries. The camera system monitors tissue light transmission (Tr) and diffuse light reflection (Rd) and calibrates employing a personalized liquid optical phantom. They tested 30 specimens from diverse organs. The optimal spectrum range for Tr was 640–680 nm, while the optimum spectrum for Rd was 720–760 nm for mutual group I (kidney vs. heart) and II (kidney vs. liver), as shown in Fig. 1.5C [168].

In another study, Mehta et al. utilized a HSI approach integrated with spectral angle mapping-based machine learning analysis to distinguish human adipose-derived stem cells from other stem cells. They identified the differentiated cells at a single molecule scale [169].

1.4 Integrating High-Throughput and Label-Free Techniques

The combination of high throughput with label-free methods has opened new avenues for modern research that offers a simultaneous approach that enhances both the precision and accuracy of biological and chemical analyses. HTS has reached a critical threshold where millions of data are now able to be screened rapidly. At the same time, label-free methods permit the study of biological processes in their natural state, undisturbed by any external probes or labels. Integrating these two techniques into one provides an in-depth analysis and understanding. Hence, it is very useful in pharmaceutical development, clinical diagnosis, and fundamental biological research.

High-throughput technologies are formed to operate large datasets, which allow the analysis of multiple samples in one run using parallel processing. Integration of these platforms with label-free techniques allows the study of complex biological interactions in real-time, thus preserving the molecular or cellular behavior of the sample under investigation. Many technologies, such as microarray, mass spectroscopy, and SPR are integrated into this comprehensive methodology. Microfluidics is one of the leading technologies, because it offers a controlled environment for the precise manipulation and monitoring of minute volumes of fluid [170]. Thus, dynamic molecular processes can be observed with high sensitivity and throughput efficiency by using integrated microfluidic systems with label-free detection methodologies such as SPR, electrical impedance, and optical-based approaches. These include fully automated systems based on the label-free sensor that are capable of detecting biomolecular interactions without the use of fluorescent or radioactive probes [35]. This is particularly the case for studies involving living cells or complex biomolecular interactions, where external labeling might interfere with normal behavior. Electrical biosensors, which include impedance or capacitance-based biosensors, can detect minimal changes in the physiological condition of a cell in high-throughput experiments. This, in turn, makes the integration of these technologies particularly effective for screening potential drug candidates or observing disease markers. The advantages of the combination of high-throughput and label-free methods are not confined to the question of data acquisition alone. In contrast, combined systems have specific benefits by providing qualitative and quantitative data within one experimental format. This aspect is particularly important in the context of drug discovery applications, as researchers often need to test large quantities of compounds with detailed information to identify cellular or molecular targets. For instance, when HTS platforms are combined with label-free imaging techniques, including holographic microscopy or phase contrast imaging, researchers can get cell morphology, behavior, and viability in real-time without photobleaching or toxic effects of fluorescent labels [171]. Other important advantages include the fact that dynamic interactions can be measured in real-time. For example, in label-free systems like SPR, biomolecular binding kinetics can be observed to investigate the speed of action of a drug on its target. Together with high-throughput robotic systems, these approaches can screen many samples simultaneously, which can be of great importance in the process of drug optimization and the time taken for discovery.

Several case studies explain the effectiveness of the integration of high-throughput techniques with label-free methodologies. For example, in cancer research, phase holographic imaging, a label-free technique, has been utilized in coordination with HTS platforms to observe the real-time effects of anticancer drugs on tumor cells. This can allow the observation of temporal effects of different drug concentrations on cancer cells without interference from dyes or labels that might skew the data. In this way, dynamic cellular behavior assessment capability shows higher resolution information on treatment response and also helps in identifying potential mechanisms of drug resistance.

Another important example is biomarker recognition and identification in biosensors for use in clinical diagnostics [172]. These biosensing methodologies without labels include QCM and electrical biosensors that are integrated with high-throughput robotic systems to screen large numbers of samples effectively. Such a platform possesses the capability for the detection of disease markers with high sensitivity and specificity, hence finding applications in early disease diagnosis or monitoring. Automation pushes the envelope of throughput even further, enabling hundreds or thousands of samples in a single experiment, where the capability to perform would be hard in terms of time by traditional methods. As technological development continues, the integration of high-throughput with label-free approaches is expected to reach further levels of sophistication and precision [173]. Methods in machine learning and AI are currently utilized to analyze large datasets generated by such systems, enabling insights into complex biological processes at higher resolutions. Moreover, device miniaturization and improvements in sensor sensitivity will enable the use of such platforms within an expanding diversity of uses, from personalized medicine to environmental monitoring. In general, this combination of high throughput with label-free methodologies provides an approach to scientific investigation that is extremely powerful and flexible. Thus, with the strengths of one complementary to the weaknesses of the other, new frontiers of rate and depth of insight into biological and chemical processes can be achieved. The combination is opening up new possibilities in drug development, diagnostics, and systems biology.

1.5 Applications in Life Sciences

The application of high-throughput and label-free methods has greatly impacted almost every aspect of the life sciences, from drug discovery and proteomics to genomics and clinical diagnostics [174]. These approaches allow faster, more sensitive, and higher-throughput analyses, which enhance research and further understanding of complex biological systems. Moreover, novel technologies like HSI and electrical detection further improve the resolution and sensitivity of these applications. Table 1.3 presents the summary of high-throughput technologies that are commonly used.

Table 1.3 Selected high-throughput technologies

	Nature of Analytes	Number of Analytes	Detected Abnormalities
Genomics			
Global genome sequencing	Genomic DNA	Unlimited	Mutations
CGH arrays	Sequences of genomic DNA	Up to 2 million probes	Copy number alterations
SNP arrays	Oligonucleotides	Up to 2 million probes	Polymorphisms, loss of heterozygosity, copy number alterations
Epigenomics			
DNA methylation arrays	Oligonucleotides covering whole genome or CpG islands +/− promoters	Up to 3,85,000 probes	DNA hyper- or hypo-methylation

	Nature of Analytes	Number of Analytes	Detected Abnormalities
Transcriptomics			
Gene expression arrays	Oligonucleotides or cDNA	Up to 3,85,000 probes	Modification of gene expression levels
miRNA expression arrays	Oligonucleotides	Up to 1,300 probes	Modification of miRNA levels
Proteomics			
Two-dimensional gel electrophoresis	Proteins, peptides	100–1,000 proteins	Modification of proteins relative abundance and activity
MALDI-TOF MS	Proteins, peptides	100–300 features	
Tandem MS	Peptide sequences	500–5,000 proteins	
Protein arrays	Lysates, proteins, peptides	Hundreds of probes	
Tissue microarrays	Tissue biopsies	Hundreds of tissue specimens	

1.5.1 Drug Discovery and Development

High-throughput and label-free analysis allow rapid screening of large libraries of chemical compounds, thereby playing a significant role in drug discovery and development. HTS platforms are commonly used for the identification and investigation of promising drug candidates that interact with specific biological targets. These platforms, coupled with label-free methods like mass spectrometry or SPR, allow researchers to investigate drug-target interactions without the need for external labels or tags that may interfere with the natural behavior of the molecule.

1.5.2 Screening of Potential Drug Candidates

HTS has been very useful in the identification of new drug candidates. This is due to the automation of testing large numbers of compounds, through which researchers can quickly evaluate the effects of those compounds on specific cellular targets or molecular pathways. The combination of phenotype assays with HTS allows the screening of chemical libraries and, therefore, offers more unbiased drug identification [175]. HTS is potentially utilized in cancer research, where the cell-based models include 3D tissue culture that screens the compounds with capabilities to target particular cellular behaviors, including growth inhibition, induction of apoptosis, and manipulation of cellular metabolism [8]. For instance, Lee et al. developed a 3D-cell spotter that improves 3D cell-based HTS (3D-HTS) of breast cancer responses to therapy. It separates drug responsiveness based on HER2 expression, discovers two novel non-EGFR-targeted effective medicines, and preserves analytic sensitivity in spite of increased resistance to medication [176].

1.5.3 Pharmacokinetics and Pharmacodynamics

Apart from preliminary screening, HTS platforms have the potential to study pharmacokinetics (PK) and pharmacodynamics (PD). These methods provide insights into the absorption, distribution, metabolism, and excretion of drugs in the body. Label-free methods, such as mass spectrometry, have been used in integration with HTS for the real-time study of drug interactions and metabolism, thus enabling the detection of metabolites and their potential toxic side effects [8, 34, 177]. Microfluidic systems allow for finer control of experimental conditions and, therefore, more detailed PK/PD studies [178].

1.5.4 Proteomics and Genomics

Proteomics and genomics depend significantly on high-throughput and label-free techniques to create large datasets for studying biological activities at the molecular level [174]. High-throughput sequencing has become a standard in genomics, enabling the quick sequencing of genomes, which is critical for personalized therapy, evolutionary studies, and disease research [179].

1.5.4.1 Protein identification and quantification

High-throughput methodologies have hugely benefited the study of proteins and proteomics [174]. Techniques like mass spectrometry can be utilized in a high-throughput manner to identify and quantify proteins in complex mixtures, providing information on cellular functions and pathways [180]. A combination of these techniques is notably useful in the understanding of disease mechanisms and biomarkers, identifying therapeutic targets [34]. By eliminating the need for fluorescent labeling, researchers can study protein interactions and post-translational modifications in their native states and obtain much more accurate and reliable results [181].

Proteomics is an approach that employs proteome examination and data processing to examine protein structure, expression, and interactions. It integrates genomes and transcriptomics by mapping sophisticated molecular mechanisms that regulate biological processes. Proteomics technologies have progressed from traditional to high-throughput procedures, boosting efficiency and the extent of coverage [182]. Notable high-throughput proteomic approaches comprise mass spectrometry, protein pathway arrays, tissue microarrays, and Luminex. Protein pathway array (PPA) is a high-throughput device for detecting antigens in specimens of proteins extracted from biopsy or tissue. It enhances protein percentage by microdissection, translates immunofluorescence cues to numerical data, and provides uses in disorders such as essential thrombocytopenia and papillary thyroid cancer [183].

1.5.4.2 DNA/RNA analysis

High-throughput sequencing has revolutionized the study of genomics by facilitating rapid and inexpensive sequencing of an entire genome. The new technology has been applied in many different types of studies, including the discovery of genetic variants that cause disease and the deciphering of complex regulatory networks [5]. In transcriptomics, RNA sequencing is among the powerful tools for studying patterns of gene expression and helps in understanding "how genes regulate in different conditions and during developmental stages." High-throughput sequencing, together with label-free approaches like CRISPR-based screening, allows the investigation of deeper functional genomics by observation of genetic functions and interactions [184]. Recent genome editing developments, consisting of ZFN, TALENs, and the CRISPR/Cas framework, have had a substantial influence on gene research in mammals. The CRISPR/Cas machinery enables adaptive protection against viruses and plasmids while simultaneously permitting straightforward programming and targeting of genomic loci. By applying this method, Zhou et al. developed a CRISPR/Cas-based lentiviral library for effectively identifying host genes required for anthrax and diphtheria toxin poisoning, as evidenced by functional testing. This robust genetic screening technique can help identify genes linked to bacterial toxicity as well as other biological operations [185].

Additionally, RNA interference allowed scientists to analyze the function of genes more efficiently, and high-throughput RNA interference (HT-RNAi) screening has culminated in the finding of physiologically significant mediators of cell function. Henderson et al. developed a strategy that screens HeLa cells injected with short interfering RNA pools in 384-well microplates to identify particular gene regulators and drug development leads [186].

1.5.5 Clinical Diagnosis

High-throughput and label-free approaches are increasingly being used in clinical diagnostics to improve diagnosis and treatment. These technologies enable the simultaneous detection of many biomarkers, offering a full assessment of the health condition of a patient.

1.5.5.1 Biomarker discovery

Biomarkers are the biological molecules that reflect a certain disease condition. The identification of biomarkers has increased considerably as a result of high-throughput and label-free techniques [187]. For instance, high-content screening using imaging approaches to monitor cell changes linked with certain disease states, as well as mass spectrometry, which allows researchers to investigate protein and metabolite expression and uncover disease-specific profiles [188]. In addition to diagnosis, these technologies enable disease progression and therapy response monitoring, making them personalized medicine tools. For instance, Wasuworawong et al. investigated metastatic protein markers in cholangiocarcinoma, employing protein expression profiles on the KKU-M213 and KKU-100 cell types. They hypothesized that S100A2 may represent a critically important protein implicated in the development of cholangiocarcinoma, as well as an intriguing biomarker and therapy candidate [189].

1.5.5.2 Disease diagnosis and monitoring

In the field of clinical diagnostics, high-throughput techniques and label-free approaches have resulted in significant increases in disease diagnosis accuracy and speed. Electrical biosensors that detect changes in electric properties during biomolecular interactions have found applications in finding biomarkers for diseases such as cancer and cardiovascular disease [190]. Many of these sensors are linked to microfluidic devices, which can examine minute amounts of biological material to facilitate rapid and cost-effective point-of-care diagnosis [191].

Specifically, DNA sequences help in disease diagnosis; DNA sequences can be employed as barcodes for HTS, particularly to store binary data. Such barcodes are applicable to single-cell sequencing, pharmaceutical screening, proteomic assessment, and cancer research. For example, single-cell sequencing is a technique that allows for concurrent DNA sequencing from multiple cells at once, decreasing manual labor, duration, and costs. Gel beads are labeled with distinctive DNA sequences encapsulated in droplets, and reverse-transcripted mRNA is used to make cDNA. This HTS technique allows for the investigation of specific cell properties as well as the identification of putative gene-regulating linkages [192, 193].

1.5.6 Role of HSI and Electrical Detection in Life Sciences

HSI and electrical detection methodologies have emerged as powerful tools within the life sciences. HSI offers the advantage of both imaging and spectroscopy, recording a large range of light wavelengths at every pixel, enabling comprehensive investigation into biological tissues without labeling. This ability is being used in a variety of applications, such as differentiating healthy from diseased tissues and analyzing the chemical composition of cells. In cancer studies, HSI has been used to delineate tumor tissues by their unique spectral signature that increases diagnostic precision in imaging [194]. On the other hand, electrical detection methodologies are based on the assessment of electrical properties as a way to identify molecular interactions [195]. These sensors possess high sensitivity, enabling them to detect minimal changes in the microenvironment, even those caused by the binding of an antibody to an antigen. Integration into high-throughput systems provides for the ability of high-speed examination of large data arrays and thus makes the latter suitable for drug screening and pathogen detection. Among the methods of electrical detection, biosensing has proved particularly effective, as it provides speedy and reliable results, which is vital in real-time diagnosis [31].

1.6 Applications in Organoids and Tissue Engineering

Organoids and tissue engineering studies have recently made great advances, especially with the use of high-throughput and label-free analytical techniques [196]. Indeed, this has allowed for increased throughput in disease modeling, drug testing, and regenerative medicine development [197]. Organoids are 3D miniature representations of organs, while biologically engineered tissue constructs have created new paths toward the study of biological mechanisms by structurally and functionally mimicking real organs in a controlled laboratory setting.

1.6.1 Organoids

Organoids are 3D-cell cultures that self-organize to mimic the structure and function of real organs. These microtissues are mostly derived from stem cells and replicate complex cell behaviors and interactions, making them highly valuable for studying organ development, disease mechanisms, and drug responses. Over the past few years, organoids have become a versatile model system for various tissues, including the brain, liver, kidney, and intestine. Their ability to mimic in vivo–like environments helps to connect the gap between traditional 2D cultures and animal models, which can eventually offer more physiologically relevant data [196].

One of the most promising applications of organoids is in disease modeling to mimic human-specific disease conditions, such as cystic fibrosis, cancer, or other diseases. By deriving organoids from patient-specific cells, researchers can study disease progression in a personalized manner [198]. For instance, patient-derived tumor organoids have been utilized to predict drug responses, which eventually help in personalized therapy approaches. This was further demonstrated in studies that employed patient-derived organoids of gastrointestinal cancers to test drug sensitivities, leading to predictive results that could be used to improve personalized therapy. In addition, organoids have been used to model genetic abnormalities, infectious diseases, and neurodegenerative disorders, thus providing a suitable, physiologically relevant platform for testing therapeutic interventions and performing HTS of drugs [196, 197].

HTS now supports various research, investigating the therapeutic potential of large libraries of compounds with the help of this organoid technology [196]. Automation of the culture and maintenance of organoids and their analysis gives the potential to HTS platforms for quickly testing the action of drug candidates on cellular activity, which is very important and provides great insight into both the efficacy and toxicity of the compounds (Fig. 1.6).

Techniques that do not require labeling (label-free), such as electrical impedance sensing and optical metabolic imaging, allow researchers to track changes in cell behavior while leaving the organoids in their natural state, which leads to more accurate and reliable data. In addition, automated systems have improved the ability to scale these experiments, which allows hundreds of organoids to run in parallel assays [199].

Additionally, applications of HTS with organoids have been numerous. For instance, arrayed colonic organoids have been used to screen for secretagogues and toxins, an application that shows how well organoids can model not only organ-specific functions but also drug responses. Other examples of drug testing, such as effective drug combinations and dosing, are determined by a systematized process, which includes the culture of pancreatic cancer organoids on microfluidic platforms [200]. The advancement in the field of drug discovery and personalized medicine is dependent on the ability of organoid systems to integrate with high-throughput platforms [175].

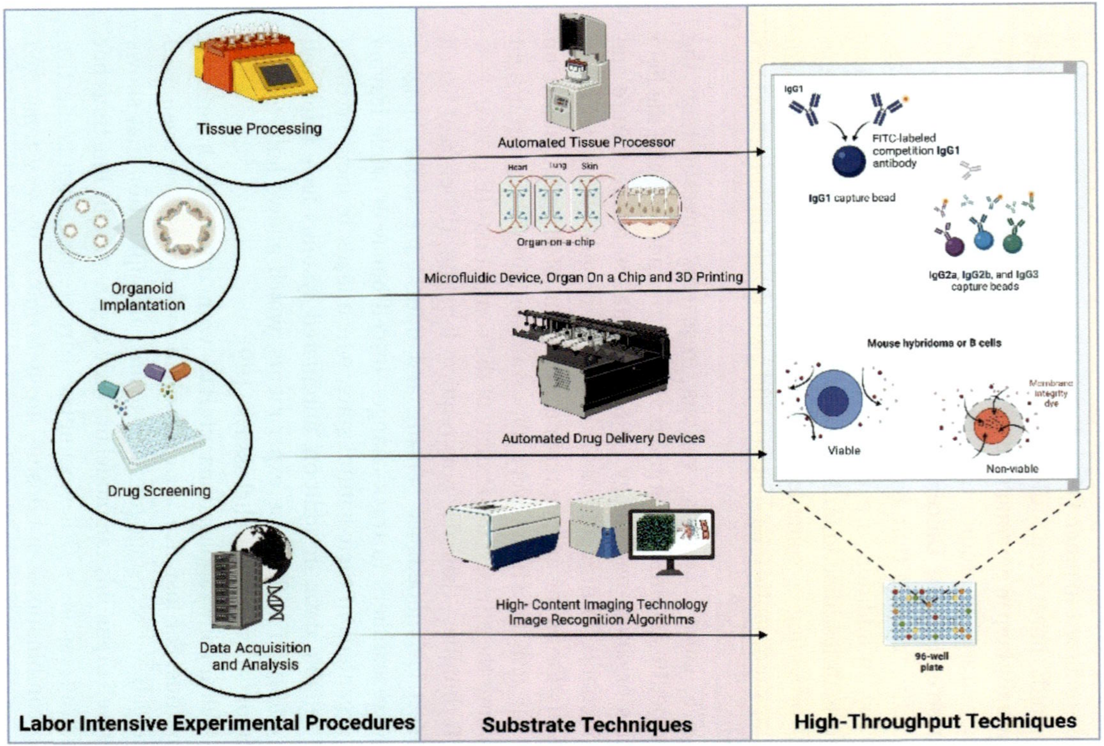

Figure 1.6 High-throughput and label-free analysis in organoid research.

1.6.2 Tissue Engineering

Tissue engineering focuses on creating functional tissue constructs by combining cells, biomaterials, and biologically active molecules, mainly for the repair or replacement of damaged tissues. Unlike traditional organoids, which form spontaneously, tissue-engineered organs or constructs are designed to promote specific cell–cell behaviors and tissue structures. These bioengineered tissues can be used to replace damaged organs, test drug efficacy, or study cell–cell and tissue behavior under various disease conditions. This flexibility in designing the biological tissue scaffolds allows for controlled cell locus and arrangement, which is important and essential in applications ranging from simple wound healing to complex organ repair [179].

In the field of regenerative medicine, tissue engineering offers great potential for the repair or replacement of damaged tissues and diseased organs. Due to recent advances in scaffold fabrication, stem cell approaches, and bioprinting, bioengineered tissues can successfully mimic the in vivo functions of native organs [181]. For example, bioengineered skin grafts, cartilage, and heart valves have been engineered for use in tissue damage, shortages, or dysfunctional organs. Furthermore, engineered tissue platforms are used to investigate cell-matrix interactions that provide valuable information regarding the mechanisms of wound healing and tissue regeneration.

High-throughput approaches have also been applied to the testing of tissue-engineered scaffolds. These techniques help in the rapid and simultaneous screening of multiple biomaterials for their suitability in supporting cell–cell proliferation and tissue development. Label-free technologies, including SPR and acoustic sensors, can detect changes in the mechanical properties of scaffolds and also identify cell attachment and growth on these matrices [201]. This methodology provides direct feedback for the improvement in scaffold design by improving tissue integration. In a particular investigation, advanced high-throughput platforms were employed to evaluate biomaterials that facilitate the differentiation of neural stem cells, thereby demonstrating the potential of these methodologies to accelerate the creation of functional scaffolds for nerve regeneration [202]. Furthermore, the integration of 3D bioprinting methods with automated high-throughput systems has enabled the fabrication of intricate tissue architectures for applications in regenerative medicine, including cartilage and liver tissues, thus putting the groundwork for prospective therapeutic interventions. Specifically, HSI has proven to be particularly useful as a key methodology in the study of organoids and tissue-engineered constructs. Unlike the more traditional approaches to imaging, HSI captures detailed spectral information across a range of wavelengths, thus allowing researchers to assess not only the external but also the internal features of samples under investigation. This is particularly useful in differentiating cell types, identifying cell heterogeneity, and detecting metabolic changes [197]. In tissue engineering, HSI has been used to assess the integration of scaffolds and the maturation of tissues, providing a noninvasive way to track the course of tissue development.

Additionally, electrical detection techniques that involve impedance sensing and electrical impedance spectroscopy are important to monitor cellular growth, differentiation, and tissue maturation. The applied methodologies are able to detect indirect changes in the electrical features associated with cell functions such as cell–cell proliferation and cell apoptosis. Coupling these sensors with high-throughput systems enables the observation, in real-time, of tissue reactions against a wide range of stimulations, providing a complete view of tissue dynamics [203].

1.7 Applications in Materials Science

The development of high-throughput and label-free analytical techniques has been one of the major factors driving the field of materials science. It enables faster and more profound research on materials across various stages of their development. It can rapidly screen and analyze a material at micro- and nanoscale dimensions, which has become necessary in industries ranging from electronics to biomaterials [204]. Recent progress in this field has allowed detailed surface characterization, quality control, and nanomaterial characterization through the combination of high-throughput methodologies with various advanced imaging and sensor technologies.

1.7.1 Characterization of Nanomaterials

Nanomaterials have unique properties like high surface area-to-volume ratios and smaller size, making them useful for several applications in catalysis, electronics, and drug delivery systems. High-throughput methodologies are thus important for efficiently characterizing these materials, especially in the research of large numbers of compositions and structures [205]. Techniques such as secondary ion mass spectrometry (SIMS), X-ray photoelectron spectroscopy (XPS), and AFM are commonly used together to evaluate surface chemistry, topography, and nanoscale molecular interactions [203]. These techniques can generate real-time data that can be utilized to discover the best compositions to improve catalytic effects, mechanical properties, and suitable structures to effectively deliver the drug. HSI is increasingly being used in materials science due to its ability to capture detailed spectral information across multiple wavelengths. Unlike traditional imaging, which might only show visible light, HSI can detect sensitive variations in material composition, allowing for the identification of contaminants, defects, or compositional differences. It has been used in the analysis of film and coating uniformity, an issue highly important in the electronics industry. Its ability to detect fine changes in the ripeness or spoilage of food makes it an ideal analysis tool for food products [206]. To study electrical properties in the material, impedance spectroscopy has become an important electrical detection technique. These techniques provide information about conductivity, capacitance, and other resistance properties of the materials which are crucial aspects in the development of sensory techniques and batteries. For the characterization of the materials, high-throughput platforms also help in the integration of electrical measurements. The application of CMOS-based microelectrode arrays has enabled the simultaneous consideration of multiple parameters, thus contributing to the development of advanced materials for energy storage and electronic applications [207, 208].

The integration of high-throughput and label-free analysis is opening new avenues in materials science. Machine learning and data analysis let this huge volume of data generated by these techniques be processed and, as a consequence, disclose a number of patterns and correlations that would otherwise be hard to find with human effort. Miniaturization of sensor technologies enables measurements with much higher resolution and localization, which is decisive for the development of next-generation materials [207]. Subsequent investigations are anticipated to prioritize the improvement of the sensitivity and rapidity of these methodologies, further decreasing expenses and broadening their relevance across various industrial domains.

1.7.2 Surface and Interface Analysis

Surface attributes and the surrounding environment together determine the performance and functionality of a material. HTS can help in the rapid testing of different surface functionalization moieties, coatings, and other surface modifications for improved interactions [209]. Label-free techniques, such as SIMS and contact angle goniometry, allow researchers to study the effect of surface modifications on adhesion and chemical reactivity without requiring any dyes or surface markers [210]. Such approaches are essential in coatings, adhesives, and biomedical implant applications, where the surface behavior directly influences product success.

1.7.3 Quality Control in Manufacturing Processes

Manufacturing entails a major aspect of quality control, where products are made to meet specified standards. The use of high-throughput systems enables the analysis of multiple parameters simultaneously, hence reducing the time needed for quality assessment. Automated systems fitted with sensors monitor the production lines continuously to ensure consistent quality from batch to batch. Examples are the use of HSI to pinpoint color, texture, and compositional defects, while electrical impedance sensors do the monitoring of structural integrity [203]. These technologies provide untold value in those industries that demand high precision, including semiconductor manufacturing and food processing.

1.8 Technological Advancements

Recent advances in high-throughput technologies, methods of label-free detection, HSI, and electrical detection have significantly driven progress in different areas of research and diagnostics. All of these new approaches enhance the speed, accuracy, and flexibility of analytical methodologies, enabling deeper insight into complex systems and creating the basis for further applications in many fields [207].

High-throughput technologies have continued to improve. The newest modernizations in the field aim further at increasing automation, reducing sample sizes, and further integrating with other systems of analytics. Of interest is the development of microfluidic platforms that allow high-speed processing of biological and chemical samples. These microfluidic devices are capable of carrying out multiple analyses in parallel—a development enhancing faster and more comprehensive analysis [211]. Fundamentally, the application of machine learning and artificial intelligence algorithms has greatly enhanced the ability to analyze the massive datasets generated from HTS, hence making the process more efficient and accurate. The application of droplet microfluidics in the field of single-cell analysis has also emerged as a powerful tool for high-throughput applications [212]. This approach allows for the separation of individual cells into microscopic droplets, each of which can then be analyzed separately for an extended analysis of cellular heterogeneity. It has been applied so far with great success in the fields of drug development, genomics, and proteomics, where novel insights are gained into cellular behavior and molecular interactions at an unprecedented resolution.

In recent years, label-free detection has seen phenomenal development, particularly in the field of electrochemical and optical-based sensors. Recent innovations have focused on sensitivity, miniaturization, and portability, which are essential for point-of-care diagnostics. Electrochemical sensors, for example, have reached a high level of sophistication, using nanomaterials for the enhancement of sensitivity and to detect biomarkers of low concentration [68]. Various methods, therefore, have been developed for performing such analyses without the use of fluorescent or radioactive labeling; these include differential pulse voltammetry, cyclic voltammetry, and electrochemiluminescence. Moreover, FET biosensors have also attracted consideration based on their capability in biomolecular interaction detection by electrical changes. Recent developments have involved graphene and various nanomaterials that enhance the sensitivity of the FET devices and enable trace amounts of analytes, including cancer biomarkers, to be detected in clinical samples [68]. Such developments help in the creation of portable and label-free diagnostic tools suitable for on-site and point-of-care applications.

From the very first applications of HSI in remote sensing, to biomedical imaging and material science, and up to food safety, it has come a long way. It has become indispensable in all the mentioned fields. The technique provides deep insight into the chemical composition of the samples and thus allows for the correct identification of all substances present in a sample, label-free [149]. It has been applied in the biomedical field to distinguish between healthy and diseased tissues based on their spectral signatures, hence finding its place as a valuable tool in noninvasive diagnostics. Furthermore, high-speed HSI systems have been developed by researchers integrated with automated high-throughput platforms for screening biological samples rapidly for different applications such as the detection of cancer and the analysis of tissue [150].

Most of the developments pertaining to electrical detection techniques have to do with integrating microelectronic devices with biological or chemical sensors. Other innovations in this respect include the use of nanomaterial-based electrodes and improvements in the sensitivity and specificity of electrical sensors [151]. More recently, through this development, it has become possible to produce a device that integrates electrochemical impedance spectroscopy with microfluidics, enabling fast label-free detection of target molecules in complex samples. These systems have applications in the identification of cancerous cells, infectious agents, and environmental pollutants, with very promising use in point-of-care diagnostics. For example, the latest developments in FET-based sensors have realized the development of biosensors for detecting variations in electrical signals due to biomolecular interactions [152]. Such sensors are extremely sensitive and capable of detecting low-abundance analytes—suitable for the early diagnosis of diseases, cancer being one of them. Coupling such sensors with high-throughput formats has enabled the rapid screening of large numbers of samples and, consequently, comprehensive studies at advanced levels of precision.

The domain of high-throughput and label-free technologies keeps evolving, driven by increasing demands for more efficient, precise, and versatile analytical tools. Among the most recent trends is the integration of artificial intelligence and machine learning within advanced high-throughput technologies for better data analytical capability. This synthesis allows for even more proficient pattern recognition, a crucial component that controls the management of large datasets that emanate from HTS.

One major trend for the future is toward small, truly portable instruments that can be used outside traditional laboratory settings. Such instruments are likely to revolutionize point-of-care medicine, environmental monitoring, and food testing by offering both speed and reliable results right at the point of need. These would probably remain the focus of further research in increasing the sensitivity and specificity of the devices while trying to bring down their cost and complexity to extend their availability for various applications.

1.9 Challenges and Solutions

Although high-throughput and label-free analytical methods have gained wide acceptance in life sciences, materials science, and diagnostics, there are challenges associated with this methodology. Such challenges must be addressed to ensure that their implementation at different scales is reliable, reproducible, and effective. The subsequent sections describe the main challenges and suggest some solutions to technical phases, data organization and management, standardization, and some issues related to HSI and electrical detection.

1.9.1 Challenges in HTS

High-throughput methods that do not require labeling need modern instrumentation and regulated experimental conditions. One of the major challenges from a technical point of view is maintaining sensitivity with high accuracy while processing large amounts of data. For example, in the case of HTS, microfluidic systems implemented need to ensure consistent fluid flow and precise manipulation of reagents, which becomes challenging at such a small scale [153]. Moreover, the surrounding environmental conditions should not affect the stability of biological samples in label-free detection. The integration of multiple sensing systems without any interference, like the combination of sensory and electrical detections, is the main technical challenge. Such integrated platforms must be carefully calibrated to ensure that signals from one approach do not mask or otherwise modify signals from another.

With higher throughput, the quantity of data produced by such systems can easily become overwhelming. Effective handling of such data is important to draw meaningful insights. However, challenges encompass batch effects, data heterogeneity, and the reproducibility of results. Data storage solutions have to handle the volume while supporting efficient retrieval and processing. These are some of the challenges that have necessitated the development of sophisticated software instruments. New-generation HSI system helps in data analysis with machine learning algorithms allowing users to identify patterns and differences that are not easily detected with traditional approaches. However, it is important that such tools should be carefully validated in order to avoid the risk of biases being introduced during data interpretation.

The area of reproducibility is one of the major challenges in the standardization of high-throughput and label-free analysis. Differences in instrument calibration, sample preparation, and surrounding conditions can easily lead to differences in results across different laboratories or even in individual experiments run under the same institute. For example, high-content imaging can be very sensitive to minute changes in light intensity, leading to variations in data that cannot be validated for reproducibility [154]. As a result of this, efforts have been made to develop guidelines on sample preparation, data acquisition, and analysis methodology. There has also been a development of several organizations to establish general practices that can universally reduce variability. In addition, well-thought-out quality control processes must be developed, which include the use of reference standards and periodic calibration of instruments to ensure repeatability. Both HSI and electrical detection techniques face separate challenges unique to their respective methodologies. For example, comprehensive spectral data requires sophisticated instrumentation to be acquired in the case of HSI. In addition, it is very challenging to maintain uniform imaging conditions within different ranges of wavelengths. The quality of the data acquired could be negatively affected by light-scattering effects, sample autofluorescence, and other complications of image processing. Moreover, the high resolution of HSI results in the generation of huge data output, which is difficult to store economically and analyze effectively.

On the other hand, specificity is the main issue that needs to be overcome by electrical detection. Various techniques, such as field-effect transistors, have significantly improved the capability of detecting low concentrations of analytes, but they cannot accurately maintain signal integrity in complex biological matrices. Various false-positive results can be obtained due to nonspecific binding or noise, hence, stringent calibration processes should be used along with the use of blocking agents to resolve the problem.

1.9.2 Proposed Solutions and Best Practices

There are several strategies that can be used to improve automation and integration like full automation of the sample handling, data acquisition, and analysis to decrease the differences and increase throughput. Careful design and calibration of these systems are required to minimize cross-interference. The application of machine learning and AI in handling large amount of data generated by high-throughput and label-free techniques has proven effective. AI can detect minute patterns in data that are crucial for early-stage diagnosis or pharmaceutical screening. Moreover, the accuracy and reliability should be tested against the already available standards. To enhance reproducibility, standards should range from sample preparation to data analysis to make the results comparable across various platforms and research laboratories. International collaborations may also contribute to the development and adoption of such protocols for easier replication in and out of various countries globally. Specifically for HSI, the development of better optical filtering and data processing algorithms can go a long way toward alleviating problems with both autofluorescence and light scattering. In electrical detection, the development of new nanomaterials with much better conductivity and lower noise properties could make signal resolution better [155]. Furthermore, regular calibration and control samples would be important in maintaining the accuracy of the methods.

1.10 Conclusion

High-throughput analyses are critical tools that provide effective, noninvasive ways for testing biological and chemical materials in a rapid mode and inexpensive manner. Recent developments in sensing technology, especially HSI, have improved diagnostic potential as they distinguish tumor tissues by comparing distinct spectral signatures of different tissues. For instance, the high-throughput mass spectroscopy technique is specifically used in drug discovery.

The usage of automation and robotics raises efficiency and uniformity in experimental circumstances, saving time and money over human screening, and additionally boosts the safety of laboratories through decreasing human encounters with harmful compounds. Additionally, machine learning and artificial intelligence methods coupled with mass spectroscopy, AFM, and multi-omics techniques aid in evaluating enormous datasets produced through such systems to get an understanding of complicated biological events at a greater precision. The combination of high-throughput and label-free methods offers unprecedented research opportunities but, at the same time, creates technical challenges. The challenges include preserving sensitivity and precision and dealing with massive volumes of information. Data storage systems must be able to manage large amounts of information while also providing efficient recovery. Standardization is critical owing to variations in calibrating instruments, preparing samples, and circumstances. HSI and electrical detection systems provide distinct obstacles, including rigorous calibration procedures and blocking compounds. These can be overcome by careful design, standardization, and use of advanced data analysis tools; these are crucial for the continued effectiveness and adoption of such methods. Automated methods, machine learning, artificial intelligence, and multinational cooperation can help with the initial stages of diagnostics and pharmacological screening. Advanced optical filtering and nanomaterials for HSI increase signal image quality, while frequent calibration and control ensure process correctness. The scientific community needs to be open to best practices and interdisciplinarity in overcoming such challenges for better and more effective research outcomes. With the further development of HTS and a label-free approach in the near future, it will become the mainstream tool in the field of clinical diagnosis and drug discovery.

References

1. Carulli, J. P., et al., *High throughput analysis of differential gene expression.* J Cell Biochem Suppl, 1998. **30–31**: p. 286–96.
2. Proll, G., et al., *Potential of label-free detection in high-content-screening applications.* J Chromatogr. 2007. **1161**(1–2): p. 2–8.
3. Zhou, S., et al., *Computer vision meets microfluidics: A label-free method for high-throughput cell analysis.* Microsyst Nanoeng, 2023. **9**(1): p. 116.
4. Fraikin, J. L., et al., *A high-throughput label-free nanoparticle analyser.* Nat Nanotechnol, 2011. **6**(5): p. 308–13.
5. Schie, I. W., et al., *High-throughput screening Raman spectroscopy platform for label-free cellomics.* Anal Chem, 2018. **90**(3): p. 2023–30.
6. Ozkumur, E., et al., *Label-free and dynamic detection of biomolecular interactions for high-throughput microarray applications.* Proc Natl Acad Sci U S A, 2008. **105**(23): p. 7988–92.
7. Schmidt, B. and A. Hildebrandt, *Next-generation sequencing: Big data meets high performance computing.* Drug Discov Today, 2017. **22**(4): p. 712–7.
8. Blasi, T., et al., *Label-free cell cycle analysis for high-throughput imaging flow cytometry.* Nat Commun, 2016. **7**: p. 10256.
9. Wildey, M. J., et al., *High-throughput screening,* in Annual Reports in Medicinal Chemistry. Goodnow R.A., editor. Academic Press; Cambridge, MA, USA: 2017. pp. 149–95.
10. Bubendorf, L., *High-throughput microarray technologies: From genomics to clinics.* Eur Urol, 2001. **40**(2): p. 231–8.
11. Manjasetty, B. A., et al., *A high-throughput approach to protein structure analysis.* Genet Eng (N Y), 2007. **28**: p. 105–28.
12. Cooper, M. A., *Label-free screening of bio-molecular interactions.* Anal Bioanal Chem, 2003. **377**(5): p. 834–42.
13. Guenther, P., et al., *Potential of label-free detection in high-content-screening applications.* Journal of chromatography. A, 2007. **1161**(1–2): p. 2–8.
14. Lausted, C., et al., *SPR imaging for high throughput, label-free interaction analysis.* Comb Chem High Throughput Screen, 2009. **12**(8): p. 741–51.

15. Wong, J. W. and G. Cagney, *An overview of label-free quantitation methods in proteomics by mass spectrometry.* Methods Mol Biol, 2010. **604**: p. 273–83.
16. Miccio, L., et al., *Perspectives on liquid biopsy for label-free detection of "circulating tumor cells" through intelligent lab-on-chips.* VIEW, 2020. **1**(3): p. 20200034.
17. Chen, Y., et al., *Label-free microfluidics for single-cell analysis.* Microchem J, 2022. **177**: p. 107284.
18. Lu, N., et al., *Label-free microfluidic cell sorting and detection for rapid blood analysis.* Lab on a Chip, 2023. **23**(5): p. 1226–57.
19. Schasfoort, R. B. M., et al., *Trends in SPR cytometry: Advances in label-free detection of cell parameters.* Biosensors (Basel), 2018. **8**(4).
20. Chung-Hang, L., et al., *Application of label-free techniques in microfluidic for biomolecules detection and circulating tumor cells analysis.* TrAC Trends Anal Chem, 2019.
21. Hu, P., et al., *Single cell isolation and analysis.* Front Cell Dev Biol, 2016. **4**: p. 116.
22. Pirone, D., et al., *Beyond fluorescence: Advances in computational label-free full specificity in 3D quantitative phase microscopy.* Curr Opin Biotechnol, 2024. **85**: p. 103054.
23. Dueñas, M. E., et al., *Advances in high-throughput mass spectrometry in drug discovery.* EMBO Mol Med, 2023. **15**(1): p. e14850.
24. Chieng, A., Z. Wan, and S. Wang, *Recent advances in real-time label-free detection of small molecules.* Biosensors (Basel), 2024. **14**(2).
25. Tang, J.-W., et al., *Label-free surface-enhanced Raman spectroscopy coupled with machine learning algorithms in pathogenic microbial identification: Current trends, challenges, and perspectives.* Interdiscip Med, 2024. **2**(3): p. e20230060.
26. Ibrahim, K. A., et al., *Label-free techniques for probing biomolecular condensates.* ACS Nano, 2024. **18**(16): p. 10738–57.
27. C, K., et al., *Emerging technology: Hyperspectral imaging.* Perspect Vasc Surg, 2004. **16**.
28. Elmasry, G., et al., *Principles and applications of hyperspectral imaging in quality evaluation of agro-food products: A review.* Crit Rev Food Sci Nutr, 2012. **52**(11): p. 999–1023.
29. Park, B. and R. Lu, Hyperspectral Imaging Technology in Food and Agriculture. 2015. 1–390.

30. Çetin, N., et al., *Using hyperspectral imaging technology and machine learning algorithms for assessing internal quality parameters of apple fruits.* Chemom Intell Lab Syst, 2022. **230**: p. 104650.

31. Bhunia, A., M. S. Kim, and C. Taitt, High Throughput Screening for Food Safety Assessment: Biosensor Technologies, Hyperspectral Imaging and Practical Applications. Woodhead Publishing Series in Food Science, Technology and Nutrition, 2014: p. 1–523.

32. Fox, S., *A history of high-throughput screening for drug discovery: A special report summarizing six comprehensive industry studies in the years 1998–2005.* HighTech Business Decisions, Moraga, CA, USA, 2006.

33. Soon, W. W., M. Hariharan, and M. P. Snyder, *High-throughput sequencing for biology and medicine.* Mol Syst Biol, 2013. **9**: p. 640.

34. Li, L., et al., *High-throughput imaging: Focusing in on drug discovery in 3D.* Methods, 2016. **96**: p. 97–102.

35. Haselgrübler, T., et al., *High-throughput, multiparameter analysis of single cells.* Anal Bioanal Chem, 2014. **406**(14): p. 3279–96.

36. Lee, W. E., et al., *High-throughput measure of mitochondrial superoxide levels as a marker of coronary artery disease to accelerate drug translation in patient-derived endothelial cells using Opera Phenix® technology.* Int J Mol Sci, 2023. **25**(1): p. 22.

37. Panusatid, C., N. Thangsiriskul, and C. Peerapittayamongkol, *Methods for mitochondrial health assessment by high content imaging system.* MethodsX, 2022. **9**: p. 101685.

38. Yong, J., et al., *Development of a high-content imaging assay for screening compound aggregation.* Anal Biochem, 2018. **559**: p. 30–3.

39. Kawase, K., et al., *Prolyl hydroxylase domain inhibitors prevent kidney crystal formation by suppressing inflammation.* Urolithiasis, 2025. **53**(1): p. 1–11.

40. Houl, J. H., et al., *Role of the ETV5/p38 signaling axis in aggressive thyroid cancer cells.* bioRxiv, 2025: p. 2025.02. 17.637322.

41. Welte, T., et al., *Gain-of-function chromatin remodeling activity of oncogenic FOXL2C134W reprograms glucocorticoid receptor occupancy to drive granulosa cell tumors.* Cancer Res, 2025. **85**(5): p. 875–93.

42. Groo, A.-C., et al., *Evidence from a mouse model supports repurposing an anti-asthmatic drug, bambuterol, against Alzheimer's disease by administration through an intranasal route.* Commun Biol, 2025. **8**(1): p. 155.

43. Welte, T., et al., *Gain-of-function chromatin remodeling activity of oncogenic FOXL2C134W reprograms glucocorticoid receptor occupancy to drive granulosa cell tumors.* Cancer Res, 2025. **85**(5): p. 875–93.
44. Huang, C.-J., et al., *Pterostilbene suppressed cell viability, induced apoptosis and autophagy of cisplatin-resistant gastric cancer cells.* Anticancer Res, 2025. **45**(2): p. 511–23.
45. Durens, M., et al., *High-throughput screening of human induced pluripotent stem cell-derived brain organoids.* J Neurosci Methods, 2020. **335**: p. 108627.
46. Liu, R., et al., *Melphalan induces cardiotoxicity through oxidative stress in cardiomyocytes derived from human induced pluripotent stem cells.* Stem Cell Res Ther, 2020. **11**: p. 1–17.
47. Zerafati-Jahromi, G., et al., *Sequence variants in HECTD1 result in a variable neurodevelopmental disorder.* Am J Hum Genet, 2025.
48. Battich, N., T. Stoeger, and L. Pelkmans, *Image-based transcriptomics in thousands of single human cells at single-molecule resolution.* Nat Methods, 2013. **10**(11): p. 1127–33.
49. Frechin, M., et al., *Cell-intrinsic adaptation of lipid composition to local crowding drives social behaviour.* Nature, 2015. **523**(7558): p. 88–91.
50. Dariolli, R., et al., *Optimized scaffold-free human 3D adipose tissue organoid culture for obesity and disease modeling.* SLAS Discov, 2025: p. 100218.
51. Sanz-Martinez, P., R. Berkane, and A. Stolz, *Function of CSNK2/CK2 selectively affects the endoplasmic reticulum and the Golgi apparatus in mtor-mediated autophagy induction.* Autophagy, 2025. **21**(2): p. 480–6.
52. Goethals, O., et al., *Identification of a broad-spectrum flavivirus inhibitor targeting NS2A, a previously unidentified target.* 2025.
53. Sandu, C., et al., *Thiostrepton interacts covalently with Rpt subunits of the 19S proteasome and proteasome substrates.* J Cell Mol Med, 2015. **19**(9): p. 2181–92.
54. Mohamed, Z.A.M.A., et al., High Content Screening: Understanding Cellular Pathway.
55. Samoto, M., et al., *Novel bone microenvironment model of castration-resistant prostate cancer with chitosan fiber matrix and osteoblasts.* Oncol Lett, 2021. **22**(4): p. 689.
56. Yang, S., et al., *Control of cell geometry through infrared laser assisted micropatterning.* JoVE (Journal of Visualized Experiments), 2021(173): p. e62492.

57. Johnston, S. T., et al., *Estimating cell diffusivity and cell proliferation rate by interpreting IncuCyte ZOOM™ assay data using the Fisher-Kolmogorov model.* BMC systems biology, 2015. **9**: p. 1–13.

58. Churko, J. M., et al., *Overview of high throughput sequencing technologies to elucidate molecular pathways in cardiovascular diseases.* Circ Res, 2013. **112**(12): p. 1613–23.

59. Moukarzel, G., et al., *Automation of biochemical assays using an open-sourced, inexpensive robotic liquid handler.* SLAS Technol, 2024. **29**(6): p. 100205.

60. Earley, J. J., et al., *Robotic automation of dideoxyribonucleotide sequencing reactions.* Biotechniques, 1994. **17**(1): p. 156–8, 160, 162–5.

61. Socea, J. N., et al., *Implementing laboratory automation for next-generation sequencing: benefits and challenges for library preparation.* Front Public Health, 2023. **11**: p. 1195581.

62. Keeney, S., *Use of robotics in high-throughput DNA sequencing.* Methods Mol Biol, 2011. **688**: p. 227–37.

63. Łącki, K.M., *High-throughput process development of chromatography steps: advantages and limitations of different formats used.* Biotechnol J, 2012. **7**(10): p. 1192–202.

64. Soleilhac, E., R. Nadon, and L. Lafanechere, *High-content screening for the discovery of pharmacological compounds: advantages, challenges and potential benefits of recent technological developments.* Expert Opin Drug Discov, 2010. **5**(2): p. 135–44.

65. Kircher, M. and J. Kelso, *High-throughput DNA sequencing–concepts and limitations.* Bioessays, 2010. **32**(6): p. 524–36.

66. Zeng, W., et al., *High-throughput screening technology in industrial biotechnology.* Trends Biotechnol, 2020. **38**(8): p. 888–906.

67. Lee, M. Y., et al., *Three-dimensional cellular microarray for high-throughput toxicology assays.* Proc Natl Acad Sci U S A, 2008. **105**(1): p. 59–63.

68. Yung-Shin, S., *Use of microarrays as a high-throughput platform for label-free biosensing.* SLAS Technol, 2015. **20**(4): p. 334–53.

69. Al Kuraya, K., R. Simon, and G. Sauter, *Tissue microarrays for high-throughput molecular pathology.* Ann Saudi Med, 2004. **24**(3): p. 169–74.

70. Wingren, C. and C. A. Borrebaeck, *High-throughput proteomics using antibody microarrays.* Expert Rev Proteomics, 2004. **1**(3): p. 355–64.

71. Zhou, Y., et al., *Microarray chip-based high-throughput screening of neurofilament light chain self-assembling peptide for noninvasive monitoring of Alzheimer's disease.* ACS Nano, 2024. **18**(28): p. 18160–75.

72. Schmitz, M. G. J., et al., *Merging modular molecular design with high throughput screening of cell adhesion on antimicrobial supramolecular biomaterials.* Macromol Rapid Commun, 2024: p. e2300638.

73. Fernandes, T. G., et al., *High-throughput cellular microarray platforms: Applications in drug discovery, toxicology and stem cell research.* Trends Biotechnol, 2009. **27**(6): p. 342–9.

74. Lee, M. Y., et al., *Metabolizing enzyme toxicology assay chip (MetaChip) for high-throughput microscale toxicity analyses.* Proc Natl Acad Sci U S A, 2005. **102**(4): p. 983–7.

75. Bailey, S. N., D. M. Sabatini, and B. R. Stockwell, *Microarrays of small molecules embedded in biodegradable polymers for use in mammalian cell-based screens.* Proc Natl Acad Sci U S A, 2004. **101**(46): p. 16144–9.

76. Gopalakrishnan, S. M., et al., *A cell-based microarrayed compound screening format for identifying agonists of G-protein-coupled receptors.* Anal Biochem, 2003. **321**(2): p. 192–201.

77. Fernandes, T. G., et al., *On-chip, cell-based microarray immunofluorescence assay for high-throughput analysis of target proteins.* Anal Chem, 2008. **80**(17): p. 6633–9.

78. Bailey, S. N., et al., *Microarrays of lentiviruses for gene function screens in immortalized and primary cells.* Nat Methods, 2006. **3**(2): p. 117–22.

79. Yoshikawa, T., et al., *Transfection microarray of human mesenchymal stem cells and on-chip siRNA gene knockdown.* J Control Release, 2004. **96**(2): p. 227–32.

80. Falsey, J. R., et al., *Peptide and small molecule microarray for high throughput cell adhesion and functional assays.* Bioconjug Chem, 2001. **12**(3): p. 346–53.

81. Soen, Y., et al., *Detection and characterization of cellular immune responses using peptide-MHC microarrays.* PLoS Biol, 2003. **1**(3): p. E65.

82. Anderson, D. G., et al., *Biomaterial microarrays: Rapid, microscale screening of polymer-cell interaction.* Biomaterials, 2005. **26**(23): p. 4892–7.

83. Anderson, D. G., S. Levenberg, and R. Langer, *Nanoliter-scale synthesis of arrayed biomaterials and application to human embryonic stem cells.* Nat Biotechnol, 2004. **22**(7): p. 863–6.

84. Flaim, C. J., S. Chien, and S. N. Bhatia, *An extracellular matrix microarray for probing cellular differentiation.* Nat Methods, 2005. **2**(2): p. 119–25.

85. Flaim, C. J., et al., *Combinatorial signaling microenvironments for studying stem cell fate.* Stem Cells Dev, 2008. **17**(1): p. 29–39.

86. Soen, Y., et al., *Exploring the regulation of human neural precursor cell differentiation using arrays of signaling microenvironments.* Mol Syst Biol, 2006. **2**: p. 37.

87. Phillippi, J. A., et al., *Microenvironments engineered by inkjet bioprinting spatially direct adult stem cells toward muscle- and bone-like subpopulations.* Stem Cells, 2008. **26**(1): p. 127–34.

88. Hung, P. J., et al., *Continuous perfusion microfluidic cell culture array for high-throughput cell-based assays.* Biotechnol Bioeng, 2005. **89**(1): p. 1–8.

89. Wada, K., et al., *Live cells-based cytotoxic sensorchip fabricated in a microfluidic system.* Biotechnol Bioeng, 2008. **99**(6): p. 1513–7.

90. Khetani, S. R. and S. N. Bhatia, *Microscale culture of human liver cells for drug development.* Nat Biotechnol, 2008. **26**(1): p. 120–6.

91. Yamazoe, H. and H. Iwata, *Cell microarray for screening feeder cells for differentiation of embryonic stem cells.* J Biosci Bioeng, 2005. **100**(3): p. 292–6.

92. Orner, B. P., et al., *Arrays for the combinatorial exploration of cell adhesion.* J Am Chem Soc, 2004. **126**(35): p. 10808–9.

93. Doh, J. and D. J. Irvine, *Immunological synapse arrays: patterned protein surfaces that modulate immunological synapse structure formation in T cells.* Proc Natl Acad Sci U S A, 2006. **103**(15): p. 5700–5.

94. Ruiz, A., et al., *Micro-stamped surfaces for the patterned growth of neural stem cells.* Biomaterials, 2008. **29**(36): p. 4766–74.

95. Derda, R., et al., *Defined substrates for human embryonic stem cell growth identified from surface arrays.* ACS Chem Biol, 2007. **2**(5): p. 347–55.

96. Nakajima, M., et al., *Combinatorial protein display for the cell-based screening of biomaterials that direct neural stem cell differentiation.* Biomaterials, 2007. **28**(6): p. 1048–60.

97. Chin, V. I., et al., *Microfabricated platform for studying stem cell fates.* Biotechnol Bioeng, 2004. **88**(3): p. 399–415.

98. Ochsner, M., et al., *Micro-well arrays for 3D shape control and high resolution analysis of single cells.* Lab on a Chip, 2007. **7**(8): p. 1074–77.

99. Khademhosseini, A., et al., *Co-culture of human embryonic stem cells with murine embryonic fibroblasts on microwell-patterned substrates.* Biomaterials, 2006. **27**(36): p. 5968–77.

100. Moeller, H. C., et al., *A microwell array system for stem cell culture.* Biomaterials, 2008. **29**(6): p. 752–63.

101. Cordey, M., et al., *Enhancing the reliability and throughput of neurosphere culture on hydrogel microwell arrays.* Stem Cells, 2008. **26**(10): p. 2586–94.

102. Ashton, R. S., et al., *High-throughput screening of gene function in stem cells using clonal microarrays.* Stem Cells, 2007. **25**(11): p. 2928–35.

103. Cimetta, E., et al., *Micro-bioreactor arrays for controlling cellular environments: Design principles for human embryonic stem cell applications.* Methods, 2009. **47**(2): p. 81–9.

104. Gómez-Sjöberg, R., et al., *Versatile, fully automated, microfluidic cell culture system.* Anal Chem, 2007. **79**(22): p. 8557–63.

105. Kim, L., et al., *Microfluidic arrays for logarithmically perfused embryonic stem cell culture.* Lab on a Chip, 2006. **6**(3): p. 394–406.

106. Zhong, J. F., et al., *A microfluidic processor for gene expression profiling of single human embryonic stem cells.* Lab Chip, 2008. **8**(1): p. 68–74.

107. Newell-Price, J., A. J. Clark, and P. King, *DNA methylation and silencing of gene expression.* Trends Endocrinol Metab, 2000. **11**(4): p. 142–8.

108. Seligson, D. B., et al., *Global histone modification patterns predict risk of prostate cancer recurrence.* Nature, 2005. **435**(7046): p. 1262–6.

109. Ambardar, S., et al., *High throughput sequencing: An overview of sequencing chemistry.* Indian J Microbiol, 2016. **56**(4): p. 394–404.

110. Boldrini, B., et al., *Hyperspectral imaging: A review of best practice, performance and pitfalls for in-line and on-line applications.* Journal of Near Infrared Spectroscopy, 2012. **20**: p. 438–508.

111. Roberts, J., et al., *A short update on the advantages, applications and limitations of hyperspectral and chemical imaging in food authentication.* Appl Sci, 2018. **8**: p. 505.

112. Liu, Y., H. Pu, and D.-W. Sun, *Hyperspectral imaging technique for evaluating food quality and safety during various processes: A review of recent applications.* Trends Food Sci Technol, 2017. **69**.

113. Lai, X., L. Wang, and F. A. Witzmann, *Issues and applications in label-free quantitative mass spectrometry.* Int J Proteomics, 2013. **2013**: p. 756039.

114. Wang, M., et al., *Label-free mass spectrometry-based protein quantification technologies in proteomic analysis.* Brief Funct Genomics, 2008. **7**(5): p. 329–39.

115. Bloom, E., *Mycotoxins in indoor environments. Determination using mass spectrometry.* 2008. Doctoral Thesis (compilation), Division of Medical Microbiology, Lund University.

116. Neilson, K. A., et al., *Less label, more free: Approaches in label-free quantitative mass spectrometry.* Proteomics, 2011. **11**(4): p. 535–53.

117. Hendriks, T. F. E., et al., *Integrating rapid evaporative ionization mass spectrometry classification with matrix-assisted laser desorption ionization mass spectrometry imaging and liquid chromatography-tandem mass spectrometry to unveil glioblastoma overall survival prediction.* ACS Chem Neurosci. **16**(6): p. 1021–33.

118. Wood, R. W., *On a remarkable case of uneven distribution of light in a diffraction grating spectrum.* Proc Phys Soc Lond, 1902. **18**(1): p. 269.

119. Homola, J., *Present and future of surface plasmon resonance biosensors.* Anal Bioanal Chem, 2003. **377**(3): p. 528–39.

120. Das, S., R. Devireddy, and M. R. Gartia, *Surface plasmon resonance (SPR) sensor for cancer biomarker detection.* Biosensors (Basel), 2023. **13**(3).

121. Karami, P., et al., *Colorimetric immunosensor for determination of prostate specific antigen using surface plasmon resonance band of colloidal triangular shape gold nanoparticles.* Spectrochim Acta A Mol Biomol Spectrosc, 2019. **222**: p. 117218.

122. Thurmond, R. L., et al., *Kinetics of small molecule inhibitor binding to p38 kinase.* Eur J Biochem, 2001. **268**(22): p. 5747–54.

123. Adamczyk, M., J. A. Moore, and Z. Yu, *Application of surface plasmon resonance toward studies of low-molecular-weight antigen-antibody binding interactions.* Methods, 2000. **20**(3): p. 319–28.

124. Loo, J. F., et al., *An assay using localized surface plasmon resonance and gold nanorods functionalized with aptamers to sense the cytochrome-c released from apoptotic cancer cells for anti-cancer drug effect Determination.* Micromachines (Basel), 2017. **8**(11).

125. Boozer, C., et al., *Looking towards label-free biomolecular interaction analysis in a high-throughput format: A review of new surface plasmon resonance technologies.* Curr Opin Biotechnol, 2006. **17**(4): p. 400–5.

126. Hiroshi, A., M. C. Robert, and M. Brandon, *MicroRNA detection on microsensor arrays by SPR imaging measurements with enzymatic signal enhancement.* Biosens Bioelectron, 2019. **142**: p. 111565.

127. Hartmann, B., et al., *Profiling native pulmonary basement membrane stiffness using atomic force microscopy.* Nat Protoc, 2024. **19**(5): p. 1498–528.

128. Jaydeep, R., et al., *AI guided measurement of live cells using AFM.* IFAC-PapersOnLine, 2021. **54**(20): p. 316–21.

129. Rade, J., et al., *Deep learning for live cell shape detection and automated AFM navigation.* Bioengineering (Basel), 2022. **9**(10).

130. Huff, J. L., et al., *Label-free protein and pathogen detection using the atomic force microscope.* J Biomol Screen, 2004. **9**(6): p. 491–7.

131. Länge, K., *Bulk and surface acoustic wave sensor arrays for multi-analyte detection: A review.* Sensors (Basel), 2019. **19**(24).

132. Byung-Su, J., H. Jeung-Soo, and L. Duk-Dong, *Fabrication of polymer SAW sensor array to classify chemical warfare agents.* Sens Actuators B: Chem, 2007. **121**(1): p. 47–53.

133. Piro, L., et al., *Flexible SAW microfluidic devices as wearable pH sensors based on ZnO nanoparticles.* Nanomaterials (Basel), 2021. **11**(6).

134. Länge, K., et al., *A surface acoustic wave biosensor concept with low flow cell volumes for label-free detection.* Anal Chem, 2003. **75**(20): p. 5561–6.

135. Zeng, Y., et al., *Foodborne pathogen detection using surface acoustic wave biosensors: A review.* RSC advances, 2024. **14**(50): p. 37087–103.

136. Angelos, N. and G. Electra, *Portable surface acoustic wave device platform coupled with a paper-based capillary fluidics for real-time biosensing applications.* Sens Actuators A: Phys, 2024. **378**: p. 115814.

137. Xu, Y., et al., *A review of impedance measurements of whole cells.* Biosens Bioelectron, 2016. **77**: p. 824–36.

138. Nahid, M.A., et al., *An evaluation of the impact of clinical bacterial isolates on epithelial cell monolayer integrity by the electric cell-substrate impedance sensing (ECIS) method.* J Microbiol Methods, 2020. **169**: p. 105833.

139. Szulcek, R., H. J. Bogaard, and G. P. van Nieuw Amerongen, *Electric cell-substrate impedance sensing for the quantification of endothelial proliferation, barrier function, and motility.* J Vis Exp, 2014. (85).

140. Yu, C., et al., *CMOS high density electrical impedance biosensor array for tumor cell detection.* Sens Actuators B: Chem, 2012. **173**: p. 903–7.

141. Saeed Khoshfetrat, P., *Stretchable micro-electrode arrays for electrophysiology*. 2015. Doctoral Thesis.

142. Castillo-Fernandez, O., et al., *High-speed counting and sizing of cells in an impedance flow microcytometer with compact electronic instrumentation.* Microfluid Nanofluid, 2013. **16**.

143. Chalklen, T., Q. Jing, and S. Kar-Narayan, *Biosensors based on mechanical and electrical detection techniques.* Sensors (Basel), 2020. **20**(19).

144. Qian, C., et al., *Dielectrophoresis for bioparticle manipulation.* Int J Mol Sci, 2014. **15**(10): p. 18281–309.

145. Il, D. and C. Young-Ho, *A continuous cell separation chip using hydrodynamic dielectrophoresis (DEP) process.* Sens Actuators A: Phys, 2005. **121**(1): p. 59–65.

146. Li, M. and R. K. Anand, *Cellular dielectrophoresis coupled with single-cell analysis.* Anal Bioanal Chem, 2018. **410**: p. 2499–2515.

147. Zhao, K., et al., *Continuous cell characterization and separation by microfluidic alternating current dielectrophoresis.* Anal Chem, 2019. **91**(9): p. 6304–14.

148. Asami, K. and K. Sekine, *Dielectric modelling of cell division for budding and fission yeast.* J Phys D: Appl Phys, 2007. **40**(4): p. 1128.

149. Nerguizian, V., et al., *The effect of dielectrophoresis on living cells: Crossover frequencies and deregulation in gene expression.* Analyst, 2019. **144**(12): p. 3853–60.

150. Bolognesi, C., et al., *Digital sorting of pure cell populations enables unambiguous genetic analysis of heterogeneous formalin-fixed paraffin-embedded tumors by next generation sequencing.* Sci Rep, 2016. **6**(1): p. 20944.

151. Wang, Q., et al., *Microfluidic dielectrophoresis illuminates the relationship between microbial cell envelope polarizability and electrochemical activity.* Sci Adv, 2019. **5**(1): p. eaat5664.

152. Schulte, J., et al., *DEPArray™ single-cell technology: A validation study for forensic applications.* Forensic Sci Int Genet, 2024. **70**: p. 103026.

153. Makowski, M. S. and A. Ivanisevic, *Molecular analysis of blood with micro-/nanoscale field-effect-transistor biosensors.* Small, 2011. **7**(14): p. 1863–75.

154. Estrela, P., et al., *Field effect detection of biomolecular interactions.* Electrochimica Acta, 2005. **50**(25): p. 4995–5000.

155. Lee, W., et al., *Nonthrombogenic, stretchable, active multielectrode array for electroanatomical mapping.* Sci Adv, 2018. **4**(10): p. eaau2426.

156. Sang, S., et al., *Progress of new label-free techniques for biosensors: A review.* Crit Rev Biotechnol, 2016. **36**(3): p. 465–81.

157. Bergveld, P., *Development, operation, and application of the ion-sensitive field-effect transistor as a tool for electrophysiology.* IEEE Trans Biomed Eng, 1972. **19**(5): p. 342–51.

158. Bergveld, P., *Thirty years of ISFETOLOGY: What happened in the past 30 years and what may happen in the next 30 years.* Sens Actuators B: Chem, 2003. **88**(1): p. 1–20.

159. Kaisti, M., *Detection principles of biological and chemical FET sensors.* Biosens Bioelectron, 2017. **98**: p. 437–48.

160. van der Schoot, B. H. and P. Bergveld, *ISFET based enzyme sensors.* Biosensors, 1987. **3**(3): p. 161–86.

161. Caras, S. and J. Janata, *Field effect transistor sensitive to penicillin.* Anal Chem, 1980. **52**(12): p. 1935–7.

162. Shen, M. Y., B. R. Li, and Y. K. Li, *Silicon nanowire field-effect-transistor based biosensors: From sensitive to ultra-sensitive.* Biosens Bioelectron, 2014. **60**: p. 101–11.

163. Chen, K.-I., B.-R. Li, and Y.-T. Chen, *Silicon nanowire field-effect transistor-based biosensors for biomedical diagnosis and cellular recording investigation.* Nano Today, 2011. **6**(2): p. 131–54.

164. Kim, K. H., et al., *Ultrasensitive stress biomarker detection using polypyrrole nanotube coupled to a field-effect transistor.* Micromachines (Basel), 2020. **11**(4).

165. Samuel, V. and J. R. Korukonda, *A review on label free biosensors.* Biosens Bioelectron: X, 2022. **11**: p. 100216.

166. Sun, D.-W., *Hyperspectral imaging for food quality analysis and control.* 2010.

167. Lodhi, V., D. Chakravarty, and P. Mitra, *Hyperspectral imaging system: development aspects and recent rrends.* Sens Imaging, 2019. **20**.

168. Aref, M. H., et al., *Custom hyperspectral imaging system reveals unique spectral signatures of heart, kidney, and liver tissues.* Spectrochim Acta A: Mol Biomol Spectrosc, 2024. **305**: p. 123363.

169. Mehta, N., et al., *Multimodal label-free monitoring of adipogenic stem cell differentiation using endogenous optical biomarkers.* Adv Funct Mater, 2021. **31**(43): p. 2103955.

170. Wu, Y., et al., *Review on the application of hyperspectral imaging technology of the exposed cortex in cerebral surgery.* Front Bioeng Biotechnol, 2022. **10**: p. 906728.

171. Rozenberg, J. M., et al., *Molecules promoting circulating clusters of cancer cells suggest novel therapeutic targets for treatment of metastatic cancers.* Front Immunol, 2023. **14**: p. 1099921.

172. Aćimović, S. S., et al., *Superior LSPR substrates based on electromagnetic decoupling for on-a-chip high-throughput label-free biosensing.* Light Sci Appl, 2017. **6**(8): p. e17042.

173. Akgonullu, S., E. Özgür, and A. Denizli, *Recent advances in quartz crystal microbalance biosensors based on the molecular imprinting technique for disease-related biomarkers.* Chemosensors, 2022. **10**: p. 106.

174. Weisser, H., et al., *An automated pipeline for high-throughput label-free quantitative proteomics.* J Proteome Res, 2013. **12**(4): p. 1628–44.

175. Pegoraro, G. and T. Misteli, *High-throughput imaging for the discovery of cellular mechanisms of disease.* Trends Genet, 2017. **33**(9): p. 604–15.

176. Lee, S.-Y., H. J. Hwang, and D. W. Lee, *3D-cell spotter cooperated high-throughput screening (HTS) for estimating targeted drugs in breast cancer cells.* Adv Ther, 2023. **6**(2): p. 2200161.

177. Guangshan, L., M. F. Ann, and F. Ye, *Label-free profiling of ligands for endogenous Gpcrs using a cell-based high-throughput screening technology.* SLAS Technol, 2006. **11**(4): p. 181–7.

178. Zhou, S., et al., *Computer vision meets microfluidics: a label-free method for high-throughput cell analysis.* Microsyst Nanoeng, 2023. **9**: p. 116.

179. Yokoyama, S., *Protein expression systems for structural genomics and proteomics.* Curr Opin Chem Biol, 2003. **7**(1): p. 39–43.

180. Zhang, Z., et al., *High-throughput proteomics.* Annu Rev Anal Chem (Palo Alto Calif), 2014. **7**: p. 427–54.

181. Cui, M., C. Cheng, and L. Zhang, *High-throughput proteomics: A methodological mini-review.* Lab Invest, 2022. **102**(11): p. 1170–81.

182. Wilhelm, M., et al., *Mass-spectrometry-based draft of the human proteome.* Nature, 2014. **509**(7502): p. 582–7.

183. Zhang, D.Y., et al., *Proteomics, pathway array and signaling network-based medicine in cancer.* Cell Div, 2009. **4**: p. 20.

184. Bonini, A., et al., *A label-free impedance biosensing assay based on CRISPR/Cas12a collateral activity for bacterial DNA detection.* J Pharm Biomed Anal, 2021. **204**: p. 114268.

185. Zhou, Y., et al., *High-throughput screening of a CRISPR/Cas9 library for functional genomics in human cells.* Nature, 2014. **509**(7501): p. 487–91.

186. Henderson, M. C. and D. O. Azorsa, *High-throughput RNAi screening for the identification of novel targets*, in *Target Identification and Validation in Drug Discovery: Methods and Protocols*, J. Moll and R. Colombo, Editors. 2013, Humana Press: Totowa, NJ. p. 89–95.

187. Riedmaier, I. and M. W. Pfaffl, *Transcriptional biomarkers–high throughput screening, quantitative verification, and bioinformatical validation methods.* Methods, 2013. **59**(1): p. 3–9.

188. Menzel, C., et al., *High-throughput biomarker discovery and identification by mass spectrometry.* Comb Chem High Throughput Screen, 2005. **8**(8): p. 743–55.

189. Janvilisri, T., et al., *High-throughput screening for biomarker discovery.* Dis Markers, 2015. **2015**(1): p. 108064.

190. Beiram Zadeh, Z., et al., *Label-free detection of cardiac biomarkers: A review on microfluidic electrochemical biosensors.* ACS Appl Bio Mater, 2023. **6**(7): p. 2622–2635.

191. Rapp, B. E., F. J. Gruhl, and K. Länge, *Biosensors with label-free detection designed for diagnostic applications.* Anal Bioanal Chem, 2010. **398**(6): p. 2403–12.

192. Erlich, Y. and D. Zielinski, *DNA Fountain enables a robust and efficient storage architecture.* Science, 2017. **355**(6328): p. 950–4.

193. Klein, A. M., et al., *Droplet barcoding for single-cell transcriptomics applied to embryonic stem cells.* Cell, 2015. **161**(5): p. 1187–1201.

194. Aref, M., A.-B. Youssef, and Y. El-Sharkawy, *Novel approach exploiting the hyperspectral imaging system for breast cancer therapy and diagnosis.* Arch Breast Cancer, 2020: p. 189–201.

195. Mahlein, A. K., et al., *Hyperspectral sensors and imaging technologies in phytopathology: State of the art.* Annu Rev Phytopathol, 2018. **56**: p. 535–58.

196. Lampart, F. L., D. Iber, and N. Doumpas, *Organoids in high-throughput and high-content screenings.* Front Chem Eng, 2023. **5**.

197. Zuo, J., et al., *High-throughput solutions in tumor organoids: From culture to drug screening.* Stem Cells, 2025. **43**(1).

198. Choi, K. G., et al., *Utilizing organoid and air-liquid interface models as a screening method in the development of new host defense peptides.* Front Cell Infect Microbiol, 2020. **10**: p. 228.

199. Bengtsson, A., et al., *Organoid technology for personalized pancreatic cancer therapy.* Cell Oncol (Dordr), 2021. **44**(2): p. 251–60.

200. Mao, W., et al., *Spectroscopic techniques for monitoring stem cell and organoid proliferation in 3D environments for therapeutic development.* Adv Drug Deliv Rev, 2023. **201**: p. 115074.

201. Serafini, C. E., et al., *Non-invasive label-free imaging analysis pipeline for in situ characterization of 3D brain organoids.* Sci Rep, 2024. **14**(1): p. 22331.

202. Demircan Yalcin, Y. and R. Luttge, *Electrical monitoring approaches in 3-dimensional cell culture systems:Toward label-free, high spatiotemporal resolution, and high-content data collection in vitro.* Organs-on-a-Chip, 2021. **3**.

203. Yang, L., et al., *High-throughput methods in the discovery and study of biomaterials and materiobiology.* Chem Rev, 2021. **121**(8): p. 4561–677.

204. Hook, A. L., et al., *High throughput methods applied in biomaterial development and discovery.* Biomaterials, 2010. **31**(2): p. 187–98.

205. Xu, X., et al., *The new era of high-throughput nanoelectrochemistry.* Anal Chem, 2023. **95**(1): p. 319–56.

206. Yesiloz, G., M. S. Boybay, and C. L. Ren, *Label-free high-throughput detection and content sensing of individual droplets in microfluidic systems.* Lab on a Chip, 2015. **15**(20): p. 4008–19.

207. Macchia, E., et al., *Large-area interfaces for single-molecule label-free bioelectronic detection.* Chem Rev, 2022. **122**(4): p. 4636–99.

208. Carey, T. R., et al., *Developments in label-free microfluidic methods for single-cell analysis and sorting.* Wiley Interdiscip Rev Nanomed Nanobiotechnol, 2019. **11**(1): p. e1529.

209. Guselnikova, O., et al., *Label-free surface-enhanced Raman spectroscopy with artificial neural network technique for recognition photoinduced DNA damage.* Biosens Bioelectron, 2019. **145**: p. 111718.

210. Vervoort, N., et al., *Recent advances in analytical techniques for high throughput experimentation.* Anal Sci Adv, 2021. **2**(3-4): p. 109–27.

211. Halai, R. and M. A. Cooper, *Using label-free screening technology to improve efficiency in drug discovery.* Expert Opin Drug Discov, 2012. **7**(2): p. 123–31.

212. Singh, D. K., et al., *Label-free, high-throughput holographic screening and enumeration of tumor cells in blood.* Lab on a Chip, 2017. **17**(17): p. 2920–32.

Chapter 2

High-Throughput Analysis of Stem Cells

Mohan K. Dey, Sreyashi Das, Kirti Agrawal, Ram V. Devireddy, and Manas R. Gartia
Department of Mechanical and Industrial Engineering,
Louisiana State University,
Baton Rouge, LA 70803, USA
mgartia@lsu.edu

2.1 Introduction

2.1.1 Definition of High-Throughput Analysis

High-throughput analysis refers to the use of automated and parallel processes to analyze large volumes of data or samples, with the goal of generating meaningful results in a short amount of time. This term is commonly used in various fields such as genomics, proteomics, and drug discovery, where large amounts of biological or chemical

High-Throughput Imaging Methods: Cells and Tissues
Edited by Manas Ranjan Gartia and Ram V. Devireddy

ISBN 978-981-5352-08-5 (Hardcover), 978-1-003-77417-4 (eBook)
www.jennystanford.com

DOI: 10.1201/9781003774174-2

information need to be processed and analyzed efficiently. This type of analysis has revolutionized many areas of research and is becoming increasingly important in the development of new treatments and technologies. Modern high-throughput analysis often involves the use of high-performance computing systems, machine and deep learning algorithms, automated sample processing, and data analysis pipelines [1]. The focus of modern high-throughput analysis is on integrating multiple data sources, reducing complexity, and improving the accuracy and speed of data analysis. Figure 2.1 demonstrates the progression of techniques for single-cell optical image processing and analysis [2]. The goal of this type of analysis is to uncover insights, patterns, and trends in the data that may not be visible to the human eye, and can be used to improve decision-making, identify potential problems, and develop predictive models, as shown in Fig. 2.1. For example, Fig. 2.2 presents a demonstration of an example of high-throughput analysis by constructing and implementing an osteogenic convolutional neural network (OCNN) to demonstrate how automated imaging, computational modeling and biological validation may be used to construct a robust framework of analyzing the osteogenic differentiation process rapidly and with high accuracy. During the model development phase (Fig. 2.2A), the data in the confocal laser scanning microscopy (LSCM) are inputted and the results are processed by the OCNN architecture where the network is trained to identify the important cell image features, and its accuracy is measured based on predictive measures (ROC curves). Osteogenic biomarker validation (Fig. 2.2B) is the next step which uses traditional biological assays such as polymerase chain reaction (PCR), flow cytometry and immunofluorescence staining of markers such as RUNX2 to validate that computational outputs reflect biological markers of osteogenesis. Lastly, morphological analysis and comparison (Fig. 2.2C) are conducted by first extracting cellular features through image analysis packages like Cell Profiler, but then statistical evaluation and ROC curve analysis are performed to make sure that OCNN prediction does not vary with the morphological features under various experimental settings. In Fig. 2.2D it also marks the wide scope of the applications of the OCNN approach. The model can be used to predict cellular responses to various drugs, cytokines, and surfaces on the material, in high throughput, by combining imaging data with the deep learning. When discussing this, the analysis of the effect of different concentrations of compounds on osteogenic differentiation can be used to generate dose-relevant predictions and the classification models can be improved to work better in a variety of conditions of the experiment. This directly applied phase shows how OCNN can be used as more than a validation and analysis platform but also a predictive platform that can expedite the drug screening process, biomaterial evaluation phase, and the identification of novel treatment approaches.

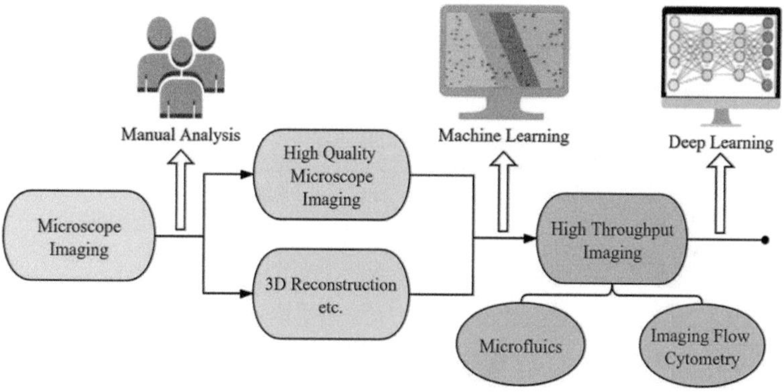

Figure 2.1 The development of single-cell optical image analysis techniques. Initially, experts manually edited and evaluated optical images of cells. Machine learning is used to automate the analysis of single-cell optical images as optical image quality has improved and single-cell difficult tasks like 3D reconstruction have emerged. Big data is brought by high-throughput technologies like microfluidics and image flow cytometry. Nowadays, single-cell optical image research makes extensive use of deep learning. View the color figure at wileyonlinelibrary.com [2].

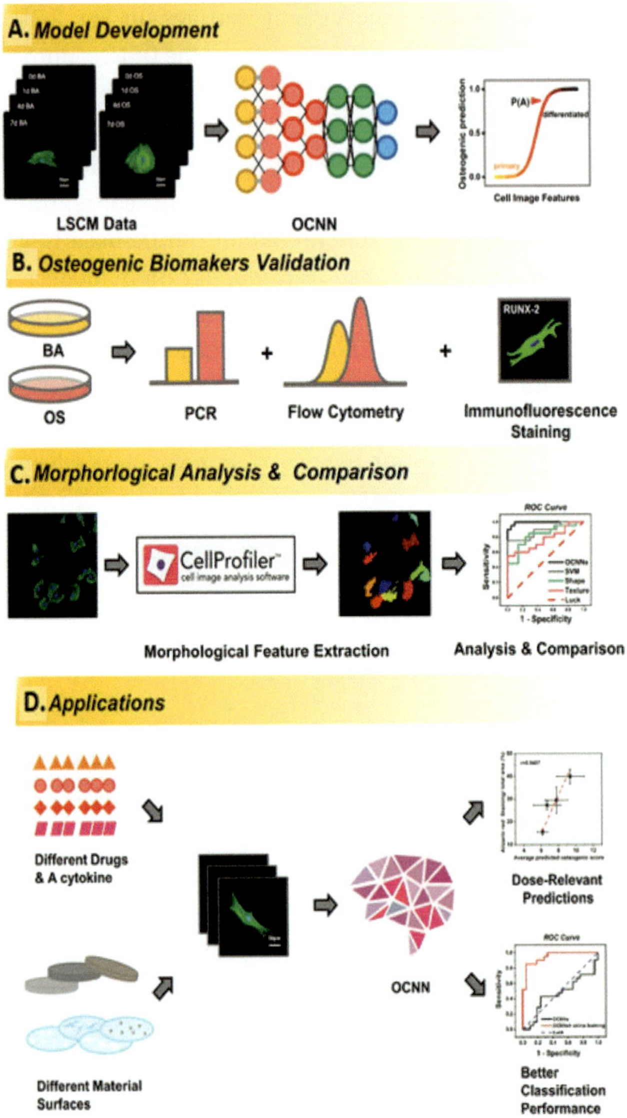

Figure 2.2 Osteogenic convolutional neural network (OCNN) training, validation, comparison, and application processes and results are shown schematically in the flow chart [3]. (A) Model development phase showing input of confocal laser scanning microscopy (LSCM) images into the OCNN architecture and evaluation via ROC curves. (B) Biological validation of osteogenic biomarkers (e.g., RUNX2) using PCR, flow cytometry, and immunofluorescence staining. (C) Morphological analysis and feature extraction using image analysis tools and statistical comparison to ensure prediction accuracy. (D) Application of OCNN for high-throughput prediction of cellular responses to compounds, cytokines, and biomaterial surfaces, supporting drug screening and biomaterial evaluation.

2.1.2 Stem Cells Morphology and Gene Expression

Stem cells are unique in that they can distinguish into different types of cells and divide indefinitely, meaning they can produce an indefinite number of daughter cells. This unique ability of stem cells to self-renew and generate differentiated progeny makes them a valuable source for tissue engineering and in vitro drug discovery. Morphologically, they are characterized by their undifferentiated state, which is characterized by a round nucleus, with a prominent nucleolus, and a small cytoplasm. Due to their properties, stem cells are a promising tool to understand normal and diseased cellular processes. The morphology of stem cells can be studied using light microscopy, transmission electron microscopy (TEM), scanning electron microscopy (SEM), and immunofluorescence microscopy. Light microscopy is used to view the general morphology of stem cells, such as cell size, shape, and the presence of cell processes, such as dendrites and axons. TEM and SEM are used to identify the ultrastructure of stem cells, such as the cell wall, nucleus, and organelles. Immunofluorescence microscopy is used to identify specific proteins and markers on the cell surface, which could be used to recognize various types of stem cells. To further study the characteristics of stem cells, high-throughput analysis can be used to study their gene expression profiles. This involves sequencing the mRNA from the cells to identify the genes that are expressed. This analysis provides a comprehensive view of the gene expression patterns in the stem cells, providing insight into their differentiating capacity. By studying the morphology and gene expression of stem cells, researchers can gain insight into the normal and diseased cellular processes that are regulated by stem cells. This information can then be used in developing new therapies for diseases and to upgrade our knowledge of normal and diseased cellular processes. Additionally, this approach can be used to identify novel genes that are associated with specific stem cell functions, such as self-renewal, differentiation, and proliferation. Using deep learning-based methods, as shown in Fig. 2.3, high-throughput analysis can be employed to extract morphology and intensity information from individual cells.

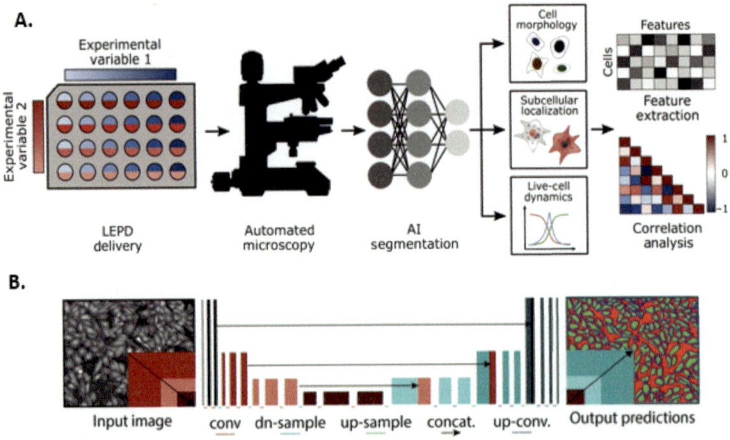

Figure 2.3 Image analysis workflow. (A) Combinatorial delivery experiments, automated imaging of each well, cell, and nuclei identification using deep learning algorithms, extraction of cellular features (morphology, subcellular localization, and dynamics), feature postprocessing, and correlation analysis make up the experimental and analysis workflow of the multi-well LEPD system. (B) The FCN's architectural design for segmenting cells and nuclei in an image. Convolution, down-sampling, up-sampling, concatenation, and up-convolution operations are used in 20 hidden layers to transform the input image into a three-class (external, interior, and border) probability map for each pixel. Based on changes in the network's dimensions (height and width) as the encoder and decoder portions move up and down, respectively, layers are color-coded. Each layer line's thickness varies in direct proportion to the layer's width. There are 7,787,523 trainable parameters in the network for a 520 × 696 grayscale image [4].

2.1.3 Overview of High-Throughput Analysis of Stem Cells

To fully utilize the potential of stem cells, researchers need to be able to study them at a high-throughput scale, i.e., the ability to process large amounts of data quickly and efficiently. However, the study of stem cells is complicated by the circumstances that they are very difficult to analyze, due to their small size, their ability to rapidly divide, and the complexity of their environment. As a result, traditional methods of stem cell analysis are often time-consuming and labor-intensive, making them unsuitable for large-scale studies. In recent years, high-throughput analysis has emerged as an attractive alternative to traditional methods of stem cell analysis. High-throughput analysis combines the power of modern computing technology with powerful laboratory techniques to analyze large numbers of stem cells quickly and accurately in a single experiment. This approach has been used to study a wide variety of stem cell characteristics to measure various biological parameters such as gene expression, cell proliferation, cell differentiation, and protein synthesis. Researchers can study thousands of cells simultaneously, thereby providing a comprehensive understanding of stem cell biology.

One of the most used high-throughput technologies for stem cell analysis is microarray-based gene expression analysis. The expression levels of thousands of genes can be simultaneously measured using this approach in a single sample. This information can be used to identify genes that are specifically expressed in stem cells, which can then be used to develop markers for stem cell identification. Another widely used technology for high-throughput analysis of stem cells is flow cytometry. This method allows for the simultaneous analysis of multiple parameters in thousands of cells in a single experiment. Flow cytometry is particularly useful for analyzing cell proliferation, apoptosis, and cell surface markers, which are important indicators of stem cell behavior. A third technology commonly used for high-throughput analysis of stem cells is high-content imaging by machine learning algorithms. Machine learning algorithms have recently emerged as a promising tool for high-throughput analysis of stem cells. These algorithms can analyze vast amounts of complex data and identify patterns and relationships that might be difficult to detect using traditional methods. This method allows for the simultaneous analysis of multiple cellular parameters, such as cell morphology, protein expression, and cell signaling pathways. High-content imaging can be used to visualize stem cell behavior in response to different stimuli and can provide insights into the underlying mechanisms of stem cell differentiation and self-renewal. In addition to these commonly used technologies, other high-throughput approaches, such as mass cytometry, transcriptomics, and proteomics, are increasingly being used to study stem cells. Mass cytometry allows for the simultaneous measurement of multiple parameters in individual cells, making it useful for analyzing complex biological systems such as stem cells.

Transcriptomics and proteomics provide a global view of the genetic and protein landscape of stem cells, respectively, which can be used to understand the underlying mechanisms of stem cell behavior.

In conclusion, high-throughput analysis of stem cells is a rapidly growing field that is revolutionizing our understanding of stem cell biology. The use of automated technologies and platforms is enabling researchers to study stem cells at a large scale, providing a comprehensive understanding of stem cell behavior and opening new avenues for stem cell-based therapies.

2.1.4 Purpose of This Chapter

The purpose of this chapter on high-throughput analysis of stem cells is to provide an overview of the field, including its recent developments, applications, and potential impact on stem cell biology and disease treatment. This chapter will cover the various high-throughput technologies and platforms used to study stem cells, such as microarray-based gene expression analysis, flow cytometry, high-content imaging, and machine learning algorithms. The chapter will also discuss the advantages and limitations of these technologies, and how they can be integrated to provide a more complete understanding of stem cell biology.

Additionally, this chapter will explore the recent developments in the field of high-throughput analysis of stem cells, including the integration of machine learning algorithms and the use of high-throughput techniques for drug discovery and development.

2.2 High-Throughput Analysis Techniques

2.2.1 Microarray Analysis

The process of screening many molecular compounds against markers in search of potential "hits" is known as high-throughput screening (HTS). HTS is distinguished by its simple setup, speed, low cost, and exceptional performance. It comprises an operational robotic platform, an exceptionally sensitive screening framework, an accurate diagnostic model, a large compound library, and an input/output and data processing system. Microarray analysis stands out among HTS techniques, allowing for the analysis of over 100,000 specimens per day [5]. The advancement of genome research and associated methodologies has resulted in improved comprehension of genetic data and its functional aspects. This has facilitated in-depth studies of genome sequences. While a thorough examination of microarray usage in biology and medicine is beyond this discussion's scope, notable publications cover these topics extensively, such as those by Conzone and Pantano and van Berkum and Holstege [6]. Label-based fluorescence sensing is used in current microarray methods, which may cause interference and lead to erroneous measurements in single-molecule binding interactions. Tag photobleaching, quenching, and the time-consuming and expensive nature of fluorophore labeling for many proteins interfere with the precise and quantitative assessment of biomolecular interactions [7]. Microarrays, on the other hand, have the perk of simultaneously estimating multiple variables, which speeds up the testing of large libraries of active drug ingredients and biomolecular interactions. On a single microscope slide, these arrays provide extensive label-free detection. A dual filter image processing technique is described to improve highly multimode detectors' precision, durability, and signal-to-noise ratio [8]. Microarrays are classified into transcriptomic and tissue microarrays (TMAs). The ability to collect numerous DNA sequences on a small surface, typically a glass slide, is at the heart of gene microarray innovation. Traditional methodologies such as reverse transcriptase-polymerase chain reaction (RT-PCR) and Northern blot allowed for examining just a few genes per test, whereas "global expression profiling" allows for examining thousands of genes, overcoming feature selection limitations. mRNA, the biomolecule that transports genetic data from the nucleus to the cytoplasm for protein biosynthesis, is used as a diagnostic biomarker to assess genetic expression [9]. Fluorescent dyes have been used to tag cDNA, allowing complementary sequence binding and improving our understanding of the molecular mechanism's underneath alterations in gene expression. DNA chips, which are used to identify DNA fragments via fluorescent tags and their interactions with DNA probe arrays, enable the measurement of fluorescence emission via a computer by enacting a light beam through the target DNA fragments [8]. The investigation of gene expression in a sample is not based on assumptions, and it is frequently referred to as "discovery-type research." cDNA from stem cells is combined with a chip to investigate gene expression, and changes in gene deletion or amplification can be evaluated through the integration of fluorescently tagged DNA from a differentiated specimen onto these chips. TMA components are made by transferring paraffin-embedded tissues to pre-cored pits in a receiver paraffin frame, which allows them to be used in immunocytochemistry (IHC) or in situ hybridization experiments [8].

Chang et al. used interference lithography in HTS to form a large-scale, low-cost microarray on a standard microscope slide. They successfully manufactured up to 1.5 million detection pixel intensities on a single glass surface at a density of 816 sensors per mm^2. Figure 2.4 shows this capacity is equivalent to state-of-the-art fluorescent dye microarrays. Chen et al. investigated the upregulation of Dsg2 expression during the transition of mouse-induced pluripotent stem cells (miPSCs) to cancer stem cells (CSCs) and its relationship to activating the Wnt/-catenin signaling pathway. The researchers used DNA microarray technology to study gene expression. They used an unintended tag to create Cy3-labeled cDNAs, which were then mixed with a cellular marker DNA microarray. After the fluorescence emission intensities were collected using GenePixR Pro 5.1 software, the images were analyzed and adjusted to a control RNA. Genes on the Mouse Cell Surface Gene expression differences between miPSCs, miPS-LLCcm, miPS-P19cm, and miPS-P19c cells were assessed using Oligo Microarray (Fig. 2.4A). The microarray analysis revealed that several genes were upregulated in transformed cells versus miPSCs. Dpep1, Dpagt1, H2-Eb1, and Dsg2 were found to be significantly more expressed in miPS-LLCcm and miPS-P19cm cells than in miPS-P19c cells [10].

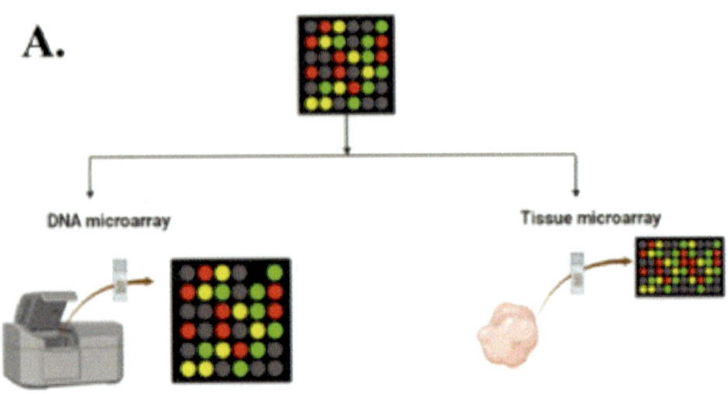

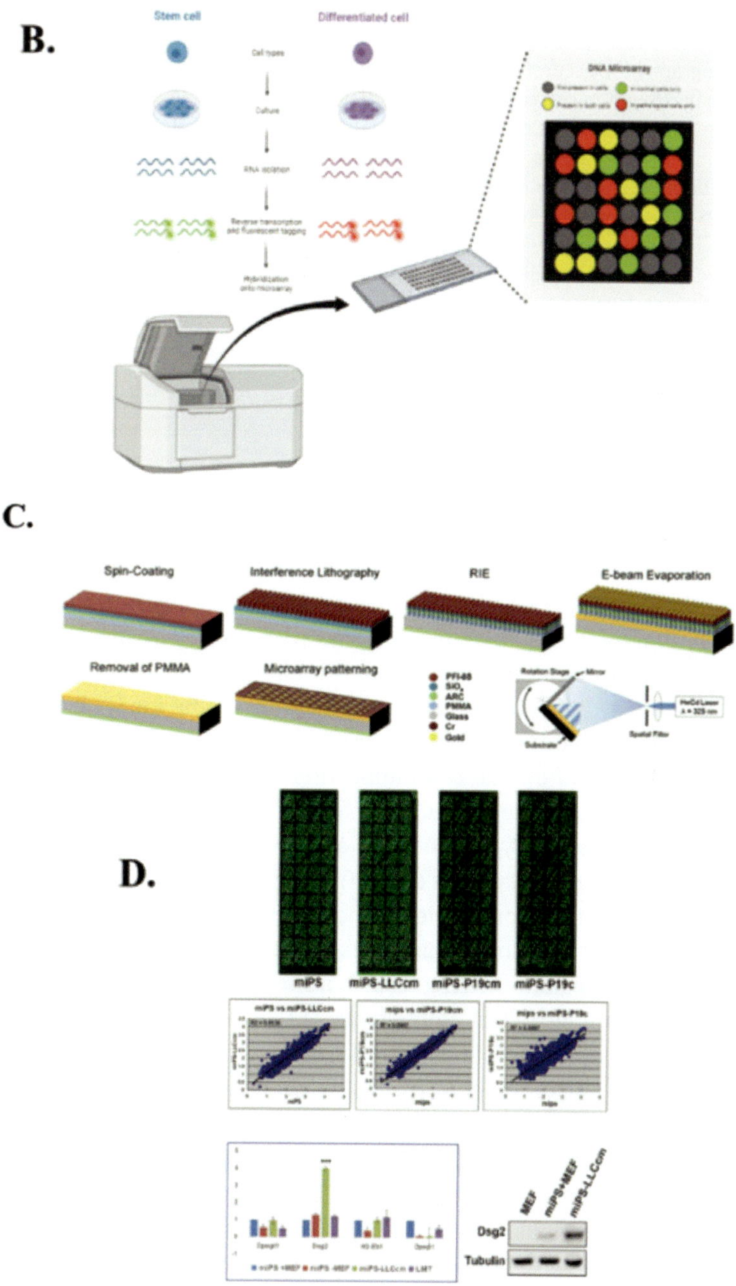

Figure 2.4 (A) Schematic representation of different types of microarrays. (B) The process of distinction of diverse genetic regulation between stem cells and

differentiated cells through DNA microarray. (C) The visual illustration provides an overview of the fabrication procedure which employed interference lithography for accurate structure sensing pixel arrays onto a conventional 75 mm by 25 mm microscope glass slide [7]. (D) Representation of transformation of miPSCs to CSCs, the expression of the Dsg2 gene was found to be upregulated. Comparative analyses using Mouse Cell Surface Genes Oligo Microarray were conducted to identify genes with differential expression between miPSCs and miPS-LLCcm, miPS-P19cm, and miPS-P19c cells. The mRNA expression levels of the four differentially expressed genes identified in the microarray analyses were validated using qRT-PCR in both miPSCs and the derived cells. The protein expression level of Dsg2 in miPS-LLCcm cells was compared to that in miPSCs using Western blot analysis [10].

2.2.2 Flow Cytometry

A well-developed framework for measuring various cell fluorescence parameters is flow cytometry. Thanks to recent developments, high-volume tissue samples can now be processed up to 30 times faster. Medical screening effectiveness can be raised by using computational techniques and multiplexing bioassays with colored beads. The real-time analysis of molecular signaling components made possible by this assay system makes flow cytometry appealing for HTS. With a low limit of detection (LOD) as low as 10–100 pM, it enables precise measurement of optical communication from particles like cells or beads, facilitating the screening of tens of thousands of compounds [11]. Flow cytometry is known as a potent measurement tool that admits for the simultaneous quantification and analysis of numerous optical biomarkers of physiochemical expression or response over time. A multiplexed bead-based suspension component that allows for up to 100 ligand-binding tests in a single batch has been created to take advantage of this competence [11].

To identify, record, and sort cells using flow cytometry, they are passed in a single file before a beam. Fluorescent dyes are used to label cells before analysis, and when they are excited by a beam of light, different wavelengths of light are emitted from the cells. The quantity of the fluorescently labeled cells can then be measured to determine their variety and number [12]. While a number of strategically positioned sensors are used to measure the Forward and Side Scatter, a laser is pointed at a fluid stream that is carrying the cells. The fluorescently labeled cells are stimulated by the laser, which causes them to emit light at higher wavelengths that are then picked up by the system's detectors [12]. The software then processes the information through a flow cytometer-connected device, as shown in Fig. 2.5A. Although commercial robotic sample handling processes are only capable of handling two specimens per minute, flow cytometry is an HTS tool. Over the past 200 years, numerous specimen-processing devices have been created to get around this restriction. The rotary multiport flow injection nozzle is one such tool that can carry out up to 10 assays per minute. Additionally, to achieve endpoint test efficiency at rates greater than 1 sample per second. A flow cytometer and an autosampler are combined in the HyperCyt® system to enable effective sample handling shown in Fig. 2.5B. A peristaltic pump successively draws sample particle suspensions from each well as the autosampler moves across a multi-well microplate, producing a series of samples that have been separated by bubbles. Compared to commercial alternatives, this system offers faster sample delivery and data acquisition. Each microplate serves as a single data file, and samples are continuously delivered to the flow cytometer, where the data is collected in a continuous stream. The time-resolved data is analyzed using a proprietary program called FCSQuery that was created by one of the authors and contains periodic gaps representing the passage of air bubbles. Due to its low particle carryover and lack of inter-sample probe rinsing, the HyperCyt® system offers impressive throughput rates in endpoint assays [11]. One notable application is the search for small molecule ligands for the human formyl peptide receptor (FPR) by combinatorial chemical library screening. Compounds are assessed for their capacity to prevent the binding of a fluorescently tagged peptide ligand to the FPR on cell membranes using a representative 96-well assay. The median fluorescence intensity of the peptide bound to cells in each well is calculated using the data that was gathered, which was obtained in less than 3 minutes. The assay's response range is established by control wells containing buffer or blocking peptide that are not labeled. Compounds that inhibit peptide binding by 70% or more are thought to be promising leads for additional research. In this homogeneous assay, fluorescent peptide, compounds, and cells are all added one after the other without any intermediate washing procedures. As a result of the assay, ligands with inhibition constants (Ki) that vary from nanomolar to tens of micromolar have been successfully identified [11].

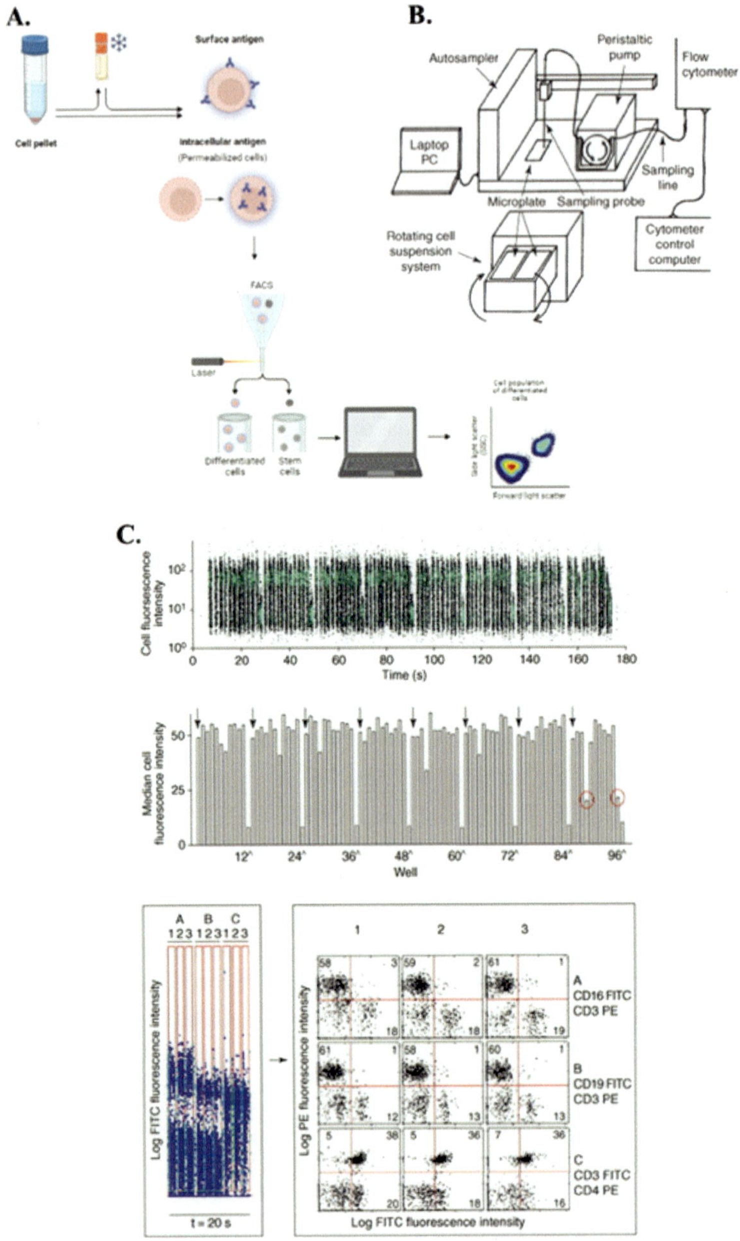

Figure 2.5 (A) Schematic representation of the working principle of flow cytometry. (B) The schematic representation of HyperCyt® system, which collects specimens from the wells of microplates using an autosampler and

sampling probe, is intended for endpoint assays. After that, a flexible sampling line transports these samples to the flow cytometer. A peristaltic pump runs continuously, switching between fluid samples and air bubbles, to maintain a constant flow of samples. Furthermore, a rotating cell suspension system rotates the microplates every so often to ensure the particles suspended during any necessary pre-analysis incubation steps, such as cells or beads, are distributed evenly [11]. (C) Application of HyperCyt® for HTS through time-resolved cell fluorescence measurements from a 96-well microplate in a search for small-molecule FPR ligands. Each data cluster represents the fluorescent ligand binding to cells from a different well. A blue-green color indicates high cell density. The median fluorescence intensity of the cells in each well is determined by the data shown in (A). In control wells, fluorescent ligand binding is completely blocked. Red circles indicate the compounds that block peptide binding by at least 70%. Immunophenotyping, time, and output for one parameter [11].

Hahne et al. created the R package flowCore, which contains a collection of adaptable and open-source computational methods to make it easier to analyze these complicated datasets. These techniques have proven to be very helpful for organizing information, streamlining the analytical process, and analyzing various datasets [13]. This article introduces "flowCore," a crucial package created as part of the Bioconductor project for processing high-throughput flow cytometry data. It provides a full set of flexible tools for data preprocessing, quality assurance, visualization, and analysis. The authors highlight its numerous features, such as data import, cell population gating, sub-setting, compensation for spectral overlap, normalization, transformation, and statistical analysis. The importance of quality control measures is highlighted, including outlier and artifact removal [13]. The article discusses how users can combine the analysis of flow cytometry data with the use of genomic and proteomic tools, thanks to flowCore's seamless integration with other Bioconductor packages. The practical application of flowCore functions is demonstrated with the help of examples and code fragments. As a result, flowCore is positioned as a reliable and approachable package that enables effective and repeatable analysis of high-throughput flow cytometry data within the R programming environment. Its integration with the Bioconductor ecosystem further improves its capacity for thorough data analysis. The flowCore framework is made to import and handle FCS files containing raw data and their metadata as shown in Fig. 2.6. The Gating Markup Language Candidate Recommendation, a norm created in cooperation with the ISAC Data Standards Task Force, is incorporated. Through this integration, flowCore can easily interact with other FCM tools that adhere to the standard and integrate into already-existing workflows. Adherence to these standards is crucial for new flowCore-based methods to be more widely accepted and included in the accepted procedures for FCM data analysis [13].

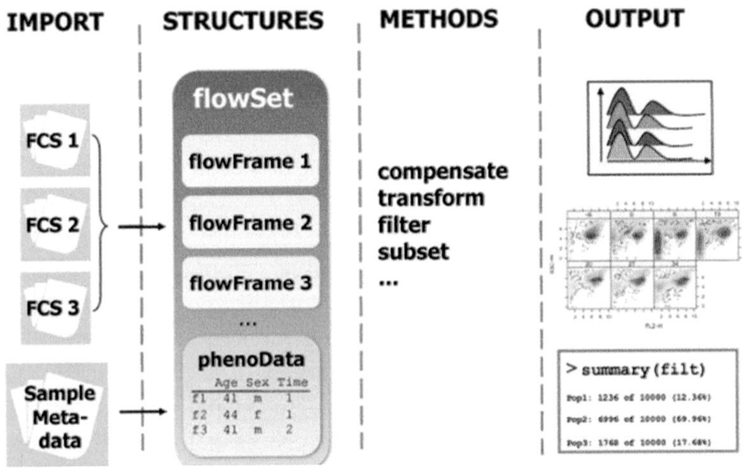

Figure 2.6 Schematic representation of the flowCore framework [13].

2.2.3 Mass Spectrometry

Mass spectrometry (MS) is a procedure of analysis that divides ionized particles based on variations between their charge-to-mass proportions. It comprises four components: the source of ion, mass analyzer, detector, and computer as shown in Fig. 2.7A [14]. There are numerous variances on the prevalent methods for the mass analyzer and ion source, and deciding on a suitable pairing is critical. For several decades, MS has served as a critical component in the field of biomedical sciences and drug discovery [14]. For the discovery of new drugs, and determination of disease biomarker, rapid detection and screening techniques are essential. MS is a very practical method of fast and direct examination of a vast range of samples with little preparation. In recent years, HTS has advanced greatly and increased the efficiency and use of MS in biomedical research. This subsection emphasizes how MS methods like desorption electrospray ionization (DESI), direct analysis in real-time (DART), and extractive electrospray ionization (EESI) are used in HTS [14]. These methods can work with a variety of compounds and complex matrices, and they are essential for drug discovery. Since they have a high throughput, are highly sensitive, and can identify molecular structures, MS-based methods are frequently used in HTS. Still, its application has been restricted by speed. Implementing matrix-assisted laser desorption ionization (MALDI)-MS to execute greater throughput screening has proven efficient [15]. Still, the drawbacks include several transfer phases of the reaction mixture and possibly crucial impact of the matrix. Agilent RapidFire^{TM}-MS, combined with controlled surface handling and an easy sample-manufacturing process, minimized the processing duration to seconds, facilitating HTS to get to MS completion points despite a breach [14]. HTS has succeeded with hyphenated MS methods like gas chromatography/mass spectrometry (GC/MS) and liquid chromatography/mass spectrometry (LC/MS). However, ambient mass spectrometry (AMS), which can analyze samples directly in the open air with little preparation, has drawn more attention. Spray/jet-based techniques (DESI), electric discharge-based techniques (DART), and gas/heat/laser-assisted techniques (EESI) are the different types of AMS methods. AMS methods have been successfully used in HTS applications, such as drug screening, forensic identification, food safety screening, pollutant detection, and quality control [16].

Cooks developed the ambient ionization technique known as DESI for MS in 2004 [17]. It uses a charged spray to ionize and desorb analytes on the sample surface, which the mass spectrometer then analyzes. Without the need for an additional matrix, DESI offers convenience, simplicity, and the ability to analyze both small and large molecules. It has the potential for HTS, and DESI has been combined with a 96-sample plate to make it possible to analyze lots of samples at once. DESI parameters like voltage, temperature, and solvent composition have been optimized to achieve selectivity, sample rate, and matrix tolerance. Yet when working with complicated specimens, DESI encounters difficulties with sensitivity and selectivity [17, 18]. Reactive DESI methods have been developed to cope with this problem by forming easily ionized adducts through chemical reactions [17, 18]. Nitrogen or helium plasma is used in the 2003-developed DART ion source to produce excited-state atoms and ions [19]. It has been applied to high-throughput drug, pesticide, and food additive screening. For quantitative analysis, improvements like modifications and sample pretreatment have been investigated. Enhancing sensitivity and linearity is done by combining DART with a triple-quadrupole MS. Food safety inspections make use of selective ionization. However, DART-MS has drawbacks like poor sensitivity, a constrained mass range, and difficult quantitative analysis. By using internal standards and parameter optimization, researchers are working to overcome these constraints [16]. ESI is a cutting-edge method for analyzing trace substances in complicated matrices. Two sprayers are used to nebulize the sample solution, which is followed by the creation of charged droplets and the extraction of the desired compounds as shown in Fig. 2.7B [20]. For quick in situ analysis, nano-EESI has been developed, and it has applications in pesticide residue detection, food quality control, and pharmaceutical screening [21]. Rapid screening on biological surfaces and detecting explosives are made possible by neutral desorption (ND) devices, which allow for the gentle sampling of volatile and non-volatile analytes from various surfaces. These developments have shown respectable drive limits and relative deviations from standard, proving successful surface-specific sampling and analysis [21].

MALDI, created by Koichi Tanaka and improved by Franz Hillenkamp and Michael Karas, has become one of the commonly utilized techniques to analyze an extensive range of biomolecules, coming from small metabolite to proteins and DNA, ionized using traditional ionization process methods [22, 23]. A scaffold, which includes 2,5-dihydroxybenzoic and sinapinic acid, is employed to safeguard biomolecules while enabling the vaporization process and ionization which is also important for acquiring useful mass spectra [22, 23]. Imaging mass spectrometry (IMS) can integrate the two features, ranging from detecting mass to visualizing elements and determining both biochemical processes and positioning unidentified compounds at the same time [14]. The time-of-flight analyzer, which divides pulsating ions based on their flight duration and establishes their molecular weight, is the prevalent analyzer for IMS [22, 23]. Molecular structure assessment is possible with tandem MS through comparing the molecular weights and the fragmented ions. A method called IMS makes it possible to see how molecules and structures are distributed in space centered on their molecular masses. IMS has made recent strides that have enhanced resolution beyond what the human eye can see. IMS retains positional information to efficiently analyze large datasets by obtaining mass spectra from various spectra acquired during MS analysis. Sample processing techniques like on-tissue digestion preserve the positional information of biomolecules, which are frequently used for preparing thin tissue sections on conductive materials. The spray-droplet method and nanoparticle-assisted laser desorption/ionization (nano-PALDI), two techniques that have been developed to improve sensitivity and resolution, are essential to effective IMS. The atmospheric pressure IMS (Ion Mobility Spectrometry) β prototype shown in Fig. 2.7C exhibits exceptionally fine stage and spot diameter control. Both the laser optic and observation optic axes are independently controlled by it. IMS, as opposed to traditional MS, does not require the separation and purification of target molecules, protecting vital positional data like the distribution and localization of these molecules. IMS efficiently depicts the distribution of target molecules within tissue sections by examining mass spectra gathered through MS analysis [14].

At first, proteins or peptides were the main subjects of IMS studies. However, the study of detecting and observing small organic molecules has seen a noticeable increase. For instance, phospholipids, glycolipids, and cholesterol are all very plentiful and easily ionizable in living organisms [24]. It is essential to comprehend the arrangement and biological functions of phospholipids, especially phosphatidylcholines (PCs), which are significant cell membrane constituents whose biological activities are determined by two fatty acids [24]. The pattern of distribution of PCs in the mouse brain and retina, both of which have nine different membrane layers with a combined width of 150 μm, has been revealed by IMS analysis [25]. It has been determined that some signals can be PCs through MS/MS analyses. Based on these layers' fatty acid composition, our research successfully divided them into the following three categories: Segment B includes the inner segment and outer nuclear layer, segment C includes the outer plexiform layer, and segment A includes the pigment epithelium and outer segment as shown in Fig. 2.7D. Docosahexaenoate (DHA), which is predominately found in segments A and B, is a key target of lipid peroxidation, a process that plays a role in a number of diseases. Rhodopsin and DHA-PC play a significant part in the phototransduction in the retina in vivo. The outcome is intriguing and shows how effective IMS is [25].

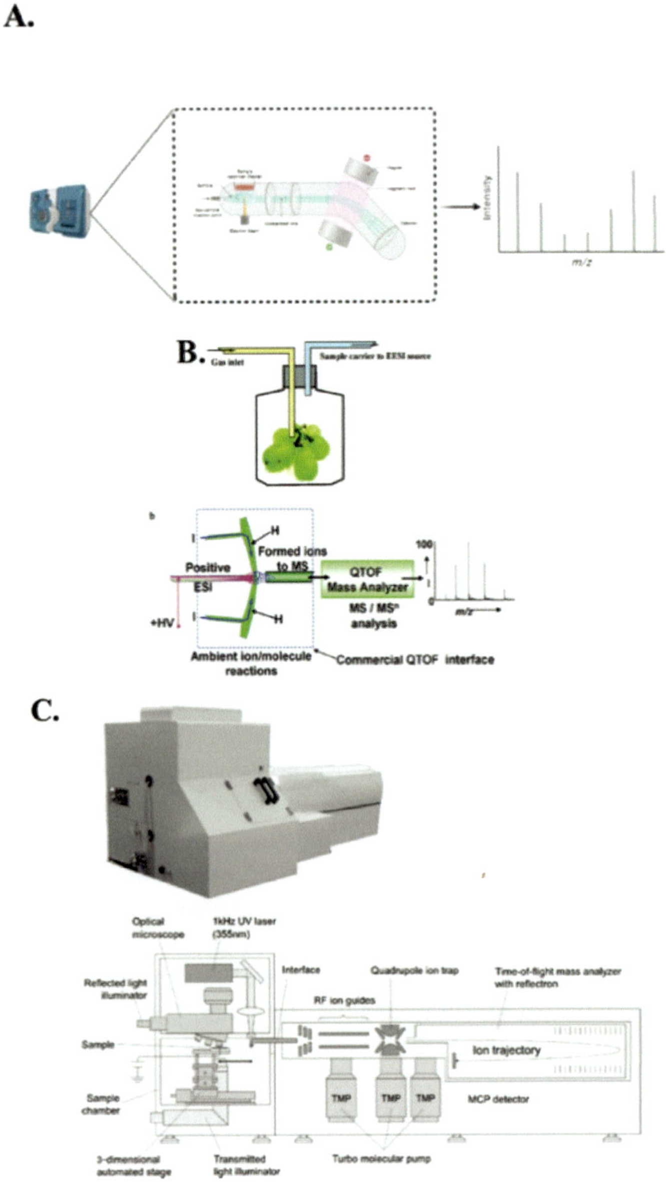

D.

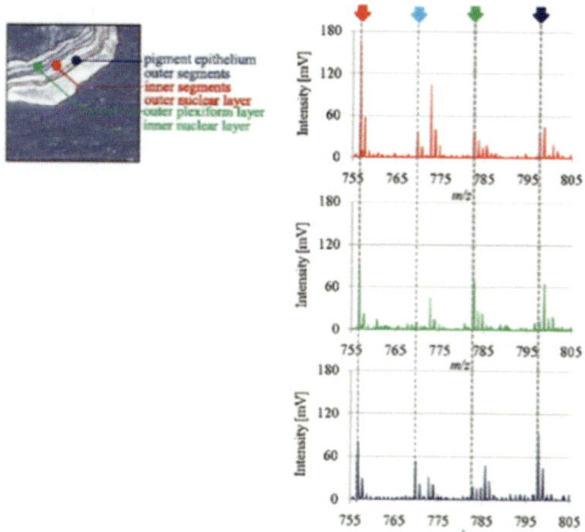

Figure 2.7 (A) Schematic representation of the working principle of MS. (B) Diagram showing the typical configuration of an EESI (electrospray-enhanced ionization) [20]. (C). Overview and design representation of the atmospheric pressure IMS prototype [14]. (D) The distribution of phosphatidylcholine (PC) molecular species can be seen in a mouse retinal section. Multiple regions can be roughly differentiated in an optical image of the retinal part, and three different colored dots are shown. In the retinal section, the blue dot (segment A), the green dot (segment B), and the red dot (segment C) represent the measurement areas where spectra between m/z 755 and 805 are compared. The green dot had the highest signal intensity at m/z 782.7 (PC diacyl-(16:0/18:1)), while the red dot had the highest signal intensity at m/z 756.5 (PC diacyl-(16:0/16:0)). The strongest signal strengths in the blue dot were observed at m/z 797.4 [PC diacyl-(18:0/22:6)] and m/z 769 [PC diacyl-(16:0/22:6)] [25].

2.2.4 Next-Generation Sequencing by Machine Learning Algorithm

Since its inception, DNA sequencing technology has made remarkable strides, moving from the manual sequencing pioneered by Frederick Sanger to the introduction of next-generation sequencing (NGS) platforms [14]. This evolution resembles semiconductor technology's revolutionary development. Sequencing speed has increased exponentially, following the path predicted by Moore's Law for the semiconductor sector [26]. By revolutionizing our ability to understand the basic structure of life, this discovery has given rise to new fields like comparative genomics, bioinformatics, systems biology, and synthetic biology. Sequencing technology is advancing rapidly, enabling thorough genome analysis and personalized medicine. NGS, also known as high-throughput sequencing (HTS), is a fast and low-cost DNA sequencing technology that generates enormous quantities of sequence data. Its advancement has transformed many fields, such as genomics, transcriptomics, epigenetics, and metagenomics, enabling investigators to read entire genomes, identify gene variation, determine gene expression levels, and investigate gene-regulating interactions [27]. NGS is an inexpensive, high-through substitute to conventional sequencing methods. HTS technologies have completely transformed research in the medical field. NGS or second-generation (2G) machines have rapidly evolved into third generation (3G) and fourth generation (4G) machines. Clinicians can transition from conventional healthcare to a genetic sequence-based era of preventative measures and therapeutic choices by DNA sequencing, the mapping process, and contrasting in both healthy and unhealthy disease states. High-throughput sequencing (HTS) has rapidly evolved with the emergence of cyclic-array sequencing. This technology is used in platforms such as Roche's 454, Illumina's Genome Analyzer, ABI's SOLiD, and Helicos' Heliscope, which enable the sequencing of millions of short sequences at once, generating different types of sequence data [27]. HTS platforms offer immense potential for genomic variant discovery, and bioinformatics experts have developed algorithms for this purpose. NGS has revolutionized genetic diagnostics but also presents difficulties with data management, interpretation, and counseling.

The workflow for NGS platforms is shown in Fig. 2.8A and includes the Illumina Genome Analyzer, Roche 454 Genome Sequencer, Life Technologies SOLiD System, Polonator G.007, and Complete Genomics [27]. They involve several steps and use sequencing-by-synthesis. To make a single-stranded template library, genomic DNA is broken up and connected with adapter sequences. This library is clonally amplified using bridge PCR, emulsion PCR, or in situ polonies after being immobilized on a solid surface, such as slides or beads shown Fig. 2.8B [28, 29]. The surface-mounted array of amplified DNA clusters is subjected to cyclic modification by polymerase or ligase enzymes [28]. During the cyclic chain extension process, an

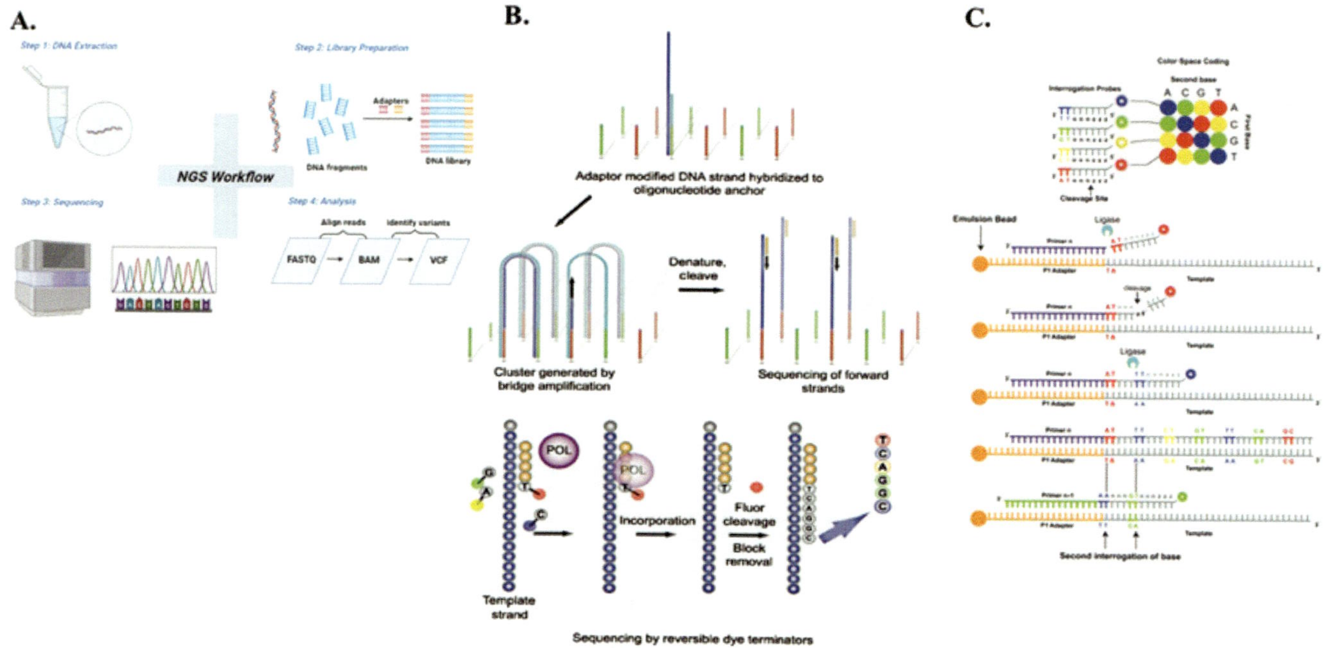

Figure 2.8 (A) Schematic illustration of the workflow of NGS. (B) Representation of the working mechanism of Illumina Genome Analyzer sequencing [29]. (C). Using a particular technique for sequencing, SOLiD color-space coding interrogates the P1 adapter and template with probes that correspond to 16 combinations. Before cleaving the final three variable bases, fluorescence is captured after annealing and ligating the probe. The next probe is sequenced through annealing, ligation, and extension steps after the cleaved probe's 5′ end is phosphorylated. The extended product is then separated from the adapter/template and sequenced again using a primer that corresponds to step (n-1) from the previous process. The known adapter sequence and the color-space coding can be combined with progressively compensate primers to determine the template sequence through deconvolution [29].

optical event is detected by a microscopic system, and pictures are taken. Computer algorithms are used to analyze the array images and derive DNA fragment sequences. To amplify single-stranded library fragments, the Illumina Genome Analyzer uses a process called "bridge amplification." This procedure uses asymmetrical adaptors to create oligo-bridges on an oligo-derived flow-cell surface. Following PCR thermal cycles, thousands of DNA copies are produced and gathered in one spot on the surface. Illumina's GA sequencing platform uses sequencing with fluorescently labeled nucleotides and reversible terminators. Each cycle combines Four labeled nucleotides with DNA polymerase to enable DNA chain extension. To ascertain the order of each cluster, a fluorescence image is acquired. Although the platform has a high throughput, its read length is constrained by optical signal decay and dephasing [28]. Sequencing chemistry mistakes can lead to signal decay and decrease the precision of the sequence. The number of errors rises with longer DNA strands [28].

The Roche 454 Sequencer uses emulsion PCR to create amplicons for sequencing. This method combines single-stranded template DNA and paramagnetic beads coated with DNA primers to form separate micro-reactors for PCR [30, 31]. Each bead becomes coated with thousands of copies of the same DNA sequence following several cycles of thermocycling. Following enrichment, these beads are transferred to a picotiter plate, which is a component of the sequencing flow cell and put there. Pyrosequencing, the sequencing technique used, uses pyrophosphate chemistry to produce an optical signal for detection [31]. The release of pyrophosphate during nucleotide incorporation causes a series of biochemical reactions that are involved in this procedure. The complementary base on the template DNA is revealed by the presence or absence of a light burst, which denotes the incorporation of the corresponding nucleotide. Pyrosequencing, with a capacity of up to 500 base pairs, offers speed and read length advantages. It eliminates the requirement for extra chemical steps above and beyond the typical DNA polymerase process, decreasing the possibility of mistakes. Pyrosequencing is susceptible to insertion-deletion errors and has limitations in homopolymer regions. Due to the need for a particular set of enzymes for pyrophosphate detection chemistry, the platform is relatively expensive [30, 31]. Like the 454 Sequencer, the SOLiD system amplifies DNA using emulsion PCR and paramagnetic beads.

The amplified beads are gathered and attached to a glass substrate after the emulsion is broken, creating a disorganized array. SOLiD uses ligation-based sequencing-by-synthesis in contrast with earlier platforms and encompasses a dual-base encoding scheme to identify mistakes [32]. Fluorescently labeled degenerated octamer probes are ligated to the prolonging chain during ligation cycles in different colors, denoting different bases at various positions. The octamer is chemically cleaved to remove the final three bases and the fluorescent label after each ligation, followed by a fluorescent image's capture shown in Fig. 2.8C [29, 32]. The extended chain can be examined at various points by performing successive rounds of ligation. Until the entire sequencing region is covered, the process is carried out by repeatedly performing the annealing and ligation cycles with one base shorter primer. This method is automated and managed by computers despite its apparent complexity. Each base is measured twice, which increases accuracy by identifying sequencing mistakes. However, a notable drawback is the sequence reads' relatively short length, which is partially due to ensemble disruptions [32].

Sequencing-by-ligation is a technique used by the NGS platform, the Polonator G.007 [33]. It uses single-base probes instead of dual-base encoding, in contrast to other platforms. A series of ligation reactions between a universal primer and nanomer probes on emulsion PCR amplified DNA clusters are required for the sequencing procedure. Degenerate oligonucleotides that have been fluorescently labeled are included in each pool of nanomer probes, with the fluorescence color indicating the base being examined. A fluorescent image is taken after each cycle of ligation, and the extended primer-probe chain is then denatured to restart the process. Up until all query positions are looked at, the process is repeated with ligations in between the primer and the following pool of degenerate oligoprobes. This system has the benefit of preventing sequencing errors from building up because subsequent ligations are not required after each reset. The range of query positions from each primer location is nevertheless constrained by this restriction, leading to a shorter read length. To lessen this, the library sequence's multiple anchor locations can be used to increase the coverage. The Polonator is more reasonably priced than other next-generation systems that are available commercially, and it also serves as an open-source platform that enables end users to potentially alter and enhance the sequencing process and chemistry [33]. The sequencing-by-ligation method used by Complete Genomics boosts DNA cluster density while using less reagent. A circular DNA template is produced by inserting numerous adaptors, and this circular DNA template is then amplified using circular PCR to produce DNA nanoballs (DNBs), which self-assemble onto a planar substrate. DNBs offer a denser array by better utilizing three-dimensional space [27, 34].

The NGS platforms mentioned earlier have greatly increased sequencing speed and decreased costs. They can generate billions of bases of sequence data each day per instrument at the cost of only a tiny fraction of a cent per kilobase. Except for the Roche 454 Sequencer, most platforms still struggle to produce short sequence reads. The dephasing problem in DNA sequencing clusters, where the optical signal drifts out of phase, is the main cause of this restriction. Single-molecule sequencing, a promising approach that eliminates the combined effect of multiple molecules, is one potential remedy for this issue.

2.3 Applications of High-Throughput Analysis in Stem Cell Research

2.3.1 Identification of Stem Cell Markers

Stem cells could be the ultimate solution to humanity's prayers for healing and regeneration to restore those worn out through aging and disorder. Embryonic stem (ES) cells are derived from the inner cell mass of a premature blastocyst [35]. Yet, somatic cell nuclear transfer (SCNT) methods allow for the generation of ES cells from a patient's genomic DNA [36]. Nevertheless, arguments for early embryo development following SCNT methods are premature since the more developed embryo only attained the six-cell phase [37]. Due to anomalies, reproductive cloning of human beings, which enables the generation of embryos via SCNT methods, is prohibited in most nations, and laws governing utilizing ES cells differ worldwide [38]. Research on stem cells promises immortality, and regenerative therapy, including OsteoCel and Chondrogen, has been making significant progress , as shown in Fig. 2.9. The pinnacle of synthetic biology is the emergence of intricate organs like the kidney, liver, and heart outside the person's body [39].

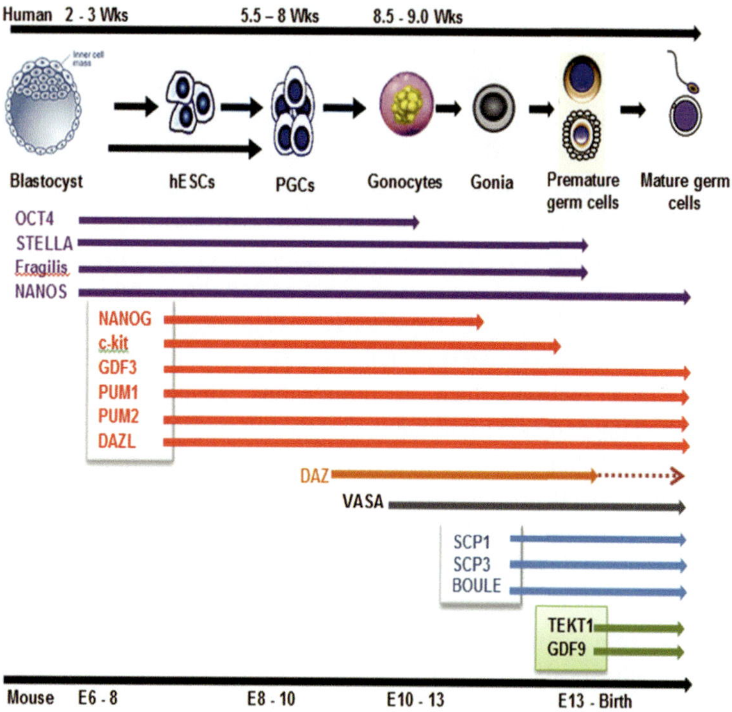

Figure 2.9 Illustration of the progression of the expression of the germ cell marker as time passes. To aid in germline differentiating in a living organism, specific germ cell markers are produced at times throughout every phase of the growth of germ cells. A totipotent zygote is created after fertilization due to the detection and drive of primordial germ cells (PGCs), crucial for germline development during the embryonic stage. The periods where the indicators are overexpressed are denoted by arrows [39].

Scientists employ genes and the protein molecules produced by those genes as stem cell markers to separate and distinguish stem cells [40]. Additionally, functional assessments, which have become the accepted norm for identifying and treatment, can be used to identify stem cells. The significance of stem cell markers for recognizing and classifying stem cells is discussed in this subsection [41]. Several marker profiles may exist according to the stem cell populace's location, species, and intended embryonic development [42]. The core cell mass of a blastocyst is where ES cells (ES cells) are produced. ES cells can generate into multiple cell types. They are pluripotent, which enables them to differentiate into nearly all cell lineages except those that generate a placenta or an embryo [43]. ES cells are a good choice for therapeutic purposes since they preserve plasticity and can remain pluripotent and divide continuously [44]. However, using ES cell-based therapeutics poses difficulties because biomarkers that detect ES cells also indicate tumor stem cells [45]. Oct4 and Nanog homeodomain transcription variables are crucial for maintaining pluripotency as in ES cells and for the initial growth of the embryo. Other variables, including Sall4, Dax1, Essrb, Tbx3, Tcl1, Rif1, Nac1, and Zfp281, have been discovered in recent investigations. In ES cells, these variables simultaneously control to create an intricate transcriptional regulation system [46–50].

Cellular features such as dimensions, shape, activity from metabolism, and sensitivity alter due to differentiation. It is crucial to show cell shape, lineage-specific markers, transcription patterns, and functional traits particular to the intended kind of cell to ensure effective differentiation. High-throughput screenings (HTS) are employed to classify novel molecules which embed procurement and differentiation actions in a variety of stem cells, including self-renewal, osteogenic differentiation, neurogenic response, and neural specification, setting the stage for HTS assays to be used routinely and broadening the plethora of molecules accessible for controlling stem cell fate. Various HTS techniques, including flow cytometry, microarray, NGS, and MS, have identified distinct stem cell and lineage-specific markers. To study human neurogenesis, the researchers employed time-dependent metabolomics, LC-MS-based measurement, and metabolomics. Investigators discovered 17 significantly different metabolites and metabolic pathways, including purines and pyrimidine metabolism, pantothenate and coenzyme A, fatty acid, and amino acid metabolism, during the early stages of neurons. These results affect other in vitro neurological disorder models and advance our neurodevelopment knowledge. Following the N-acetyl aspartic acid (NAA) levels, researchers chose representative time points and analyzed the samples using LC-MS. Day 0 and day 18 samples clustered according to the results, indicating that human neurodevelopmental metabolites differed significantly. A time-dependent metabolomics study was carried out to determine which metabolites contributed to the variations between various time points. Tyrosine was significantly consumed during the growth of neural cells, and PCA found 17 metabolites that were significantly altered during neurodevelopment. Alpha-ketoisovaleric acid, spermidine, 4-phosphopantothenoylcysteine, spermine, and uridine are a few metabolites that exhibit variation patterns that are like those of NAA and may be used as markers for the investigation of neural growth. Figure 2.10A shows the heat map used to visualize the shifts in metabolite intensities. Tyrosine was found in the culture medium and significantly used during the growth of neural cells, according to an analysis of the medium at each time point. However, the cells also generated other metabolites discharged into the medium. Alpha-ketoisovaleric acid, spermidine, 4-phosphopantothenoylcysteine, spermine, and uridine showed an identical pattern of intensity variation compared to NAA among these metabolites. As can be seen in Fig. 2.10B, these metabolites may also be used as markers for researching neural growth [51].

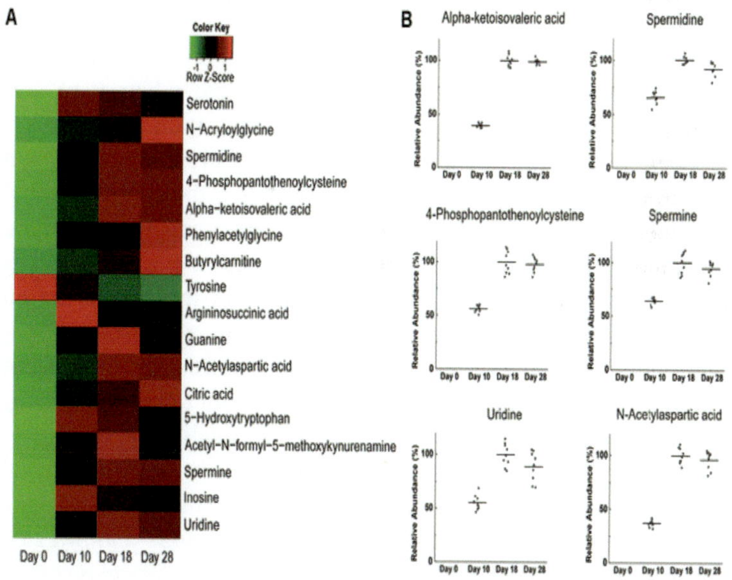

Figure 2.10 (A) The heat map displays the 17 metabolites the time-dependent metabolomics assessment determined to have undergone substantial modifications. (B) Of these metabolites, six might be used as markers for neurodevelopment. The z-score turned raw data of the metabolites are represented in the heat map employing a color scale that ranges from red to green. The heat map was produced using R with gplots. Red color indicates a rise in metabolism intensity, while a fall in intensity is indicated by green color [51].

2.3.2 Differentiation Tracking and Analysis of Stem Cells Signaling Pathways

Stem cell differentiation encompasses a transition from emergence to specialization, which results in changes in cellular morphology, hyperpolarization, metabolic functions, and signal sensitivity which commits to developmental lineages and acquiring specialized roles through signal transduction and alteration of gene expression. Takahashi and Yamanaka's research defined the development of iPSCs in adult cell populations via external expression of transcription factors such as Oct4, Sox2, Kruppel-like factor-4 (Klf4), and c-myelocytomatosis (c-Myc) were found to be important in the process of self-renewal and differentiation [52]. In responding to internal and external stimulation, numerous kinases signaling pathways influence the core stemness and ensure the preservation of pluripotency. Stem cell therapy has helped us comprehend many initially unfathomable concerns. Totipotent stem cells are primitive forms capable of generating a complete embryo, whereas ES cells generate into the three germ layers. Due to their infinite potential to regenerate themselves and competence to differentiate, adult stem cells are the first preference for stem cell-based therapies. Although treatments based on stem cells are innovative and are revamping the globe, they also have downsides, including tumorigenicity, uncontrolled proliferation, and differentiation into inappropriate lineages—especially in the case of embryonic stem cells (ESCs). The Wnt and TGF-signaling pathways are required for renewal and differentiation. So real-time tracking of their differentiation through many signaling pathways can be achieved through HTS.

Li et al. [53] created an engineered human pluripotent stem cell (hPSC), developed an HTS platform, and executed small-molecule testing that recognized two compounds (Ro1138452 and SR2211) against the prostacyclin receptor (IP) and the retinoic acid receptor-related orphan receptor g (RORg), which have been significant for the treatment of demyelinating pathologies like Multiple Sclerosis. The promising marker and numerous pathways, such as PLP1 (Proteolipid protein 1), and MBP (Myelin basic protein) associated with the central nervous system, which they may modulate, were discovered through single-cell transcriptomic assessment. Creating a report system for identifying and purifying specific cell categories originating from human ES cells (hESCs) is the sublime of this work. Secreted

NanoLuc (secNLuc) reporter and Super-fold green fluorescent protein (sfGFP) genes were inserted to PLP1 and MBP genes using CRISPR-Cas9 genome editing. The third and final reporter system, known as PTt-P1-MsNL, consists of three reporters that enable sequential sampling and quantifiable readout of MBP expression: PDGFRA-P2A-tdTomato-P2A-Thy1.2, PLP1-sfGFP, and MBP-P2A-secNLucb. Using the PTt-P1-MsNL reporter cell line, it is possible to identify and enrich adult oligodendrocytes by the expression of PLP1-sfGFP, temporally purify cells of the oligodendrocyte lineage that are positive for PDGFRA and measure the amount of MBP expression over time which is a stem cell differentiation marker. By expressing PLP1-sfGFP, this cutting-edge reporter cell line promotes the temporal purification of oligodendrocyte lineage cells that are positive for PDGFRA, the detection and enrichment of adult oligodendrocytes, and quantitative monitoring of MBP expression throughout time [53].

Merkert et al. [54] used cystic fibrosis (CF) disease-specific iPSCs to establish and perform an organotypic HTS by employing a halide-sensitive YFP subtype, allowing for automatically generated quantitative assessment of CF Transmembrane Conductance Regulator (CFTR) activity in the iPSC-derived intestinal epithelium, in addition to the identity of a succession of substantiated principal hits that boost CFTR component from a library compounds. The use of isogenic induced pluripotent stem cell (iPSC) lineages as controls in disease modeling is emphasized by this work. The correct subclone stayed more comparable to the parent cell line versus unrelated iPSC lines, according to principal component analysis (PCA). A dTomato-fluorescent protein with a nuclear membrane localization signal (dTomatonuc) was inserted into the CFTR locus of CFTR. Phe508del iPSCs and the repaired equivalent to track CFTR-expressing cells throughout iPSC development. The incorporation of dTomatonuc into the mutant allele of the heterozygously corrected CF iPSC line was confirmed by allele-specific PCR testing. Multiple reporting cell lines spanning diverse CFTR genotypes were chosen for subsequent investigation. Creating intestinal epithelia that express CFTR required the development of a sequential differentiation technique. Each cell line showed the capacity to produce definitive endoderm (DE), but the efficiency of the CFTR. Phe508del/Tom and CFTRcorr/Tom cell lines was somewhat lower. Analysis showed that different cell lines had variable proportions of positive cells, with CFTRKO/Tom and CFTRwt/Tom clones showing larger percentages. In the differentiating cultures, the expression of dTomatonuc mRNA corresponded with the expression of CFTR mRNA and red fluorescence, demonstrating the emergence of differentiation and acting as a marker [54].

2.3.3 Disease Modeling

A crucial method for comprehending the root causes of health conditions and identifying innovative therapies is disease modeling. It offers new perspectives on long-standing issues, tests cutting-edge approaches, and makes it possible to gather significant information [55]. To influence preventive methods, the distribution of resources, and studies into the underlying factors and dissemination of diseases, modeling of data and statistical evaluation are crucial. Researchers are hoping that investigations into stem cells will contribute to an improved comprehension of how diseases develop [56]. Investigators may better understand the progression of diseases and ailments by observing stem cells evolve into cells found in muscles of the heart, neurons, bones, and other systems and tissues. It is possible to direct stem cells to differentiate into particular cells that may be employed in individuals to regenerate and restore tissues that were harmed or impacted by the disease [57]. Knowing the causes of diseases and creating efficient treatments requires scientific modeling of human ailments. hPSCs have become an important tool for addressing the shortcomings of experimental animals in some disorders. ES cells or patient-specific iPSCs can produce disease models [58]. Human PSCs are suitable for examining the interactions between genotype and phenotype and acquiring large numbers of cells for drug testing and cell-based treatments, because they have benefits like primary cell line features, endless self-renewal, and the capability to be separated into various kinds of cells. The creation and use of human PSCs in disease simulation and the creation of drugs are examined in this overview difficulties. The discipline of disease demonstration has undergone a revolution thanks to the application ES cells, iPSCs, and genome editing strategies [35, 59]. The aid of patient-derived iPSCs, cellular reprogramming, and genome editing methods have made it possible to model a variety of hereditary disorders. This is a priceless asset for researching disease mechanisms and creating individualized treatments [60]. Around the world, iPSC storage facilities have been set up to offer cell lines for various genetic ailments [61]. Activating transcription factors allowed for the transformation of somatic cells into pluripotent forms in mice and, ultimately, in human cells [62–65]. This discovery has fundamentally changed how PSCs are studied and used. The iPSCs, which consist of reprogrammed cells, were used to create prototypes of human genetic diseases soon after they were identified. Furthermore, an important breakthrough in this field was made possible by the newest achievement in human SCNT, which resulted in the production of SCNT-ES cells from cells from patients shown in Fig. 2.11 [66].

HTS is the fast testing of hundreds to hundreds of thousands of specimens for biological function using computerized instruments at the model living thing, cellular, route, or molecule level. Identifying substances having biological action is a frequent practice in the biotech industries. HTS requires computerized assay layouts, robotic sample use, and automatic information processing. Investigators frequently employ HTS of stem cells to uncover substances and variables preserving stemness and to gain information about the fundamental causes of the stem cell's fate. Microtiter plates can only be used to a certain degree due to the accessibility and versatility of stem cells, necessitating the usage of smaller devices. This section introduces the role of HTS in stem cell-based disease modeling and emphasizes current developments in creating tiny HTS systems.

Zebrafish have gained popularity as models for researching cardiovascular function and disease because of their non-intrusive testing, small stature, transparency, high fertility, rapid growth, and cost-effectiveness. Zebrafish serve as a model organism for biomedical research, because they have many genes in common with human beings, including 82% of the referred human disease genes [67]. Zebrafish are a top model for biomedical research due to their strong conservation and simplicity of genetic tampering. Zebrafish positively affect genetic engineering in biomedical research, including accurate regulation of the expression of genes and the ability to conduct gain-of-function and loss-of-function examinations

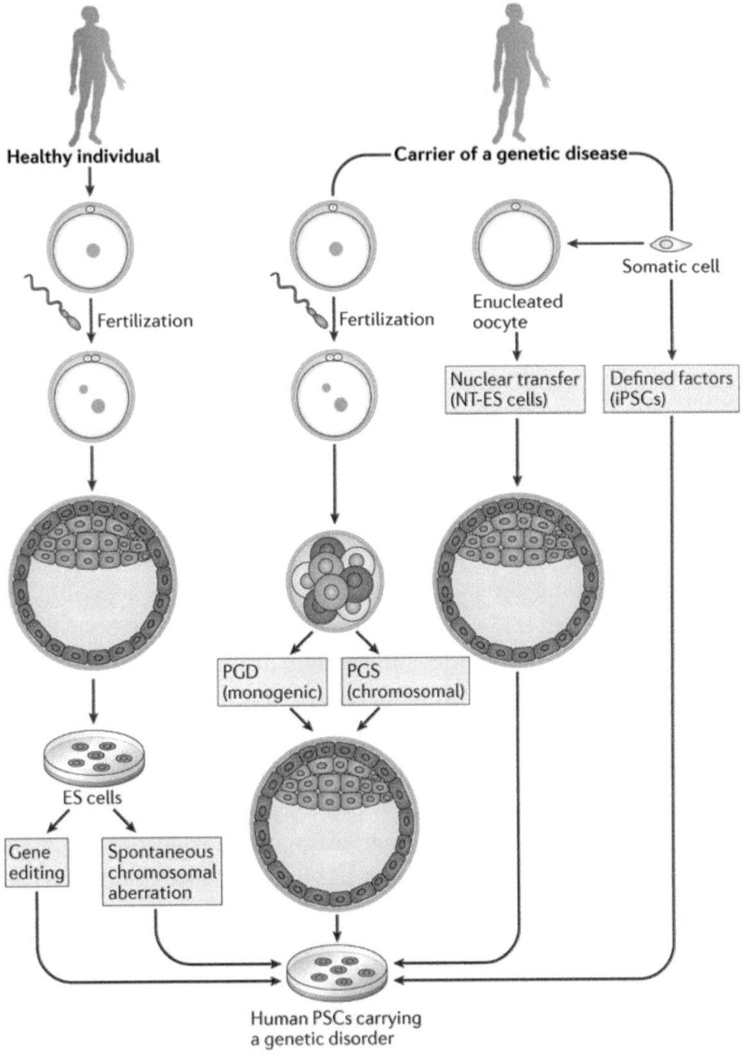

Figure 2.11 Obtaining PSCs from healthy people or people with a genetic disorder is possible. Spontaneous chromosomal alterations may be employed to model chromosomal conditions, while specific mutations can be by genes introduced into ES cells. Pre-implantation genetic diagnosis or screening for genes can also produce disease-specific ES cells. Nuclear transfer or iPSCs can reprogramme somatic cells from patients into PSCs [66].

more easily. Widespread fluorescently labeled transgenic lines have been produced, and CRISPR-Cas9 has replaced traditional gene editing techniques [68]. The partly duplicated genome of zebrafish has benefits as well as drawbacks. The zebrafish cardiovascular system can be seen using various imaging methods, including in situ hybridization and sophisticated fluorescence microscopy. New avenues for cardiovascular research have been made possible through endothelial cell calcium behavior. Numerous cardiovascular development, function, and disease aspects have been investigated and modeled using zebrafish. They have offered a beneficial understanding of heterotaxia syndrome, which may interfere with the body's left-right patterning and result in congenital heart defects (CHDs) [69, 70]. Zebrafish mutants have identified key genes involved in cardiac asymmetry with phenotypes like those of human patients, and candidate genes linked to heterotaxia have been confirmed. CRISPR-Cas9-mediated mutagenesis has improved our understanding of how these genes function. Additionally, zebrafish models have clarified the mechanisms that underlie CHDs linked to mutations in the ciliary protein CFAP53 [71]. Zebrafish are crucial to understanding the intricate connections and pathways that regulate cardiac morphogenesis because of the intricate nature of CHD genetics. Unique heart diseases, known as cardiomyopathies, are defined by alterations in the structure of the myocardium. Dilated cardiomyopathy of the heart (DCM), hypertrophic cardiomyopathy (HCM), and arrhythmogenic cardiomyopathy (ACM) are the three main groups. By identifying zebrafish counterparts for human disease-associated genes and validating candidate genes, zebrafish studies have made an important contribution to comprehending these illnesses. Zebrafish models, for instance, have provided light on the function of the TTN gene in DCM by demonstrating that the location of reducing alterations throughout the gene affects the degree of severity of the condition [72]. Additionally, zebrafish research has shed light on the responsibilities of the BAG3 and ILK genes in DCM and ACM. Accordingly, Zebrafish have also been employed in drug discovery. In general, zebrafish models have been extremely useful for bridging the gap between genetic research and translational medicine, enabling the identification of therapeutic targets, the screening of potential compounds, and the evaluation of drug efficacy and toxicity in a cost-effective and high-throughput manner.

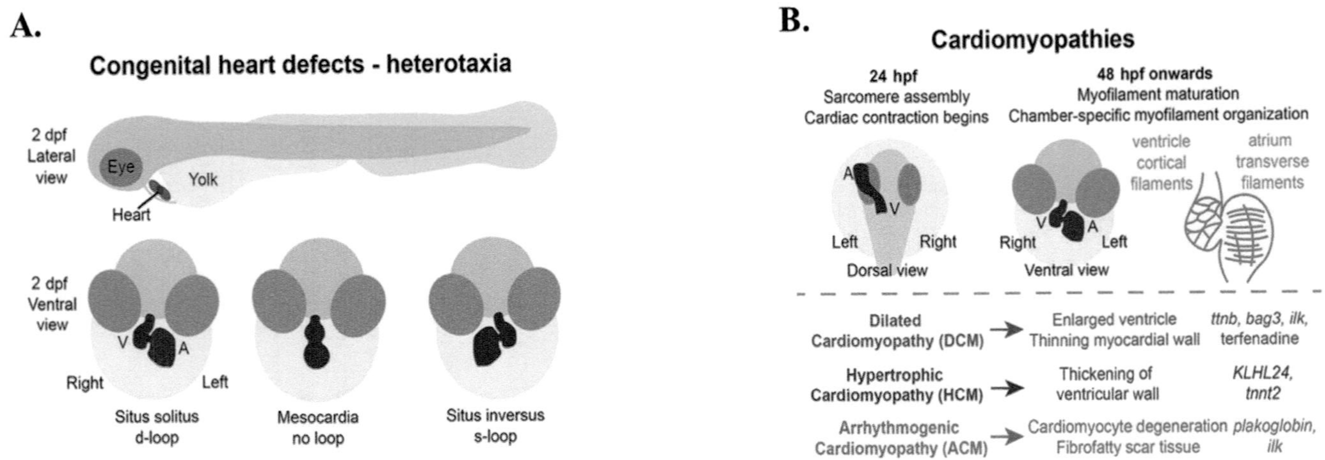

Figure 2.12 (A) CHD is frequently studied using zebrafish as models. When zebrafish embryos are two days post-fertilization, their hearts have undergone some initial morphological changes and are situated ventrally and posterior to the head over the yolk. This stage allows for the differentiation of the atrium and ventricle and the visualization of structures like the atrioventricular canal and outflow tract. Researchers can evaluate the phenotypes of heterotaxia by looking at the direction in the heart loops by using zebrafish models as their test subjects. The heart typically undergoes a d-loop under left-right patterning, or situs solitus. The embryonic laterality can be disrupted, resulting in reversed (s-loop) or absent (no-loop) looping. (B) Heart contractility develops 24 hours after fertilization in zebrafish models, which have been utilized for investigating cardiomyopathy. By 48 hours, the myofilaments of the atrial and ventricular cardiomyocytes are organized differently. Zebrafish mutations or models with altered expression of genes can be used to study the different types of cardiomyopathies caused by interruptions in sarcomere assemblage and functioning [73].

Despite significant advancements in treatment methods, cardiovascular illnesses remain the leading cause of mortality globally. The zebrafish system offers promising opportunities for drug discovery, since it combines the benefits of in vitro and mammalian systems. Eight chemicals found in zebrafish that have been identified in humans have advanced to clinical studies, proving the zebrafish's relevance as a translation model due to its great genetic and functional similarity to humans. The zebrafish model serves as an ideal tool for studying human cardiac failure due to its mix of genetic and pharmacologic models, cutting-edge imaging tools, and HTS capacities. Notably, it has already demonstrated potential in discovering substances that improve heart function in particular zebrafish cardiac models, providing a solid foundation for more research [73]. These panels combined emphasize the importance of the use of zebrafish as an effective model system in the study of structural and functional cardiac diseases. Figure 2.12A points out their usefulness in the investigation of congenital defects such as heterotaxia and Fig. 2.12B illustrates their usefulness in the discovery of cardiomyopathies mechanisms. These models connect developmental biology to translational research, which allows both the understanding of mechanics and therapeutic discovery.

2.3.4 Drug Development

Drug development aims to bring new pharmaceutical medication to the market. It entails several steps, including preclinical studies, approval by regulators, clinical studies, and submitting a new drug application to obtain regulatory authorization [74]. This procedure typically takes over 10 years, starting with identifying a lead compound in drug discovery and ending with approving a new drug or vaccine. The drug development pipeline has three main categories: preclinical development, clinical trials, and approved drugs (Fig. 2.13). Before clinical trials, drugs undergo laboratory testing in preclinical development. To be listed as a preclinical program, the program must have received at least $2 million from the Foundation or an award that includes payments based on the drug reaching clinical trials [75]. For a drug to be listed in Phases 1–3 on the pipeline, it must be tested within the Therapeutics Development Network and/or be funded by the Foundation. If a sponsor chooses not to continue developing a drug, it will be moved to the discontinued section of the pipeline [76, 77].

HTS entails rapidly testing large chemical databases for compounds with beneficial therapeutic properties. HTS is a technique that tests thousands of compounds quickly using automated equipment. It is frequently used to identify potential drug candidates that can react with a specific biological target, such as a protein or an enzyme. HTS technology includes robotics, microfluidics, and imaging systems capable of analyzing data from large samples. HTS has played a critical role in developing several successful drugs due to its affordable price and time effectiveness. This subsection describes the different techniques used in HTS, such as biological and cell-based assays, and how they have revolutionized the discovery of drugs by allowing for the rapid screening of millions of compounds.

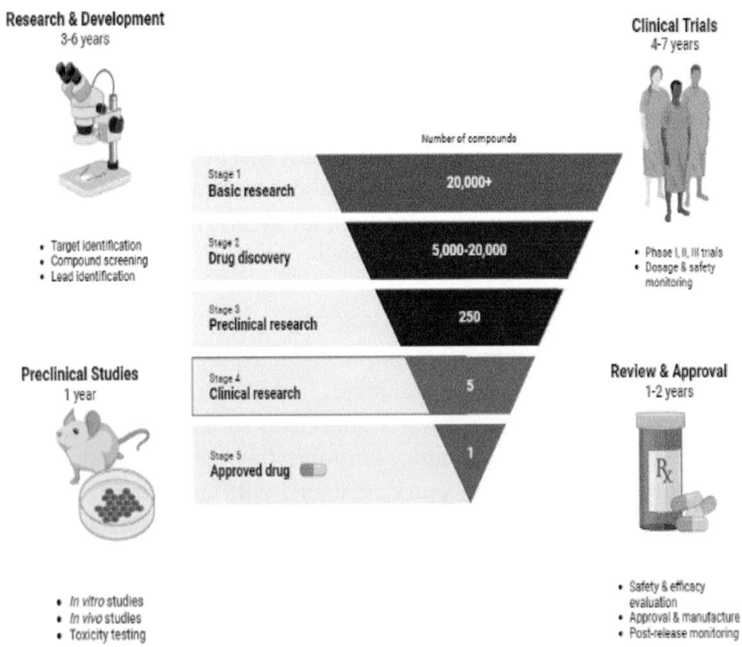

Figure 2.13 Drug development pipeline.

Lung cancer metastases exhibit epithelial-mesenchymal transition (EMT) that is linked to a poor prognosis [78]. A therapeutic approach to enhance patient outcomes is targeting EMT signaling pathways. Ishikawa et al. used A549 cells from humans with lung cancer and particular EMT markers to create a luciferase reporter cell lines system. Transforming growth factor 1 (TGF-1) persuaded EMT, and a collection of 2,350 molecules was investigated for possible inhibiting compounds [79]. The chosen substances experienced follow-up tests to validate their efficacy in repressing cell invasion and EMT. Avagacestat, GDC-0879, and levothyroxine, four substances that had favorable effects on the expression of EMT indicators and substantially decreased cell intrusiveness, were recognized as potential candidate substances. The study used a methodical approach and drugs of interest that could prevent EMT [79]. Employing TGF-induced cell lines, investigators examined the inhibitory impact of potential medications on EMT. According to Western blotting, Avagacestat suppressed E-cadherin decreased, while GDC-0879 and levothyroxine-suppressed vimentin increased. Both markers were unaffected by any substances, and cladribine had no impact. As these markers are essential for adherence and cytoskeletal changes during EMT, controlling E-cadherin and vimentin expression may regulate the invasion of cells as shown in Figs. 2.14A, 2.14B, and 2.14C. The researchers used a Matrigel invasion assay to assess how candidate drugs affected invasion caused by TGF-β1. Three groups of A549 lung cancer cells were created: control, TGF-β1 alone, and TGF-β1 paired with each potential treatment. The candidate drugs were incubated for 48 hours before the cells were gathered and seeded onto upper chambers covered in Matrigel. The cells were then incubated for a further 22 hours. The results showed that TGF-β1 increased cell invasion, whereas GDC-0879 or l-thyroxine administration substantially decreased the cells' invasion ability (Fig. 2.14D) [79].

The combination of in vitro and mammalian systems, as well as its genetic makeup or functional similarity to humans, make the model of zebrafish promising for drug discovery [80]. Yet, it has constraints, including the inability to anticipate compound exposures at higher levels and to do HTS for compounds that are water-soluble. Eight drugs found in zebrafish have advanced to clinical studies notwithstanding these restrictions, proving the relevance of this model system for translational research [81]. By utilizing HTS capacities, genetic and pharmacological designs, cutting-edge imaging tools, and the zebrafish model, it is possible to research human heart failure. Fully automated screening assays made possible by technological improvements now allow for the automated or semiautomatic evaluation of cardiovascular

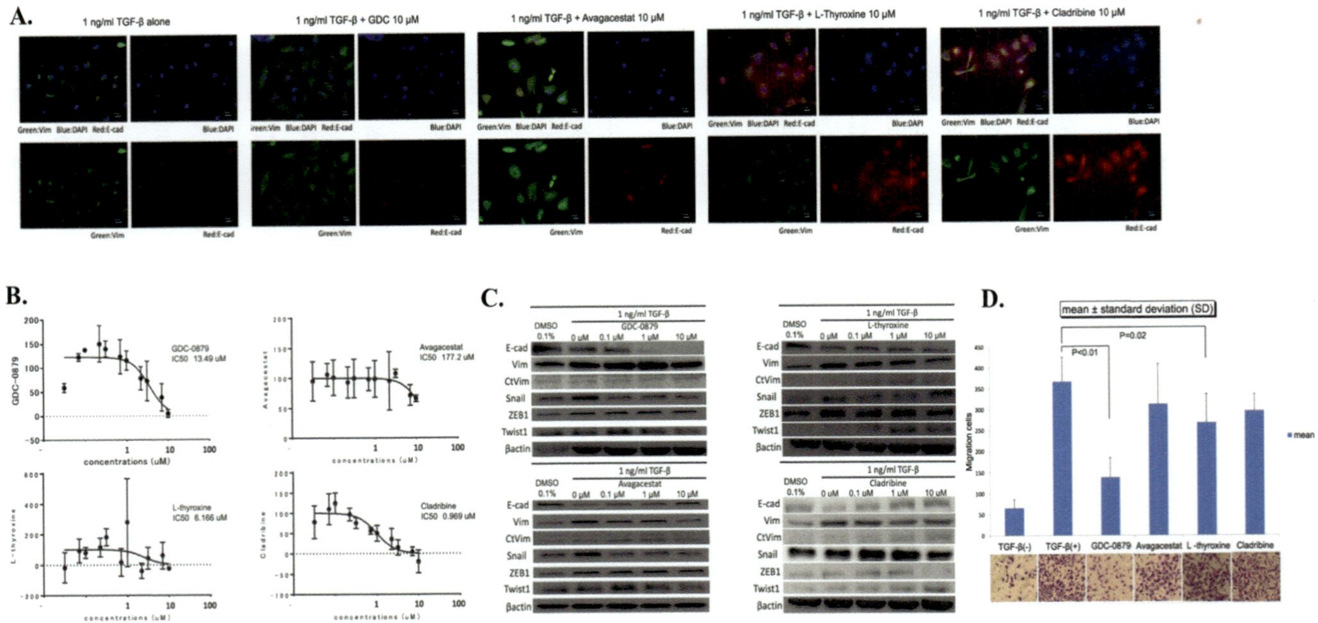

Figure 2.14 (A) An immunofluorescence image shows Vimentin and E-cadherin. E-cadherin is indicated by Alexa 568 (red) stains, vimentin by Alexa 488 (green), and the nucleus by DAPI (4′,6-diamidino-2-phenylindole) (blue). The scale bar is 184.52 m. (B) A549 lung cancer cell viability after 48 hours of treatment with drug candidates at 10 μM. Cladribine and levothyroxine (l-thyroxine), two of the candidate medications, both had IC50 values under 10 μM. (C) In A549 cells injected with TGF-β1 and various concentrations of the drug targets, Western blots of E-cadherin, vimentin, Ct vimentin, SNAIL (snail family zinc finger), ZEB1 (zinc finger E-box binding homeobox 1), TWIST1 (twist family bHLH transcription factor 1), and β-actin were performed. Avagacestat was added, inhibiting EMT in a way dependent on concentration. (D) The impact of each medication on TGF-β1-mediated cell invasion was investigated using Matrigel invasion assays. Levothyroxine (l-thyroxine) and GDC-0879 therapy substantially lowered the ability of A549 cells to invade after being exposed to TGF-β1 (1 ng/ml) [79].

parameters in zebrafish using 96-well or 384-well format [82]. Zebrafish may be utilized for both cardiotoxicity investigations and the development of drugs, making it a useful model for evaluating cardiac toxicity. Cardiotoxicity, which is brought on by irregularities in conduction, arrhythmias, and a decrease in cardiac contractility, is an important adverse effect of numerous medications that might restrict their clinical usage [73]. A zebrafish small-molecular screen indicated that 22 of the 23 investigated substances induce bradycardia and atrioventricular block, proving the viability of the zebrafish model [73]. For instance, the high effectiveness of the chemotherapeutic drug doxorubicin is constrained by the cardiac impairment caused by cumulative dose-related myocardial injury, which results in cardiac failure. A zebrafish model of doxorubicin-induced cardiomyopathy is being used to identify the cytochrome P450 family, CYP1, as a feasible therapy for human cardio protection [74]. In conclusion, the zebrafish model is a superb tool for drug discovery and cardiac research.

2.3.5 Tissue Engineering

Tissue engineering is a branch of medical engineering that uses cells, engineering, materials, and biochemical and physicochemical factors to restore, maintain, improve, or replace biological tissues. One common approach is using cells on tissue scaffolds to create new tissue that can be transplanted into the body. Tissue engineering is frequently used to repair or replace tissues like bone, cartilage, blood vessels, bladder, skin, and muscle, but it can also be used to perform specific biochemical functions. Although the terms tissue engineering and regenerative medicine are frequently used interchangeably, regenerative medicine typically emphasizes the use of stem cells or progenitor cells.

2.3.6 Understanding the Role of Epigenetic Modifications

Addressing epigenetic mechanisms, essential for gene expression and differentiation of cells, can help with optimal stem cell programming and improve the management of stem growth and differentiation [83, 84]. Epigenetic factors influence the molecular processes through which stem cell development into numerous cell lineages is quite interesting. Research is still being conducted on these pathways and how they might be used therapeutically [85–87]. By modulating gene expression, epigenetic processes are essential to modulating stem cell lineage [88, 89]. Cells have distinct gene expression profiles, although sharing an identical genetic makeup, determining how they differentiate into different kinds of cells. This cellular information is transferred throughout cell division from a single generation of cells to the next [90–94]. A stem cell's distinct pattern of expression of genes determines which cell lineages it will eventually develop into. In neurons, epigenetic processes and memory recall are retained indefinitely, limiting subsequent cell division shown in Fig. 2.15A Conrad Hal Waddington's "epigenetic landscape" theory demonstrates how diverse epigenetic changes direct the differentiation strategies of stem cell populations. He first described the epigenetic framework in the first decade of the 1900s. It shows the numerous pathways that affect the stem cell's fates. The pluripotent cells' journey throughout Waddington's landscape is depicted by troughs and crests, which stand in for genes that have experienced epigenetic alterations shown in Fig. 2.15B.

DNA-binding proteins called histones are altered via methylation, acetylation, ubiquitination, and phosphorylation, all of which control the expression of certain genes. Histone alterations, DNA methylation, and histone phosphorylation affect how chromatin is structured and how genes are regulated [96]. These epigenetic mechanisms determine the specialization and destiny of stem cells during their growth and development. When the methyl groups are inserted into cytosine residues in DNA, 5-methylcytosine (5-mC) is created [96]. This epigenetic change is known as DNA methylation. Although its purpose is complicated and reliant on CpG density, it is frequently linked to gene suppression [96]. Deamination can help maintain heterochromatic memories, long-term suppression of genes, silence transposons, and encourage mutations [97]. Comparable to somatic cells, hPSCs exhibit pervasive hypermethylation, and their ability to survive depends on persistent DNA methylation [98]. In hPSC lines, abnormal DNA methylation has been found, including alterations in promoters of genes and non-coding areas as well as lingering somatic DNA methylation [98]. Multiple hPSC lines' DNA methylation analyses have uncovered

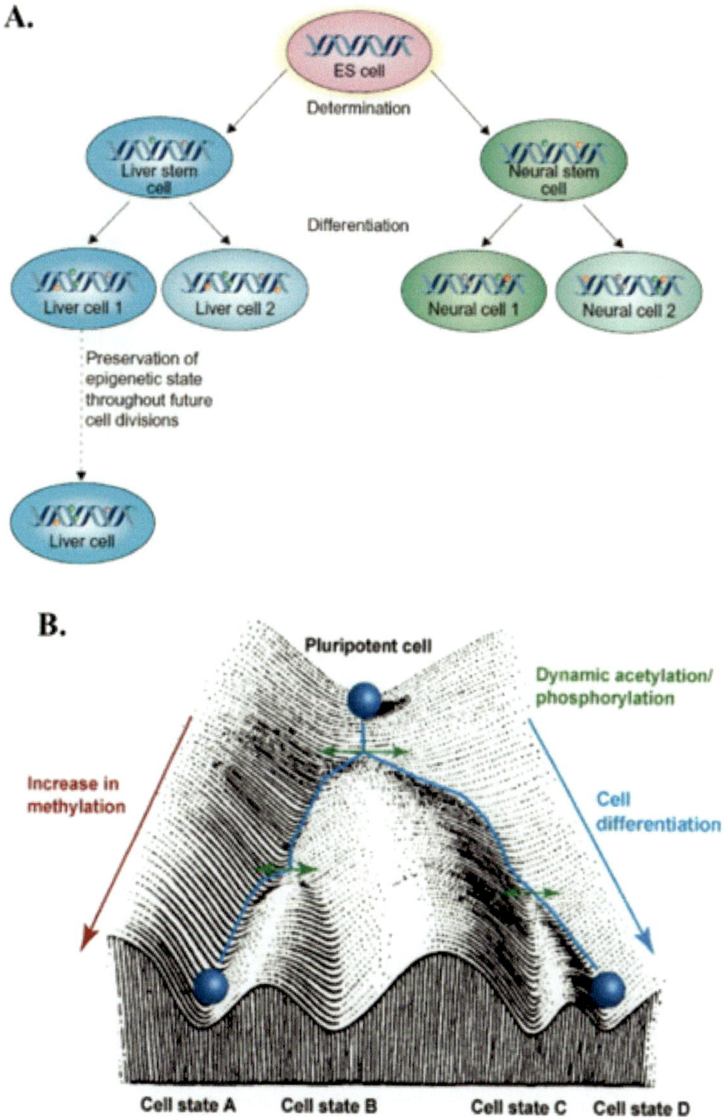

Figure 2.15 (A) The illustration depicts epigenetic imprinting at the cellular level. When liver cells and neurons are differentiated from the precursors of cells of an embryo, epigenetic information remains intact [83]. (B) Waddington's epigenetic landscape provides as a paradigm to show how different epigenetic alterations convert pluripotent ES cells to specific cell types [95].

unique trends that set them apart from somatic cells. The cells in question have aberrant DNA methylation abnormalities, numerous in malignancies, including aberrant hypermethylation of genes that suppress tumors. The hypermethylation of the gene TSPYL5, the hypermethylation and decreased activity of the antioxidant gene catalase (CAT), and the resulting increase of methylation along with gene silence are examples of specific gene changes that have been discovered [99]. Additional research is required to investigate ways to optimize culture settings to maintain stable DNA methylation. Reducing cellular stress and enhancing cell culture procedures may decrease the selection for aberrant DNA methylation in hPSCs [99]. To comprehend the pattern of distribution, particularity, stability, and consequences of DNA methylation, multiple resolutions can be used to investigate the process [100]. The regional evaluation was initially performed using methylation-sensitive enzymes and PCR amplification [101]. To discriminate between methylation and unmethylated CpGs, the bisulfite process was created [102, 103]. Genome-wide analysis at single nucleotide resolution is required to perform an objective screening of aberrant DNA methylation [104]. Because whole-genome sequencing is getting cheaper, whole-genome bisulfite sequencing (WGBS) has grown more widely available [105]. Investigating dynamical methylation alterations in individual living cells is now possible because of a newly developed DNA methylation reporter. An in-depth comprehension of the DNA methylation abnormalities in hPSCs can be obtained by integrating these techniques with RNA measurement [106].

HTS is used to detect these epigenetic aberrations in clinical settings. Cancer especially is connected to abnormal expression of genes, and epigenetic dysregulation is a cancer trigger [107]. Histones, such as monocytic leukemia zinc finger protein (MOZ), are crucial for the structure of chromatin and gene regulation, and cancer is linked to abnormalities in the MYST genes [108]. Leukemias with poor prognoses are associated with MOZ fusion protein mutations, which are also essential for healthy hematopoiesis and leukemia growth. Recently, dysregulation of histone acetyltransferases (HATs) has been related to numerous cancers [109–111]. Drugs that regulate chromatin-modifying enzymes, such as HATs inhibitors, are becoming more popular. Falk et al. created a uniform method for determining the acetylation state of synthesized histone proteins

[112]. They found antagonists from various chemical categories and prioritized verified hits over assay-interfering substances. This new HTS measures HAT activity and may lead to a new category for leukemia treatments. To gauge MOZ's acetylation efficiency on substrate peptides, investigators developed a homogenous AlphaScreen test. They made advantage of a lysine acetylation site-containing biotinylated precursor peptides from the N-terminal sequence of a core histone. An acetyl-specific antibody was used to identify acetylated residues of lysine after the biotinylated peptides were coupled to streptavidin-coated AlphaScreen donor beads. The antibody in question allows to be tested on a variety of substrates peptides since it is receptive to acetyl-lysine but does not react with nearby sequences of amino acids shown in Fig. 2.16A. With a linear correlation of 2 and 20 nM, the test signal strength was directly associated with HAT concentration in the enzyme. At 100 nM, the optimum intensity was attained. Yet, nonspecific signals rose at concentrations above 100 nM because of proteins binding to detection beads, reducing the signals-to-background ratios. The optimal working level of the enzyme being used for the experiment was determined as 10 nM shown in Fig. 2.16B. Researchers meticulously determined the AcCoA concentration of the substrate to improve the first screening experiment. Its seeming Michaelis-Menten constant, that was found to be 15 μM, was chosen as the level of concentration that they would utilize. By making this decision, the test's signal is expected to be strong and the potential of discovering drugs that attach in the AcCoA site is increased. Additionally, there is a fair chance of discovering noncompetitive blockers. Utilizing substrate peptide from histone H3 and H4 allowed researchers to gauge the MOZ enzyme's substrate specificity. In comparison to the H3 peptide, the H4 peptide continuously exhibited a greater signal-to-background ratio. A bell-like dosage response curve was shown in titration tests with the H4 peptide at a fixed MOZ dosage and continuous reaction duration, with the highest S/B reached at 50 nM peptide. The experiment could withstand up to 0.4% DMSO without suffering any appreciable signal loss. When MOZ levels were maintained between 7 and 52.9 nM, response progression graphs maintained linear for up to 120 minutes, which was shown to be the ideal range for the duration of reaction as shown in Figs. 2.16C and D. CoASH, a byproduct of the reaction that competes with AcCoA,

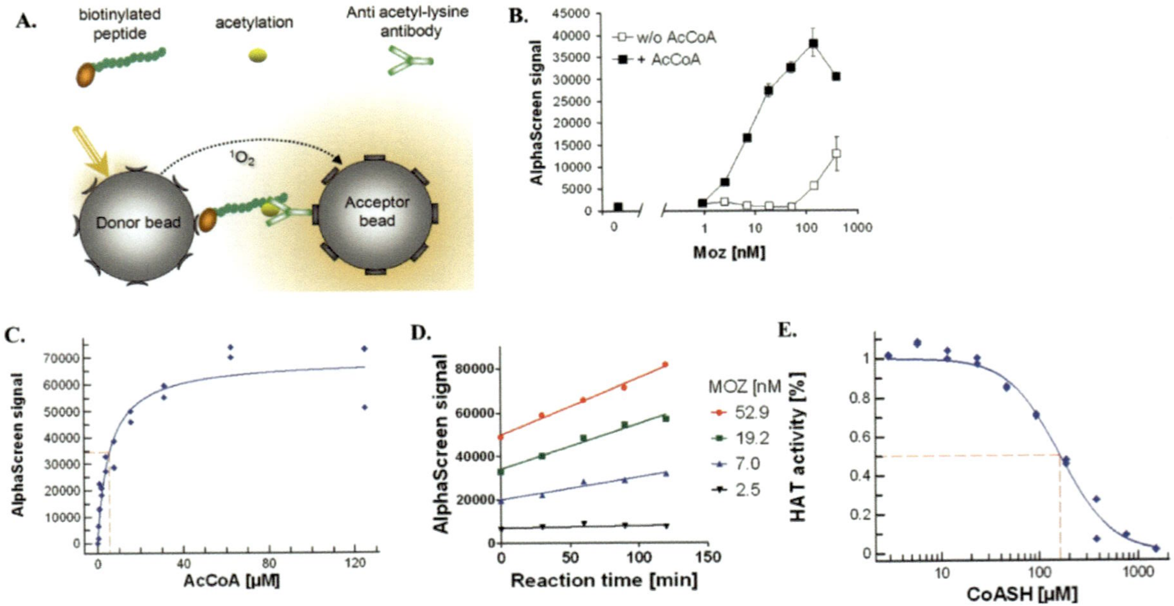

Figure 2.16 (A) The AlphaScreen HAT function assay's graphic shows the employment of an acetyl-lysine-specific antibody to draw a protein A-coated recipient bead close to a streptavidin-coated donor bead attached to a biotinylated polypeptide. Chemiluminescent emissions results from a donor bead's production of singlet oxygen when it is stimulated by light. AlphaScreen signal is decreased by inhibitors of small molecules that decrease polypeptide acetylation. (B) The assay's refinement and validation are depicted utilizing the histone H4 peptide substrates at a concentration of 50 nM. The illustration depicts the titration process of MOZ at a level of 10 µM in either with or without AcCoA. (C) AcCoA's apparent Michaelis-Menten constant (KM (app)) was established while MOZ was present at a 20 nM quantity. Depending on three tests, the mean of the findings has been determined to be 15 µM. (D) The line graph shows how the HAT response is time dependent. AcCoA was subjected to incubation with different concentrations of MOZ at a constant level of 15 µM. (E) The illustration shows the dose-response graph for CoASH, an analogous antagonist. MOZ (10 nM) and fixed AcCoA (15 µM) were used in the responses at constant concentrations [112].

was used by investigators to assess the assay's capacity to identify antagonists that block AcCoA. The investigators discovered that raising the amount of CoASH resulted in a dose-related decline in HAT action, a flat dose dependence curve, and a value for the IC50 of roughly 150 μM as shown in Fig. 2.16E. This research creates an innovative HTS for measuring HAT function and could lay the groundwork for the creation of an entirely novel category of leukemia treatments [112].

2.4 Challenges of High-Throughput Analysis of Stem Cells

2.4.1 Data Interpretation by Multi-Omics Using High-Throughput Analysis

By facilitating the creation of extensive, large-scale databases and enabling advances in numerous fields of study, which includes next generation sequencing in genomics, high-throughput RNA sequencing in transcriptomics, MS-based proteomics, high-throughput metabolomics and single-cell RNA sequencing, high-throughput technologies have certainly changed the field of medicine [113]. When discussing high throughput, the term "multi-omics data analysis" refers to the high-throughput integration and analysis of massive amounts of biological data produced by several omics technologies, as shown in Fig. 2.17. In stem cell research, this combination enables researchers to gain a thorough and systems-level understanding of biological processes or disease causes [114]. Large-scale analyses of genomic, transcriptomic, proteomic, and metabolomic information have aided in the development of targeted medicines, the identification of predictive biomarkers, and the optimization of therapeutic approaches in stem cell biology.

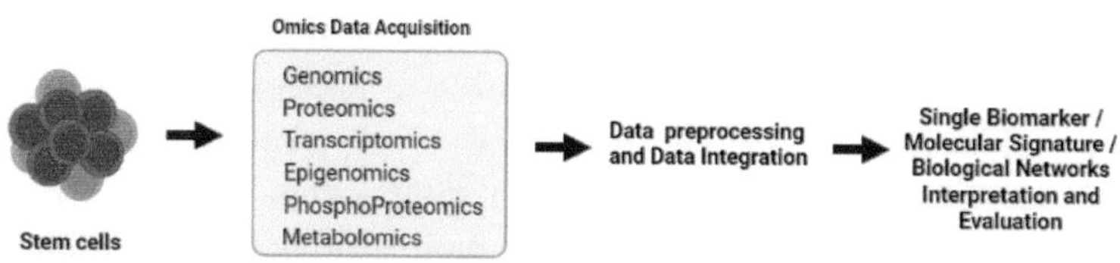

Figure 2.17 Challenges: Biological variation, integration of Omics data, and interpretation of the data.

However, there are not any strategies for integrating the resulting heterogeneous datasets without supervision. Additionally, single-cell Trio-seq and other single-cell multi-omics sequencing methods have been created [115, 116]. However, none of the methods can simultaneously evaluate many epigenome layers from the same individual cell. Researchers recently created single-cell COOL-seq technology, which has great sensitivity and coverage and can simultaneously examine chromatin accessibility, nucleosome location, DNA methylation, CNV, and ploidy from the same individual cell [117]. Multi-omics profiling is being used more frequently, and integrative analysis is required to acquire thorough understanding of biological systems. Integrating omics data utilizes various methodologies, such as kernel- or graph-based techniques, checking for weak correlations between data modalities, etc. It does, however, highlight the difficulty of interpretability and the requirement for more interpretable reconstruction of the underlying causes of sample variation [114]. The difficulties that existing approaches face Include computational scalability for larger datasets, addressing missing values, and allowing for non-Gaussian data modalities like binary or count-based features [114].

2.4.2 Navigation of Biological Variation Complexity

The integration of multi-omic data strives to convey the underlying complexity of biological systems. Due to the nonlinear and dynamic nature of biological processes, it can be difficult to comprehend and interpret the linkages and interactions across various omics layers. The latest developments in 'omics' technologies enable genomic-scale quantitative monitoring of biological components.

No single "omics" analysis can fully represent the intricacies of microbial biology, even though each one offers insightful information. To truly comprehend live microbes, it is therefore important to integrate many layers of information using a multi-'omics' approach. Although difficult, this integration of disparate "omics" information has shown its effectiveness in revealing the working ideas and dynamics of cellular systems. When it comes to reaching different biological objectives, such as cell-type categorization, regulatory network modeling, and biological process inference, there are certain potential problems. When deciding on a data integration method, it is crucial to consider the overall modeling and visualization objectives of the integrated study as well as whether the multi-omics data are matched (measured on the same cell) or unmatched (measured on distinct cells) [118]. Normalization does not reduce the impact of biological variations, reagents, personal differences, and laboratory conditions. This has been approved by a study, in which they used high-throughput data analysis on the cancer microarray data and found the batch effect, which talks about the false biological interpretation [119]. Even if the integration of massive, complex, multimodal data poses difficulties, there is a great chance of discovering novel biological insights and comprehending intricate biological mechanisms. Researchers may fully utilize multimodal data integration and improve our understanding of fundamental biological mechanisms by addressing such challenges.

2.4.3 Cost and Availability of High-Throughput Equipment

The initial cost of high-throughput equipment is frequently rather high. Examples include next-generation sequencers, robotic automation systems, and mass spectrometers. Due to the sophisticated technologies and capabilities of these instruments, their prices are frequently high. Also, to operate and maintain this equipment, a lot of capital is used. Consumables like reagents, arrays, etc., add up to the overall costs. Space and availability are also major concerns for this equipment as they require special infrastructure to fit into the working space and be available for the researchers. Currently, researchers have developed novel low-cost cloth-based microfluidic devices, such as "micro-zone plates" and "gate arrays," which offer new ways to conduct biochemical assays or regulate fluid flow, and this is made possible by the combination of low-cost cloth with low-cost high-throughput photolithography [120]. Additionally, a wider research community now has easier access to high-throughput technology because of the development of shared research resources, core facilities, and cloud-based services. The question of when to replace existing, mature technology with new practices that promise to offer outstanding benefits is a recurring conundrum. The promised benefit must be compared to the expense and risk before a final decision is made [121]. The effective integration of high-throughput data also includes computational issues, i.e., data integration, model inference, scalability, and reusability, which are also emphasized as the main activities of logic-based modeling models (Fig. 2.18).

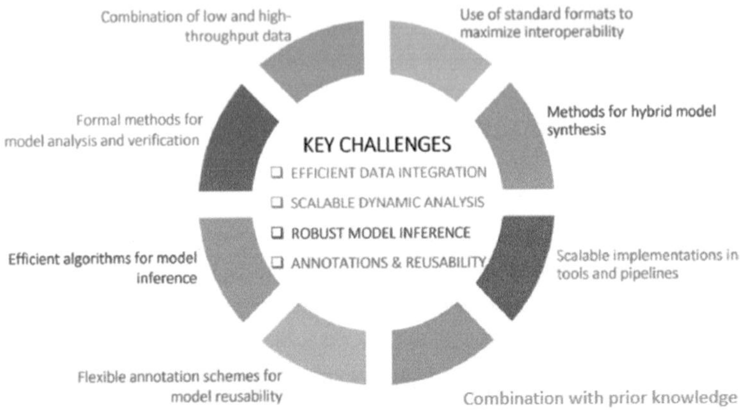

Figure 2.18 Key tasks in incorporating high-throughput information in logic-based models [122].

2.4.4 How Can Machine Learning Help to Alleviate Some of the Issues with Cell Heterogeneity and Associated Errors: Discussion?

Cell heterogeneity, or the natural variability that exists between individual cells within a population, can pose significant challenges in biomedical research and clinical applications. This variability can lead to errors and inconsistencies in experimental results, as well as difficulties in developing targeted therapies and diagnostic tools [123]. However, machine learning can help alleviate some of these issues by providing more precise and accurate analyses of heterogeneous cell populations [124]. Here are some ways that machine learning can help:

1. **Improved data analysis:** Machine learning algorithms can be trained to recognize patterns in large and complex datasets, including those generated by single-cell sequencing technologies. By analyzing these patterns, researchers can

acquire a better knowledge of the heterogeneity of unique cells within a population and identify subpopulations of cells that may have distinct biological properties or disease states [125]. This can lead to more accurate diagnosis, prognosis, and personalized treatment approaches.

2. **Cell classification and clustering:** Machine learning algorithms can be applied to classify and cluster individual cells based on their gene expression profiles, morphology, or other characteristics. This can help researchers identify distinct subpopulations of cells that may have different functions or responses to stimuli and enable them to study these cells in greater detail.
3. **Improved experimental design:** Machine learning can be used to optimize experimental design and reduce the impact of cell heterogeneity on experimental outcomes. For example, machine learning algorithms can be used to identify the optimal number of cells to sample, or to identify the most informative genes to target for sequencing.
4. **Drug discovery and development:** Machine learning can be used to identify new drug targets and develop more effective therapies for diseases. By analyzing large datasets of cell populations, machine learning algorithms can identify genes or proteins that are associated with disease states and suggest new drug targets or therapeutic strategies [126].

Overall, machine learning has the potential to significantly improve our understanding of cell heterogeneity and its impact on disease, and to enable the development of more precise and personalized treatments. However, it's important to note that these algorithms are only as good as the data they are trained on, and that careful validation and interpretation of results is still essential.

2.5 Conclusion

High-throughput analysis by machine learning is a powerful combination of automated high-throughput methods and the capabilities of machine learning algorithms. By identifying patterns and trends in large datasets that may be difficult or impossible to detect through manual analysis, machine learning algorithms can reduce the risk of errors and variability. They can also quickly analyze large amounts of data, speeding up data processing and enabling researchers to generate results more quickly. In addition, machine learning algorithms can integrate data from multiple sources and types of experiments, providing a more comprehensive view of complex biological systems. This enables researchers to identify complex relationships between different parameters, uncover novel insights, and make more accurate predictions. Furthermore, machine learning algorithms can be used to identify patient-specific biomarkers or predictors of drug response, enabling personalized medicine approaches that consider individual differences in disease susceptibility and treatment response. Overall, high-throughput analysis by machine learning has the potential to significantly improve the accuracy and efficiency of data analysis, and to enable the discovery of novel insights and personalized medicine approaches.

High-throughput analysis using machine learning is an active area of research with a lot of future potential. Researchers are focusing on developing machine learning algorithms that can effectively analyze high-throughput data, which can be complex and difficult to handle. They are also working on integrating data from multiple sources and types of experiments to provide a more comprehensive view of biological systems. Furthermore, incorporating domain knowledge into machine learning approaches is also important, as it can provide important context and constraints for high-throughput analysis. In addition, ensuring the quality of the data is crucial, and researchers are developing machine learning approaches that can help identify and correct errors or biases in the data. Real-time analysis of high-throughput data can enable more rapid decision-making and responses in healthcare and drug discovery, and machine learning can help enable this. However, machine learning models can be difficult to interpret, so researchers are working on developing approaches that provide more transparent and interpretable results. Overall, high-throughput analysis using machine learning is promising for advancing our understanding of complex biological systems, improving healthcare, and enabling more efficient drug discovery. Ongoing research in this area is likely to yield many exciting new insights and innovations in the years to come.

Acknowledgments

Manas Ranjan Gartia acknowledges the financial support from the National Science Foundation (NSF CAREER award number: 2045640) and the National Institute of General Medical Sciences of the National Institutes of Health under award number R35GM150564.

References

1. Schmidt, B. and A. Hildebrandt, *Next-generation sequencing: Big data meets high performance computing.* Drug Discov Today, 2017. **22**(4): p. 712–717.
2. Sun, J., A. Tárnok, and X. Su, *Deep learning-based single-cell optical image studies.* Cytometry A, 2020. **97**(3): p. 226–240.
3. Lan, Y., et al., *Morphology-based deep learning approach for predicting osteogenic differentiation.* Front Bioeng Biotechnol, 2022. **9**: p. 1521.
4. Patino, C.A., et al., *Multiplexed high-throughput localized electroporation workflow with deep learning–based analysis for cell engineering.* Sci Adv, 2022. **8**(29): p. eabn7637.
5. Brown, P. O. and D. Botstein, *Exploring the new world of the genome with DNA microarrays.* Nat Genet, 1999. **21**(1): p. 33–37.
6. Conzone, S. D. and C. G. Pantano, *Glass slides to DNA microarrays.* Mater Today, 2004. **7**(3): p. 20–26.
7. Chang, T.-Y., et al., *Large-scale plasmonic microarrays for label-free high-throughput screening.* Lab on a Chip, 2011. **11**(21): p. 3596–3602.
8. Zhang, L., T. Hurek, and B. Reinhold-Hurek, *Position of the fluorescent label is a crucial factor determining signal intensity in microarray hybridizations.* Nucleic Acids Res, 2005. **33**(19): p. e166–e166.
9. Sassolas, A., B. D. Leca-Bouvier, and L. J. Blum, *DNA biosensors and microarrays.* Chem Rev, 2008. **108**(1): p. 109–139.
10. Chen, L., et al., *Up-regulation of Dsg2 confered stem cells with malignancy through wnt/β-catenin signaling pathway.* Exp Cell Res, 2023. **422**(1): p. 113416.
11. Edwards, B. S., et al., *Flow cytometry for high-throughput, high-content screening.* Curr Opin Chem Biol, 2004. **8**(4): p. 392–398.
12. Adan, A., et al., *Flow cytometry: Basic principles and applications.* Crit Rev Biotechnol, 2017. **37**(2): p. 163–176.

13. Hahne, F., et al., *flowCore: A Bioconductor package for high throughput flow cytometry.* BMC Bioinformatics, 2009. **10**(1): p. 1–8.

14. Murayama, C., Y. Kimura, and M. Setou, *Imaging mass spectrometry: Principle and application.* Biophys Rev, 2009. **1**: p. 131–139.

15. Altelaar, A. M., et al., *High-resolution MALDI imaging mass spectrometry allows localization of peptide distributions at cellular length scales in pituitary tissue sections.* Int Jo Mass Spectrom, 2007. **260**(2–3): p. 203–211.

16. Li, L.-P., et al., *Applications of ambient mass spectrometry in high-throughput screening.* Analyst, 2013. **138**(11): p. 3097–3103.

17. Takats, Z., et al., *Mass spectrometry sampling under ambient conditions with desorption electrospray ionization.* Science, 2004. **306**(5695): p. 471–473.

18. Chen, H., et al., *Desorption electrospray ionization mass spectrometry for high-throughput analysis of pharmaceutical samples in the ambient environment.* Anal Chem, 2005. **77**(21): p. 6915–6927.

19. Cody, R. B., J. A. Laramée, and H. D. Durst, *Versatile new ion source for the analysis of materials in open air under ambient conditions.* Anal Chem, 2005. **77**(8): p. 2297–2302.

20. Chen, H., et al., *Differentiation of maturity and quality of fruit using noninvasive extractive electrospray ionization quadrupole time-of-flight mass spectrometry.* Anal Chem, 2007. **79**(4): p. 1447–1455.

21. Li, M., et al., *Extractive electrospray ionization mass spectrometry toward in situ analysis without sample pretreatment.* Anal Chem, 2009. **81**(18): p. 7724–7731.

22. Garrett, T. J., et al., *Imaging of small molecules in tissue sections with a new intermediate-pressure MALDI linear ion trap mass spectrometer.* Int J Mass Spectrom, 2007. **260**(2–3): p. 166–176.

23. Khatib-Shahidi, S., et al., *Direct molecular analysis of whole-body animal tissue sections by imaging MALDI mass spectrometry.* Anal Chem, 2006. **78**(18): p. 6448–6456.

24. Hannun, Y. A., *The sphingomyelin cycle and the second messenger function of ceramide.* J Biol Chem, 1994. **269**(5): p. 3125–3128.

25. Hayasaka, T., et al., *Matrix-assisted laser desorption/ionization quadrupole ion trap time-of-flight (MALDI-QIT-TOF)-based imaging mass spectrometry reveals a layered distribution of phospholipid molecular species in the mouse retina.* Rapid Commun Mass Spectrom, 2008. **22**(21): p. 3415–3426.

26. Moore, G. E., Cramming More Components onto Integrated Circuits. 1965, McGraw-Hill New York.

27. Zhou, X., et al., *The next-generation sequencing technology: A technology review and future perspective.* Sci China Life Sci, 2010. **53**: p. 44–57.

28. Fedurco, M., et al., *BTA, a novel reagent for DNA attachment on glass and efficient generation of solid-phase amplified DNA colonies.* Nucleic Acids Res, 2006. **34**(3): p. e22–e22.

29. Voelkerding, K. V., S. A. Dames, and J. D. Durtschi, *Next-generation sequencing: From basic research to diagnostics.* Clin Chem, 2009. **55**(4): p. 641–658.

30. Margulies, M., et al., *Genome sequencing in microfabricated high-density picolitre reactors.* Nature, 2005. **437**(7057): p. 376–380.

31. Ronaghi, M., et al., *Real-time DNA sequencing using detection of pyrophosphate release.* Anal Biochem, 1996. **242**(1): p. 84–89.

32. Macevicz, S. C., *DNA sequencing by parallel oligonucleotide extensions.* 1998, Google Patents.

33. Shendure, J., et al., *Accurate multiplex polony sequencing of an evolved bacterial genome.* Science, 2005. **309**(5741): p. 1728–1732.

34. Holt, R. A. and S. J. Jones, *The new paradigm of flow cell sequencing.* Genome Res, 2008. **18**(6): p. 839–846.

35. Thomson, J. A., et al., *Embryonic stem cell lines derived from human blastocysts.* Science, 1998. **282**(5391): p. 1145–1147.

36. Evans, M. J. and M. H. Kaufman, *Establishment in culture of pluripotential cells from mouse embryos.* Nature, 1981. **292**(5819): p. 154–156.

37. Ying, Q.-L., et al., *BMP induction of Id proteins suppresses differentiation and sustains embryonic stem cell self-renewal in collaboration with STAT3.* Cell, 2003. **115**(3): p. 281–292.

38. Mountford, J., *Human embryonic stem cells: Origins, characteristics and potential for regenerative therapy.* Transfus Med, 2008. **18**(1): p. 1–12.

39. Pazhanisamy, S., *Adult stem cell and embryonic stem cell markers.* Mater Methods, 2013. **3**: p. 200.

40. Nichols, J., et al., *Formation of pluripotent stem cells in the mammalian embryo depends on the POU transcription factor Oct4.* Cell, 1998. **95**(3): p. 379–391.

41. Loh, Y.-H., et al., *The Oct4 and Nanog transcription network regulates pluripotency in mouse embryonic stem cells.* Nat Genet, 2006. **38**(4): p. 431–440.

42. Zhou, Q., et al., *A gene regulatory network in mouse embryonic stem cells.* Proc Natl Acad Sci U.S.A., 2007. **104**(42): p. 16438–16443.

43. Gifford, C. A., et al., *Oligogenic inheritance of a human heart disease involving a genetic modifier.* Science, 2019. **364**(6443): p. 865–870.

44. Shi, J., et al., *PANDORA-seq expands the repertoire of regulatory small RNAs by overcoming RNA modifications.* Nat Cell Biol, 2021. **23**(4): p. 424–436.

45. Andrews, P. W., et al., *Three monoclonal antibodies defining distinct differentiation antigens associated with different high molecular weight polypeptides on the surface of human embryonal carcinoma cells.* Hybridoma, 1984. **3**(4): p. 347–361.

46. Andrews, P., et al., *Embryonic stem (ES) cells and embryonal carcinoma (EC) cells: Opposite sides of the same coin.* Biochem Soc Trans, 2005. **33**(6): p. 1526–1530.

47. Draper, J. S., et al., *Surface antigens of human embryonic stem cells: Changes upon differentiation in culture.* J Anat, 2002. **200**(3): p. 249–258.

48. Schopperle, W. M. and W. C. DeWolf, *The TRA-1-60 and TRA-1-81 human pluripotent stem cell markers are expressed on podocalyxin in embryonal carcinoma.* Stem Cells, 2007. **25**(3): p. 723–730.

49. Bühring, H. J., et al., *Novel markers for the prospective isolation of human MSC.* Ann NY Acad Sci, 2007. **1106**(1): p. 262–271.

50. Agrawal, D. K. and C. S. Boosani, *Cellular reprogramming in cardiac diseases: a feather in the hat of regenerative medicine.* J Thorac Cardiovasc Surg, 2017. **153**(2): p. 327–328.

51. Wang, X., Z. Gao, and W. Zhou, *Mass spectrometry-based quantitation combined with time-dependent metabolomics to discover metabolic features in human neurogenesis using neural constructs generated from neural progenitor cells.* Analyst, 2023. **148**(3): p. 609–617.

52. Nandan, M. O. and V. W. Yang, *The role of Krüppel-like factors in the reprogramming of somatic cells to induced pluripotent stem cells.* Histol Histopathol, 2009. **24**(10): p. 1343.

53. Li, W., et al., *High-throughput screening for myelination promoting compounds using human stem cell-derived oligodendrocyte progenitor cells.* iscience, 2023. **26**(3).

54. Merkert, S., et al., *High-throughput screening for modulators of CFTR activity based on genetically engineered cystic fibrosis disease-specific iPSCs.* Stem Cell Rep, 2019. **12**(6): p. 1389–1403.

55. Waterston, R. H., et al., *Initial sequencing and comparative analysis of the mouse genome.* Nature, 2002. **420**(6915): p. 520–562.

56. Strachan, T., S. Lindsay, and D. I. Wilson, Molecular Genetics of Early Human Development (Human Molecular Genetics). 1997.

57. Hamlin, R. L. and R. A. Altschuld, *Extrapolation from mouse to man.* 2011, Am Heart Assoc. p. 2–4.

58. Saenger, P., *Turner's syndrome.* N Engl J Med, 1996. **335**(23): p. 1749–1754.

59. Chester, N., et al., *Stage-specific apoptosis, developmental delay, and embryonic lethality in mice homozygous for a targeted disruption in the murine Bloom's syndrome gene.* Genes Dev, 1998. **12**(21): p. 3382–3393.

60. Zwaka, T. P. and J. A. Thomson, *Homologous recombination in human embryonic stem cells,* in Essentials of Stem Cell Biology. 2009, Elsevier. p. 417–422.

61. Urbach, A., M. Schuldiner, and N. Benvenisty, *Modeling for Lesch-Nyhan disease by gene targeting in human embryonic stem cells.* Stem Cells, 2004. **22**(4): p. 635–641.

62. Urbach, A. and N. Benvenisty, *Studying early lethality of 45, XO (Turner's syndrome) embryos using human embryonic stem cells.* PLoS One, 2009. **4**(1): p. e4175.

63. Eiges, R., et al., *Developmental study of fragile X syndrome using human embryonic stem cells derived from preimplantation genetically diagnosed embryos.* Cell Stem Cell, 2007. **1**(5): p. 568–577.

64. Mateizel, I., et al., *Derivation of human embryonic stem cell lines from embryos obtained after IVF and after PGD for monogenic disorders.* Human Reprod, 2006. **21**(2): p. 503–511.

65. Biancotti, J.-C., et al., *Human embryonic stem cells as models for aneuploid chromosomal syndromes.* Stem Cells, 2010. **28**(9): p. 1530–1540.

66. Avior, Y., I. Sagi, and N. Benvenisty, *Pluripotent stem cells in disease modelling and drug discovery.* Nat Rev Mol Cell Biol, 2016. **17**(3): p. 170–182.

67. Howe, K., et al., *The zebrafish reference genome sequence and its relationship to the human genome.* Nature, 2013. **496**(7446): p. 498–503.

68. Savage, A. M., et al., *tmem33 is essential for VEGF-mediated endothelial calcium oscillations and angiogenesis.* Nat Commun, 2019. **10**(1): p. 732.

69. Liang, S., et al., *Identification of novel candidate genes in heterotaxy syndrome patients with congenital heart diseases by whole exome sequencing.* Biochim Biophys Acta Mol Basis Dis, 2020. **1866**(12): p. 165906.

70. Liu, C., et al., *Rare copy number variants analysis identifies novel candidate genes in heterotaxy syndrome patients with congenital heart defects.* Genome Med, 2018. **10**(1): p. 1–13.

71. Noël, E. S., et al., *A zebrafish loss-of-function model for human CFAP53 mutations reveals its specific role in laterality organ function.* Hum Mutat, 2016. **37**(2): p. 194–200.

72. Zou, J., et al., *An internal promoter underlies the difference in disease severity between N- and C-terminal truncation mutations of Titin in zebrafish.* eLife, 2015. **4**: p. e09406.

73. Bowley, G., et al., *Zebrafish as a tractable model of human cardiovascular disease.* Br J Pharmacol, 2022. **179**(5): p. 900–917.

74. Milan, D. J., et al., *Drugs that induce repolarization abnormalities cause bradycardia in zebrafish.* Circulation, 2003. **107**(10): p. 1355–1358.

75. Paul, S. M., et al., *How to improve R&D productivity: the pharmaceutical industry's grand challenge.* Nat Reviews Drug Discov, 2010. **9**(3): p. 203–214.

76. Cummings, J. L., T. Morstorf, and K. Zhong, *Alzheimer's disease drug-development pipeline: Few candidates, frequent failures.* Alzheimer's Research Ther, 2014. **6**(4): p. 1–7.

77. Kaitin, K. I. and C. P. Milne, *A dearth of new meds.* Sci Am, 2011. **305**(2): p. 16–16.

78. Mittal, V., *Epithelial mesenchymal transition in aggressive lung cancers,* in Ahmad, A., Gadgeel, S. (eds) Lung Cancer and Personalized Medicine: Novel Therapies and Clinical Management. Advances in Experimental Medicine and Biology, vol 890. Springer, 2016: p. 37–56.

79. Ishikawa, H., et al., *A novel cell-based assay for the high-throughput screening of epithelial–mesenchymal transition inhibitors: Identification of approved and investigational drugs that inhibit epithelial–mesenchymal transition.* Lung Cancer, 2023. **175**: p. 36–46.

80. Novodvorsky, P., M. M. Da Costa, and T. J. Chico, *Zebrafish-based small molecule screens for novel cardiovascular drugs.* Drug Discov Today: Technol, 2013. **10**(1): p. e109–e114.

81. Spaink, H. P., et al., *Robotic injection of zebrafish embryos for high-throughput screening in disease models.* Methods, 2013. **62**(3): p. 246–254.

82. Burns, C. G., et al., *High-throughput assay for small molecules that modulate zebrafish embryonic heart rate.* Nat Chem Biol, 2005. **1**(5): p. 263–264.

83. Sweatt, J. D., et al., *An overview of the molecular basis of epigenetics.* 2013, *Epigenetic Regulation in the Nervous System*, 2013, Elsevier, p. 3–33.

84. Bird, A., *Perceptions of epigenetics.* Nature, 2007. **447**(7143).

85. Sell, S. and P. M. Biava, *Cancer is a problem of developmental biology: Maturation arrest and the hierarchical model of the origin of cancer from stem cells.* Stem Cells Handbook, 2013: p. 429–440.

86. Lakshmipathy, U. and C. Verfaillie, *Stem cell plasticity.* Blood Rev, 2005. **19**(1): p. 29–38.

87. Gu, W., et al., *Transplantation of bone marrow mesenchymal stem cells reduces lesion volume and induces axonal regrowth of injured spinal cord.* Neuropathology, 2010. **30**(3): p. 205–217.

88. Song, C.-H., et al., *Effect of transplantation of bone marrow-derived mesenchymal stem cells on mice infected with prions.* J Virol, 2009. **83**(11): p. 5918–5927.

89. Prockop, D. J., C. A. Gregory, and J. L. Spees, *One strategy for cell and gene therapy: Harnessing the power of adult stem cells to repair tissues.* Proc Natl Acad Sci, 2003. **100**(1): p. 11917–11923.

90. Blurton-Jones, M., et al., *Neural stem cells genetically-modified to express neprilysin reduce pathology in Alzheimer transgenic models.* Stem Cell Res Ther, 2014. **5**(2): p. 1–14.

91. Crane, A. T., J. Rossignol, and G. L. Dunbar, *Use of genetically altered stem cells for the treatment of Huntington's disease.* Brain Sci, 2014. **4**(1): p. 202–219.

92. d'Anglemont de Tassigny, X., A. Pascual, and J. López-Barneo, *GDNF-based therapies, GDNF-producing interneurons, and trophic support of the dopaminergic nigrostriatal pathway. Implications for Parkinson's disease.* Front Neuroanat, 2015. **9**: p. 10.

93. Wyse, R. D., G. L. Dunbar, and J. Rossignol, *Use of genetically modified mesenchymal stem cells to treat neurodegenerative diseases.* Int Journal Mol Sci, 2014. **15**(2): p. 1719–1745.

94. Schu, S., et al., *Immunogenicity of allogeneic mesenchymal stem cells.* J Cell Mol Med, 2012. **16**(9): p. 2094–2103.

95. Barth, T. K. and A. Imhof, *Fast signals and slow marks: The dynamics of histone modifications.* Trends Biochem Sci, 2010. **35**(11): p. 618–626.

96. Suzuki, M. M. and A. Bird, *DNA methylation landscapes: Provocative insights from epigenomics.* Nat Rev Genet, 2008. **9**(6): p. 465–476.
97. Yoder, J. A., C. P. Walsh, and T. H. Bestor, *Cytosine methylation and the ecology of intragenomic parasites.* Trends Genet, 1997. **13**(8): p. 335–340.
98. Calvanese, V., et al., *Cancer genes hypermethylated in human embryonic stem cells.* PLoS One, 2008. **3**(9): p. e3294.
99. Weissbein, U., et al., *Culture-induced recurrent epigenetic aberrations in human pluripotent stem cells.* PLoS Genet, 2017. **13**(8): p. e1006979.
100. Konki, M., et al., *Epigenetic silencing of the key antioxidant enzyme catalase in karyotypically abnormal human pluripotent stem cells.* Sci Rep, 2016. **6**(1): p. 22190.
101. Cedar, H., et al., *Direct detection of methylated cytosine in DNA by use of the restriction enzyme MspI.* Nucleic Acids Res, 1979. **6**(6): p. 2125–2132.
102. Frommer, M., et al., *A genomic sequencing protocol that yields a positive display of 5-methylcytosine residues in individual DNA strands.* Proc Natl Acad Sci, 1992. **89**(5): p. 1827–1831.
103. Tost, J. and I. G. Gut, *DNA methylation analysis by pyrosequencing.* Nat Protoc, 2007. **2**(9): p. 2265–2275.
104. Nishino, K. and A. Umezawa, *DNA methylation dynamics in human induced pluripotent stem cells.* Human Cell, 2016. **29**: p. 97–100.
105. Lister, R., et al., *Human DNA methylomes at base resolution show widespread epigenomic differences.* Nature, 2009. **462**(7271): p. 315–322.
106. Stelzer, Y., et al., *Tracing dynamic changes of DNA methylation at single-cell resolution.* Cell, 2015. **163**(1): p. 218–229.
107. Ellis, L., P. W. Atadja, and R. W. Johnstone, *Epigenetics in cancer: Targeting chromatin modifications.* Mol Cancer Ther, 2009. **8**(6): p. 1409–1420.
108. Rodenhiser, D. and M. Mann, *Epigenetics and human disease: Translating basic biology into clinical applications.* CMAJ, 2006. **174**(3): p. 341–348.
109. Voss, A. K. and T. Thomas, *MYST family histone acetyltransferases take center stage in stem cells and development.* Bioessays, 2009. **31**(10): p. 1050–1061.
110. Thomas, T. and A. K. Voss, *The diverse biological roles of MYST histone acetyltransferase family proteins.* Cell Cycle, 2007. **6**(6): p. 696–704.

111. Avvakumov, N. and J. Cote, *The MYST family of histone acetyltransferases and their intimate links to cancer.* Oncogene, 2007. **26**(37): p. 5395–5407.
112. Falk, H., et al., *An efficient high-throughput screening method for MYST family acetyltransferases, a new class of epigenetic drug targets.* J Biomol Screen, 2011. **16**(10): p. 1196–1205.
113. Hasin, Y., M. Seldin, and A. Lusis, *Multi-omics approaches to disease.* Genome Biol, 2017. **18**(1): p. 1–15.
114. Argelaguet, R., et al., *Multi-omics factor analysis—A framework for unsupervised integration of multi-omics data sets.* Mol Syst Biol, 2018. **14**(6): p. e8124.
115. Jin, W., et al., *Genome-wide detection of DNase I hypersensitive sites in single cells and FFPE tissue samples.* Nature, 2015. **528**(7580): p. 142–146.
116. Farlik, M., et al., *Single-cell DNA methylome sequencing and bioinformatic inference of epigenomic cell-state dynamics.* Cell Rep, 2015. **10**(8): p. 1386–1397.
117. Li, L., et al., *Single-cell multi-omics sequencing of human early embryos.* Nat Cell Biol, 2018. **20**(7): p. 847–858.
118. Miao, Z., et al., *Multi-omics integration in the age of million single-cell data.* Nat Rev Nephrol, 2021. **17**(11): p. 710–724.
119. Leek, J. T., et al., *Tackling the widespread and critical impact of batch effects in high-throughput data.* Nat Rev Genet, 2010. **11**(10): p. 733–739.
120. Wu, P. and C. Zhang, *Low-cost, high-throughput fabrication of cloth-based microfluidic devices using a photolithographical patterning technique.* Lab on a Chip, 2015. **15**(6): p. 1598–1608.
121. Major, J., *Challenges and opportunities in high throughput screening: Implications for new technologies.* J Biomol Screen, 1998. **3**(1): p. 13–17.
122. Hall, B. A. and A. Niarakis, *Data integration in logic-based models of biological mechanisms.* Curr Opin Syst Biol, 2021. **28**: p. 100386.
123. Zhu, Z. and C. J. Yang, *Hydrogel droplet microfluidics for high-throughput single molecule/cell analysis.* Acc Chem Res, 2017. **50**(1): p. 22–31.
124. Davatzikos, C., et al., *Precision diagnostics based on machine learning-derived imaging signatures.* Magn Reson Imaging, 2019. **64**: p. 49–61.
125. Zitnik, M., et al., *Machine learning for integrating data in biology and medicine: Principles, practice, and opportunities.* Inf Fusion, 2019. **50**: p. 71–91.

126. Catalina, M. D., et al., *The pathogenesis of systemic lupus erythematosus: Harnessing big data to understand the molecular basis of lupus.* J Autoimmun, 2020. **110**: p. 102359.

Chapter 3

High-Throughput Detection Approaches Based on Surface Plasmon Resonance

Fenni Zhang, Jinbiao Ma, Xiaoyin Liu, Jiahao Xu, Yanli Lu, and Qingjun Liu

Biosensor National Special Laboratory,
Key Laboratory for Biomedical Engineering of Education Ministry,
Department of Biomedical Engineering,
Zhejiang University, Hangzhou, 310027, P. R. China
qjliu@zju.edu.cn

Surface plasmon resonance (SPR) is a powerful analytical technique for label-free detection of molecular interactions. High-throughput approaches based on SPR have significantly contributed to fundamental biological research, clinical diagnosis, and drug discovery. In this chapter, we provide a comprehensive review of the basic principles, instrumentation, and applications of SPR-based high-throughput methods. We begin by discussing the fundamental principles and instrumentation of high-throughput SPR methods, including conventional SPR (cSPR), surface plasmon resonance

High-Throughput Imaging Methods: Cells and Tissues
Edited by Manas Ranjan Gartia and Ram V. Devireddy

ISBN 978-981-5352-08-5 (Hardcover), 978-1-003-77417-4 (eBook)
www.jennystanford.com

DOI: 10.1201/9781003774174-3

imaging (SPRi), and surface plasmon resonance microscopy (SPRM). These techniques are often combined with localized surface plasmon resonance (LSPR), microfluidics, and arrays to achieve high-throughput detection. We then introduce the latest advancements in biomolecular sensing and cell analysis using high-throughput SPR methods. These approaches have enabled the sensitive and real-time detection of biomolecular interactions, allowing for the study of complex biological processes and the identification of potential therapeutic targets. Additionally, high-throughput SPR techniques have been employed in cell analysis, providing valuable insights into cellular heterogeneity and functional characterization of individual cells. In the last part, we address the current challenges associated with high-throughput SPR and present our perspective on its potential for single-molecule and single-cell analysis in biomedical applications.

3.1 Introduction

The rapid advancements in optical technology have revolutionized the research in the study of micron-nano objects with high temporal and spatial resolution [1]. Optical detection methods, including fluorescence, chemiluminescence, colorimetry, and surface plasmon resonance (SPR) are widely used om biochemical analysis [2–5]. However, the labeling-based methods usually have drawbacks such as high biological toxicity and poor optical signal stability, which limit their applications in real-time detection [6].

Real-time tracking of target molecules is important for elucidating underlying mechanisms, and it has been a long-standing goal of the imaging community [1]. SPR, as an important optical technology, has experienced rapid development in the past 30 years due to its advantages of high sensitivity, label-free detection, and real-time monitoring. In the early 1990s, the first SPR-based biosensor was developed and applied to the study of biomolecular interactions, providing valuable information about the binding kinetics, affinity, and thermodynamics of biomolecular interactions [7]. SPR has emerged as one of the leading techniques for quantifying biomolecules and studying their interactions, and has become the "gold standard" of label-free detection. It can be combined with other techniques such as microfluidics and arrays, thus combining different advantages to achieve improved detection throughput and performance. This integration has transformed SPR into a versatile sensing platform capable of imaging individual entities ranging in size from a few nanometers to microns [8].

However, conventional SPR (cSPR) typically measures the averaged binding signals of the molecular interactions on the whole sensing chip. Heterogeneity between analytes is normally neglected due to local signal cancellation, and the sensitivity may be reduced due to the dilution of the background region [9, 10]. Besides, the single-channel cSPR analysis technology is not suitable for high-throughput and wide-range analysis of multiple vital substances. The development of multichannel, high-throughput, sensitive, and rapid SPR detection instruments has become an urgent need [11].

To address these issues, a range of advanced surface plasmon resonance imaging (SPRi) techniques have been developed to enable the imaging capabilities of SPR detection [8]. SPRi is a rapid analytical method combining SPR with a camera. SPRi retains the label-free and real-time monitoring capabilities of SPR, allowing for the measurement of molecular interactions in situ and in real time. Besides, it extends the detection capabilities by introducing imaging features. With SPRi, it is possible to track single-point resonance data across an array, enabling the analysis of individual sensing spots or regions on the chip, and overcome the limitations of averaged signals from cSPR instruments, providing improved spatial resolution and the ability to visualize and track specific binding events [12]. The combination of microfluidics with SPRi enables simultaneous measurement of multiple samples for high-throughput detection, as well as the ability to introduce reference channels to eliminate signal noise caused by disturbing factors, such as temperature drift and light source fluctuations, increasing the resolution and accuracy of the results, improving detection efficiency and reducing detection costs [13, 14]. Moreover, surface plasmon resonance microscopy (SPRM) has been developed based on the high NA microscope for high spatial resolution imaging and high-throughput detection, allowing hundreds of biomolecules to be measured in a single test. This high spatial resolution enables detailed imaging and analysis of individual binding events, expanding the capacity for high-throughput screening and analysis.

Based on these advantages, SPR-based high-throughput assays have become an important method for biochemical analysis and have been widely used in fundamental biological research [15, 16], clinical diagnosis [17], and drug discovery [18].

3.2 Typical SPR-Sensing Approaches in High-Throughput Detection

Based on the detection principle and sensing strategy, we categorize SPR techniques into cSPR, SPRi and SPRM, and the basic principles and high-throughput instrumentation strategies are discussed.

3.2.1 Detection Principle of SPR

As shown in Fig. 3.1A, SPR is an optical phenomenon that occurs as a collective oscillation of charge density at a metal-dielectric interface. When incident light passes through the substrate at an angle greater than the critical angle and is completely reflected at the metal-dielectric interface (light-dense medium propagating into the less dense medium), the total reflection at the interface not only returns the reflected light to the light dense medium, but also penetrates the medium as an evanescent wave (penetration depth ~200 nm). The evanescent wave decays exponentially to the sample solution in the vertical (z-direction) of the sensing surface:

$$\frac{\Delta I}{I} = \exp\left(-\frac{\Delta z}{L}\right) \tag{3.1}$$

where $\Delta I/I$ is the plasmonic image contrast change of samples; Δz is the vertical distance change; and L is the decay length constant of the evanescent field. Using Eq. (3.1), we can calculate the vertical positions of samples. Combined with lateral (image plane) localization, plasmonic imaging can achieve 3D tracking of samples with a nanometer resolution.

When the frequency of the evanescent wave coincides with the frequency of the free electrons (i.e., plasma) oscillating on the metal surface, the incident light may couple with the free electrons on the metal surface. The free electrons are made to resonate under certain conditions, so that the energy of the reflected light is reduced or disappears completely, at this point the angle of incidence is called the SPR angle (θ_R):

$$\sin(\theta_R) = \sqrt{\frac{\varepsilon_1 \varepsilon_m}{(\varepsilon_1 + \varepsilon_m)\varepsilon_2}} \tag{3.2}$$

where ε_1, ε_2, ε_m are the dielectric constants of the electrolyte, glass slide, and metal film, respectively. Using Eq. (3.2), the resonance angle is determined by the refractive index (RI) of the metal/dielectric interface and the solution near the surface [1]. When the RI of the metal surface changes due to molecular binding or chemical reactions, the SPR angle also changes. Therefore, the shift in the resonance angle, the change in the intensity of the reflected light, and the phase of the modulation can be used to obtain quantitative information on molecular adsorption processes and intermolecular bonding.

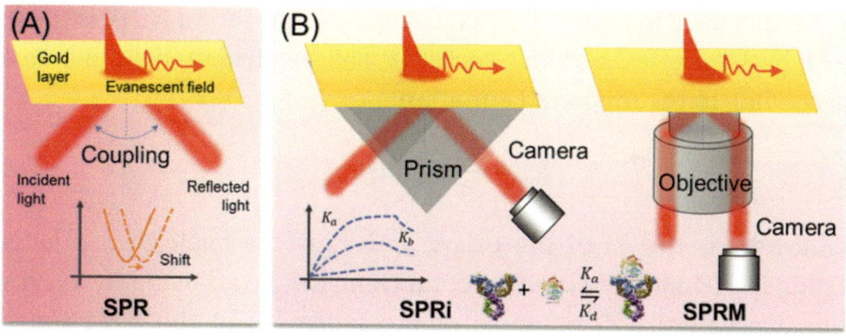

Figure 3.1 The principle of typical SPR sensing approaches. (A) SPR; (B) SPRi and SPRM.

As all substances have an index of refraction, the SPR technique maintains the natural state of the target analyte and does not require prior labeling for detection [19]. More importantly, SPR enables kinetic real-time detection, which can provide important information about molecular binding, such as binding sites, binding modes and molecular conformations by the calculation of its binding constants (k_a), and dissociation constants (k_b). It can convert undetectable biological signals into detectable optoelectronic signals. For example, when substance A with charge X_1 binds to substance B with charge X_2 to form substance C with charge X_3,

$$A^{(X_1)} + B^{(X_2)} \underset{K_b}{\overset{K_a}{\rightleftharpoons}} C^{(X_3)} \tag{3.3}$$

Using Eq. (3.3), the change in the RI during the biological binding process of substance A and substance B to form substance C can be converted into an optical signal by SPR detection. The charge change can also be accurately detected by combining it with other techniques.

However, the cSPR device only measures the averaged signals without resolving local information, which cannot meet the needs of local sensing detection and high-throughput screening [20]. To enable the imaging capability of SPR, SPRi was developed by combining a camera with a prism-SPR setup. As a coupling structure, the Otto structure requires the thickness of the dielectric between the metal and the prism to be controlled on the order of a wavelength, which is more difficult to achieve; the Kretschmann structure, as shown in Figure 3.1B, in which a thin film of a highly conductive metal (gold, silver, etc.) of suitable thickness is coated on the bevel of the prism, has a wider range of applications and is now generally used in SPRi [8]. By coupling the light through a prism with a high RI, the effective wave vector (k_e) of the incident light increases according to the following equation:

$$\mathrm{k}_e = \frac{2\pi}{\lambda} n_D \sin(\theta_R) \tag{3.4}$$

where n_D is the RI of the prism and θ_R is the angle of incidence. Using Eq. (3.4), SPR coupling conditions can be adjusted by varying the wavelength (λ) or angle of incidence (θ) of the light source [8]. However, due to its prism-based setup, the technique has only limited spatial resolution (micron scale) and magnification (one to five times) and is subject to image distortion (optical aberration effect) [9, 11].

Unlike prism-based SPRi techniques, SPRM uses a high numerical aperture (NA) oil immersion objective as the coupling component. The objective lens directs the collimated light from the laser source, reflects it out of the rear focal plane of the objective lens, and excites the surface plasmon polaritons (SPP) at a high incidence angle relative to the gold film outside the objective lens, and then finally collects the reflected SPR image. Thanks to the use of a high NA oil immersion objective, the spatial resolution and magnification of plasma imaging are significantly improved. Image distortion problems have been resolved, making SPRM a suitable tool for single-cell plasma response imaging even at the sub-cellular level. The spatial resolution of SPRM differs in two directions: in the transverse direction, where the resolution is diffraction-limited, and in the longitudinal direction, where the resolution is limited by SPR propagation.

3.2.2 High-Throughput Instrumentation of SPR

As shown in Fig. 3.2A, cSPR is often in the form of prismatic coupling. The light source is usually a composite light source or white light, which is used to provide incident light covering the applied wavelength for most detection needs. A collimating lens is placed in front of the light source to collimate the emitted light. The angle of the optical path is adjusted by simulating a specific SPR angle. Coupling the incident light at the prism allows a sensitive response to changes in the RI on the metal film carrier. When the sample flows across the surface of the metal film-coated slide, the sample to be measured will specifically bind to the pre-fixed ligand, or the micro-motion of the cell after attachment will cause the change in RI, which in turn will be collected by the reflection probe and transmitted to the spectrophotometer via optical fiber, which will decompose the received light signal and obtain the signal of the incident light at different wavelengths to obtain the SPR-sensing spectrum. The SPR-sensing spectrum is then obtained.

It has been shown that the SPR effect can only occur with P-polarized light [11]. This is because SPP places additional demands on the matching of momentum in addition to energy. As shown in Fig. 3.2B, the SPRi technology uses a single wavelength P-polarized light source. A collimating lens is placed in front of the light source to collimate the emitted light, which is then further expanded by a 4F prism, allowing the incident light to illuminate the area to be detected over a wide area, achieving a large field of view. The reflected light is captured by the camera through a zoom lens, allowing imaging at different magnifications.

As shown in Fig. 3.2C, to further improve the spatial resolution of the imaging, SPRM was developed in combination with a high NA oil immersion objective [1]. For SPRM, a P-polarized monochromatic beam is focused on the back focal plane of the objective with a certain offset from the optical axis, allowing the light to illuminate the sensing chip at a specific resonance angle. A 45° spectroscopic lens is set behind the objective lens for the reflection and transmission of the incident light. Set up a reflector to adjust the direction of the incident light to reduce the volume of the optical path. The reflected light is then collected by the same objective lens and imaged by the camera.

As shown in Fig. 32D, for high-throughput SPR detection, localized surface plasmon resonance (LSPR), microfluidics, and arrays are commonly used. A brief overview of these three high-throughput strategies is given as follows.

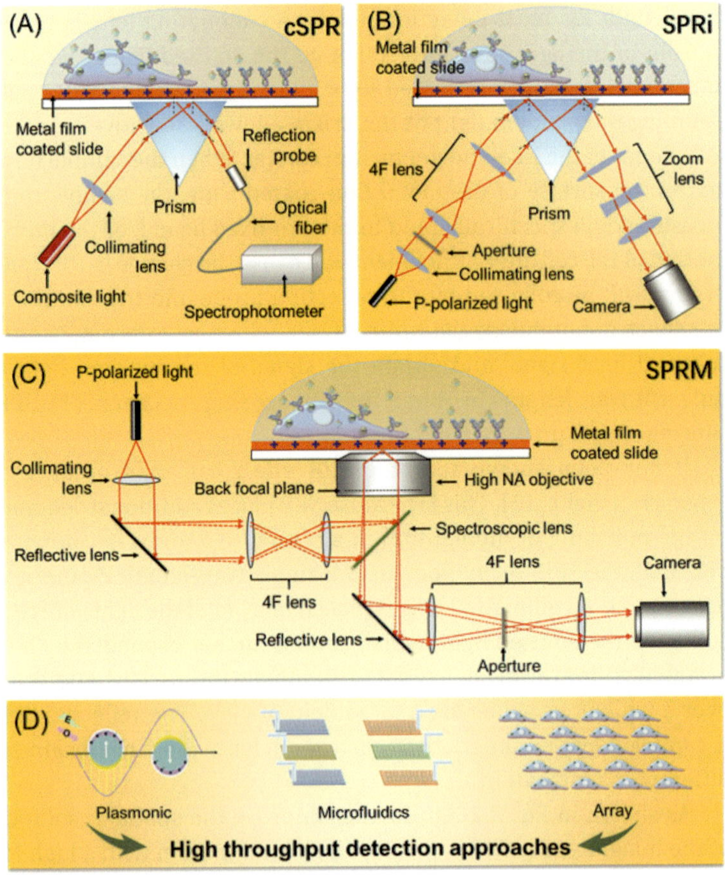

Figure 3.2 The construction of typical SPR sensing approaches in high-throughput detection.

LSPR is a nanoscale optical phenomenon. When the properties of the precious metal nanoparticles themselves are altered, the absorption intensity of the incident light by the nanoparticles is altered accordingly. By functionalizing the surface of the nanoparticles with specific molecules, they can be made to adsorb to specific molecules, and the intrinsic properties can be altered and expressed in the spectrum [21]. The main difference between SPR and LSPR is the length scale of the plasma material [22]. Both detection mechanisms are sensitive to changes within the decay length of their associated equipartition excitations. It has been shown that LSPR can be easily excited by incident light in dark-field microscopy as long as the wavelength of the LSPR matches the plasma band of the nanostructure. In combination with dark-field microscopy for tracking, analysis, and identification of nanoparticles, high spatial resolution imaging of multiple individual nanoparticles is achieved [11, 23].

Microfluidics is a key part of the SPR high-throughput approaches, which can be achieved through the preparation of multichannel chips by micro- and nano-processing. The sample transfer can be controlled by the microfluidic device and the feed system. The simultaneous analysis of different samples can be achieved by functionalizing the different channels to achieve high throughput. To prevent cross-contamination, different channels for various analyses can be made into closed chambers, helping simplify complex analytical protocols and reducing sample size, assay time, and reagent costs [13]. The application of microfluidics can reduce sample consumption, reduce the size of the assay device, improve the anti-interference capability of the biosensor assay, and make it flexible and portable for field applications, but the process of increasing the number of channels to improve the cSPR throughput is too complex and limited by the space available to improve the throughput number [24].

To fully exploit the potential of spatially resolved SPR imaging and microscopy methods, functionalized methods capable of transferring biological functions to plasma structures in a spatially resolved manner are required. In contrast to cSPR, which must successively analyze multiple species to be measured, SPRi can simultaneously record the signals generated by each array point on a microarray chip [25]. High-throughput analysis of multiple analytes to be measured in complex samples can be performed, resulting in significant savings in analysis time and sample size. The arrays are available in a variety of forms, not only with multiple different molecules immobilized in a large number of arrays, but also with plasma features that deliver a single type of biomolecule to multiple nanoscales [7]. Arrays of nanoholes [12], nanocups [26], etc., have proven to be an effective strategy for SPR to achieve high-throughput detection.

3.3 Application of SPR Sensing Approaches in Molecular Interaction Detection

Measuring molecular binding is a main task in understanding biological processes at the molecular scale, as the binding kinetic constants determine the rate of binding and dissociation of drugs and their receptors, and determining molecular binding is essential for screening drugs, detecting disease biomarkers, and understanding biological processes at the molecular level [15, 18]. SPR is widely used as an effective analytical method for studying molecular interactions, and we present here some high-throughput applications. We first present the progress of cSPR in high-throughput molecular interaction.

3.3.1 Molecular Interaction Detection with High-Throughput CSPR

Immunoassay-based methods allow high-throughput detection of molecular interactions. However, they provide endpoint results that hinder the extraction of kinetic information due to the lack of sufficient temporal resolution. Compared to other technologies, cSPR can detect molecular interactions in real time and record the entire kinetic process, which is crucial for the analysis of intrinsic mechanisms [1].

Disease diagnosis is one of the main purposes of molecular interaction detection. As important markers for medical diagnostics, miRNAs play a regulatory role in the normal function of the body and are known to be involved in a variety of biological processes (drug resistance, metabolism, proliferation, development, differentiation). They can be found in tissue samples and body fluids. As shown in

Fig. 3.3A, the SPR platforms can simultaneously detect multiple miRNAs targets, greatly improving detection efficiency for early diagnosis and prognosis of disease treatment. Different surface functionalization and signal amplification strategies can help promote the response of SPR biosensing [27].

Similarly, exosomes contain a variety of molecular contents that show potential for cancer diagnosis. However, the detection and molecular analysis of exosomes are technically challenging and usually requires extensive sample purification and labeling. A label-free, high-throughput method for the quantitative analysis of exosomes based on arrays is described by Im et al. The nano-plasma exosome assay is based on transmission SPR through periodic nanopore arrays. Each array is functionalized with antibodies to enable the analysis of exosomal surface proteins and proteins present in exosome lysates. This method consumes only a small amount of specimen, and the captured exosomes can be easily eluted from the device for downstream analysis, such as genomic analysis. In summary, these methods will facilitate comprehensive exosome analysis by generating proteomic and genetic information [28].

Recent research suggests that a turning point in SPR sensor research will be the combination of SPR strategies with other techniques to improve integration and sensitivity, with microfluidics being an ideal solution. As shown in Fig. 3.3B, Sonato et al. demonstrated a new approach to SPR biosensing based on microfluidics, SAW mixing and an innovative real-time phase-interrogating grating coupled with SPR technology, which has previously proven to be a promising technique in terms of detection sensitivity, setup scalability, and high throughput [29]. The present work can be designed into different patterns to immobilize different probes, which is expected to enable high-throughput detection of molecular interactions. As shown in Fig. 3.3C, Chen et al. developed a microfluidic-based high-throughput multi-array LSPR optical biosensor device with 480 nano-plasma-sensing sites and enabled parallel multiplex immunoassays of six cytokines in complex serum matrices throughout parallel on-chip assays (including sample and reagent loading, incubation and washing) as well as 10-fold replicate multi-analyte assays for each sample using the entire biosensor array in less than 40 min [30].

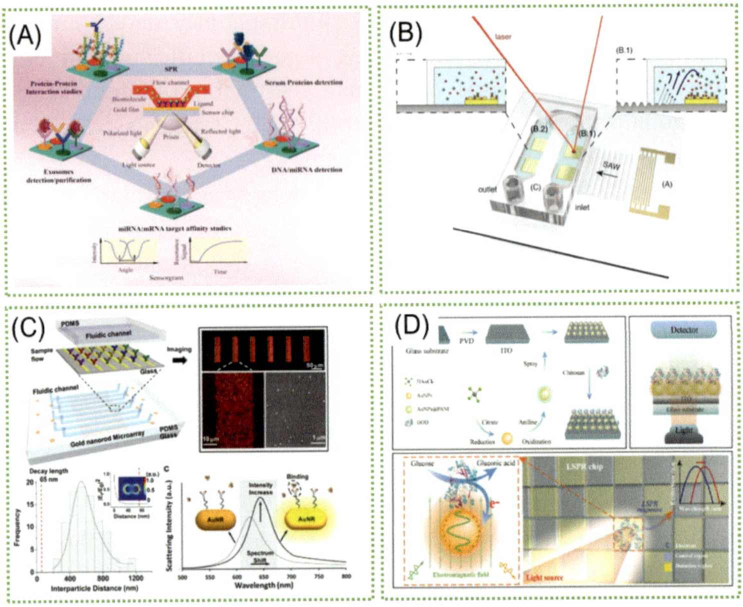

Figure 3.3 High-throughput detection applications for surfaces plasmon resonance. (A) Schematic diagram of an SPR biosensor, indicating its ability to monitor interactions between DNAs, RNAs, proteins and other ligands or cofactors (Reproduced from Jebelli et al. [27]; copyright 2020, Elsevier). (B) A surface acoustic wave (SAW)-enhanced grating coupling phase-interrogation SPR microfluidic biosensor (Reproduced from Sonato et al. [29]; copyright 2016, The Royal Society of Chemistry). (C) Microfluidic-based multi-array high-throughput LSPR chip: Schematic diagram of an LSPR microarray chip, histogram of the interparticle distance of AuNR on an LSPR microarray chip using SEM images, Principle of the LSPR microarray method (Reproduced from Chen et al. [30]; copyright 2015, Nature). (D) The bio-sensing platform for space charge density monitoring and corresponding analyte detection: fabrication process of the LSPR chip, the layered structure of LSPR chip and the detection strategy, principle of the charge density monitoring (Reproduced from Chen et al. [21]; copyright 2022, American Chemical Society).

To further improve the detection capability, electrochemical methods can be combined with cSPR detection, which can provide additional sensing information during optical measurements. As shown in Fig. 3.3D, Chen et al. proposed an LSPR biosensing platform for local charge density monitoring and corresponding analyte detection. The platform transfers electrons generated from GOD-catalyzed glucose to the nanocomposites of the LSPR chip and uses the change in the electron distribution of the nanocomposites, leading to a shift in the wavelength of the iso-exciton resonance for charge density calculation. The ability to detect spatial charge density in different regions of a single chip and the assessment of bioactive substances enables high-throughput detection [21]. With this multi-area LSPR chip, a promising approach can be provided for detecting space charge density and evaluating bio-reactive material at different locations at the chip/solution interface to achieve high-throughput detection.

3.3.2 Molecular Interaction Detection with SPR

Although cSPR enables real-time kinetic detection of molecular interaction and high-throughput detection by strategies such as LSPR or microfluidics, it lacks spatial resolution, making it difficult to further improve the local sensitivity and throughput. Moreover, single-molecule assays with heterogeneous analytical capabilities are needed to elucidate the mechanisms in biochemical processes in deeper depth. As mentioned earlier, SPRi is more suitable for high-throughput molecular detection because of its imaging capabilities. It has become a hot spot for molecular interaction detection in recent years.

By integrating SPRi with molecular microarrays, high-throughput molecular interaction detection can be achieved. As shown in Fig. 3.4A, Zhu et al. utilized SPRi in combination with antibody microarrays specific to the extracellular domains of exosome membrane proteins to detect exosomes in tumor cell culture medium quantitatively. Compared with time-consuming and inefficient methods such as electron microscopy, western blot, and ELISA, SPRi is a convenient, direct, and quantitative measurement technique for the detection of particles of around 100nm, like exosomes [31]. As shown in Fig. 3.4B, the on-demand generation of microarrays is facilitated by the use of trehalose vitrified phospholipid vesicles, allowing each element to have a unique composition for label-free and high-throughput analysis of biomolecular interactions by SPRi. This method is stable, efficient, and can be applied to various label-free sensing platforms that require lipid membranes for high-throughput analysis. The fabricated microarrays were stable over medium for long storage periods. The efficient and easily scalable nature of the method has the potential to be applied to a variety of label-free high-throughput sensing platforms [32].

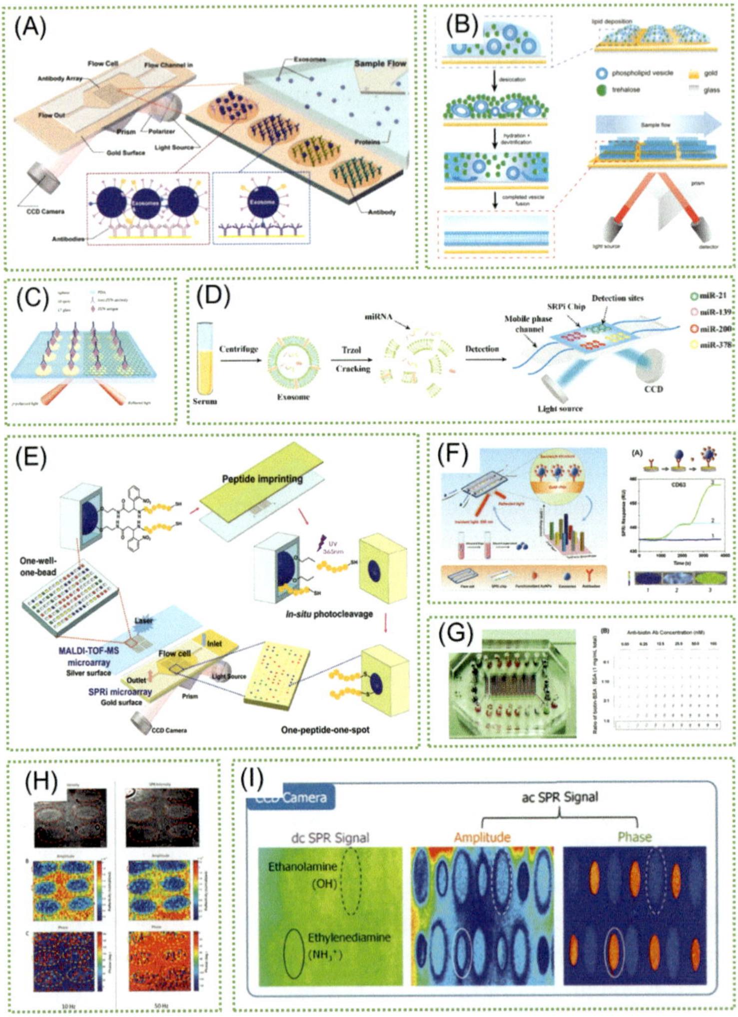

Figure 3.4 SPR imaging for high-throughput detection of molecular interaction applications. (A) Schematic diagram of SPRi in combination with antibody microarray to capture and detect exosomes in cell culture supernatant (Reproduced from Zhu et al. [31]; copyright 2014, American Chemical Society). (B) Schematic diagram of the process of vesicle deposition, desiccation, and devitrification upon hydration of the trehalose matrix on the modified SPR sensor chips (Reproduced from Hinman et al. [32]; copyright 2015, American

Chemical Society). (C) Schematic illustration of the graphene-coating enhanced SPRi chip for immunoassay (Reproduced from Wei et al. [33]; copyright 2018, Elsevier). (D) Extraction and detection of the multiplex exosomal miRNAs using SPRi biosensor (Reproduced from Wu et al. [34]; copyright 2021, Elsevier). (E) Overview of Integrated OBOC Peptide Library Screening: Peptide beads were trapped into the microwell array for MALDI-TOF-MS detection. Peptides were in situ photocleaved and imprinted onto the SPRi chips for detection (Reproduced from Wang et al. [35]; copyright 2014, American Chemical Society). (F) Working principle of plasmonic AuNPs-enhanced SPRi biosensor for multiplex analysis of NSCLC-derived exosomes via different recognition sites (Reproduced from Fan et al. [36]; copyright 2020, Elsevier). (G) The left picture is the layout of the microfluidic device. The right one is the difference between the SPR images obtained after and before the one-step immunoreaction (Reproduced from Luo et al. [37]; copyright 2008, The Royal Society of Chemistry). (H) SPR image, amplitude and phase images at 10 and 50 Hz obtained by surface impedance imaging (Reproduced from Foley et al. [38]; copyright 2008, American Chemical Society). (I) The P-EIM signals are imaged by a CCD camera, and vary with the applied voltage. The DC component of the response is the average intensity, and the ac component is comprised of a phase and amplitude (Reproduced from MacGriff et al. [39]; copyright 2013, American Chemical Society).

To improve the performance of SPRi, different materials and nanostructures have been explored for sensitivity enhancement. As shown in Fig. 3.4C, for the first time, Wei et al. demonstrated that the sensitivity of SPRi can be effectively improved by applying a single-layer graphene layer to the chip, which is suitable for high-throughput SPRi [33]. As shown in Fig. 3.4D, Wu et al. developed an SPRi biosensor for simultaneous and ultrasensitive detection of multiple exosomal miRNAs in a clinical sample using Au-on-Ag heterostructure and DNA tetrahedral framework (DTF). This high-throughput biosensor provides an accurate platform for early diagnosis of NSCLC [34]. As shown in Fig. 3.4E, Wang et al. developed a bimodal imprint microarray system for high-throughput affinity analyses by SPRi, which provides a solution for high-efficiency peptide screening [35]. As shown in Fig. 3.4F, highly sensitive, high-throughput, and low-LOD multiple feature analysis of NSCLC-related exosomes was achieved by a multichannel SPRi chip functionalizing anti-CD63, anti-EGFR, and anti-EpCAM on gold nanoparticles [36]. As shown in Fig. 3.4G, a PDMS microfluidic device containing an array of gold spots to which antigens or antibodies of interest can be attached was designed and fabricated to reduce reaction time and sample consumption, and for real-time detection of the immunocomplex formation with high sensitivity and high throughput [37]. The exploration of different materials and nanostructures has expanded the possibilities for improving SPRi performance and has opened up new avenues for high-sensitivity, high-throughput, and multiplexed biosensing applications.

Furthermore, to expand the detection capability of SPRi, the SPR-based electrochemical impedance imaging technique has been developed and used as a sensitive signal transduction method for sensors, and more information can be obtained. As shown in Fig. 3.4H, three images—the DC component and the amplitude and phase of the AC component—could be simultaneously obtained using surface impedance imaging based on the sensitive dependence of SPR on local surface charge density demonstrated by Kyle J. et al [38]. As shown in Fig. 3.4I, plasma-based electrochemical impedance microscopy can be used in the charge-based detection of small molecules. Local variations in charge density on the surface of the electrode appear as optical SPR signals. This signal consists of DC and AC components and is measured with high spatial resolution. The DC component represents the cSPR imaging signal, and the AC component represents the amplitude and phase of the surface impedance. This technique is used to detect and differentiate small molecules based on charge status, breaking the mass limitations of cSPR measurements, and providing new ways for small molecule detection [39].

3.3.3 Molecular Interaction Detection with SPRM

To further improve the spatial resolution of SPRi, SPRM was developed by coupling the SPR illumination with a high NA objective for high temporal-spatial imaging, which has been used in many applications of molecular interaction detection.

As shown in Fig. 3.5A, a time-resolved digital immunoassay based on plasmonic imaging of nanoparticles was reported, which is used to detect biomarkers with a wide dynamic range rapidly. This approach enables the detection of single-molecule binding on the sensor surface by detecting antibody-conjugated nanoparticles. SPRM enables the real-time counting of the binding events when nanoparticles interact with the antibody-captured biomarker molecules, providing dynamic information for more rapid and accurate binding detection. Real-time counting and super localization tracking of each nanoparticle enable the resolution of two binding events at a distance less than the diffraction limit, which increases the dynamic range and minimizes counting errors [40].

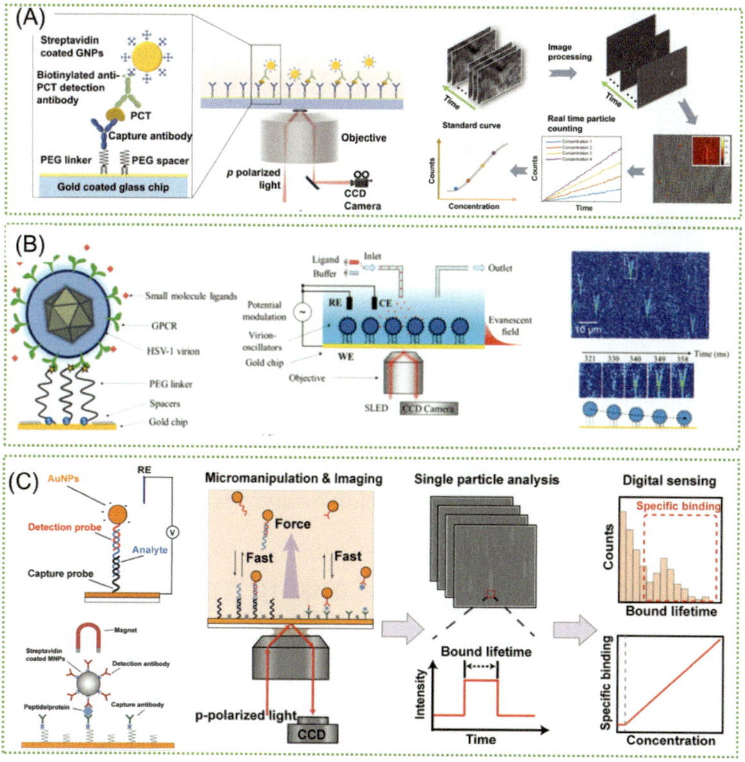

Figure 3.5 SPR microscopy for high-throughput array detection of molecular interaction applications. (A) Schematic illustration of time-resolved digital immunoassay (Reproduced from Jing et al. [40]; copyright 2019, American Chemical Society). (B) Fabricating virion oscillators and detecting ligand binding to membrane proteins with virion oscillators (Reproduced from Ma et al. [16]; copyright 2018, American Chemical Society). (C) The principle of Sensing Single Molecule under Micro Manipulation: The SSM3 actively tunes molecular binding kinetics by single-molecule micromanipulation through nanoparticles; single-nanoparticle imaging with SPRM records the dynamic binding process by a charge-coupled device (CCD) or complementary-metal-oxide-semiconductor (CMOS) image sensor. Data analysis at the single-particle level reveals the dynamic features of specific and nonspecific binding events, and digital counting of only specific events enables ultrasensitive detection of analyte concentration (Reproduced from Zeng et al. [41]; copyright 2022, Proceedings of the National Academy of Sciences of the United States of America).

By integrating SPRM with molecular manipulation assays, high-throughput single entity detection can be achieved with increased sensitivity. As shown in Fig. 3.5B, the virion oscillators displaying the receptor on the envelope are assembled on the surface of the chip through a flexible polymer, and an alternating electric field is applied to drive their oscillation. Ligand and receptor binding kinetics are measured by tracking changes in oscillation amplitude in real time with subnanometer precision. This new technique can be used to measure the binding of various ligands to different types of membrane proteins, helping to understand cell function and to screen drugs [16]. To further circumvent the limitations of mass transport and thermodynamics, as shown in Fig. 3.5C, the SPRM-based system was developed by Zeng et al. in combination with modulation such as magnetic and electric fields. By selecting low-affinity ligands to facilitate repetitive binding and dissociation, it can quantify the heterogeneous lifetime of each molecular complex to distinguish specific binding from nonspecific background noise. The limit of detection with femtomolar sublevels in buffer and spiked human serum can be achieved in 15 min by digitally counting the number of repetitive specific binding events [41].

3.4 Application of SPR-Sensing Approaches in Cell Biology

Cells are the fundamental building blocks of all living organisms, and understanding their activity is crucial for studying physiological processes. Therefore, the analysis of cellular activity has been one of the main focuses of bioanalytical chemistry [9]. Multi-cell analysis only obtains the averaged information of the dynamic signals from multiple individual cells. Single-cell behavior is neglected, and the detection sensitivity is limited due to the dilution of the background. SPR has emerged as a high-throughput application for detecting single cells.

3.4.1 Cell Biology Study with SPRi

SPRi has a large imaging field of view and can sensitively detect the molecular-binding kinetics in multiple cells simultaneously.

Detecting the binding kinetics between drug molecules and membrane protein targets is important for screening the best drug in molecularly targeted therapy of cancer. As shown in Fig. 3.6A, Zhang et al. used SPRi to map the distribution and interaction dynamics of membrane proteins in in situ environments. Quantifying the expression levels of membrane proteins on the cell membrane and the interaction kinetics of drugs is very important for cancer diagnosis and treatment. It allows direct quantification of membrane protein expression levels and ligand binding kinetics at single-cell resolution, while simultaneously examining multiple cells in real time [18].

With the proper pattern of the SPR surface, high-throughput detection of multiple channels can be achieved. As shown in Fig. 3.6B, an SPRi-based immunosensor was developed using specific monoclonal antibodies. The effects of different mixtures of SAMs on immobilized antibody density and antigen-antibody binding sensitivity were investigated. It can be applied to multi-period, multiple detections, and multichannel analysis [42]. As shown in Fig. 3.6C, deep-UV SPR enables highly sensitive autofluorescence imaging of living cells without staining. Although the typical autofluorescence of living cells is masked by background noise, deep-UV SPR enhances the autofluorescence signal [43].

As a versatile method, SPRi can be easily integrated with other techniques to enhance the detection capabilities of multi-parameter detection. As shown in Fig. 3.6D, a 2D-SPR imager is used to monitor intracellular signal transduction via local RI change in cells adhering on a gold sensor slide without an indicator reagent. Complex intracellular reactions could be monitored in real-time and characterized by the 2D–SPR imager, and it is further expected that signal transmission followed by the translocation of signaling proteins could be observed at the single-cell level with the high-resolution 2D–SPR imager [44]. As shown in Fig. 3.6E, by combining SPRi and self-sorting microwell technology, the product secretion of individual cells can be monitored, and the good position coordinates obtained from SPRi can be used to punch holes from the microwell for cell isolation [45]. As shown in Fig. 3.6F, an SPRi-impedance sensor was developed to simultaneously detect changes in RI and impedance by coating the sensor chip with comb-shaped electrodes. Multiple parameters of living cells could be detected by culturing cells on a chip. This method may enable us to clarify complex functions of living cells that affect RI and impedance, and apply this in the medical field [46].

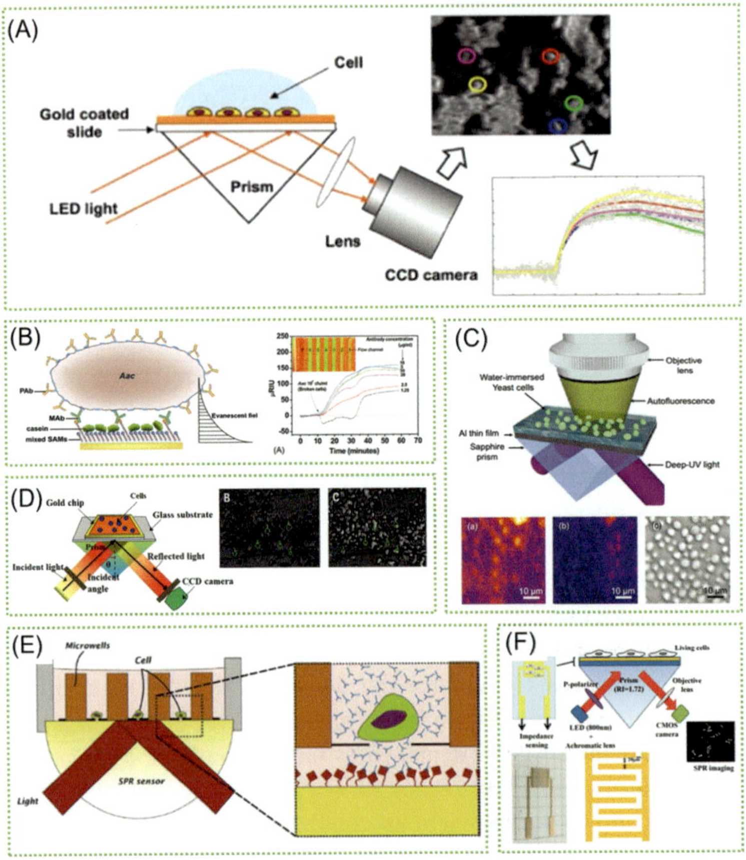

Figure 3.6 SPR imaging for high-throughput array detection of cell biology applications. (A) Schematic diagram of the SPRi setup with cells on the surface of gold-coated glass chip (Reproduced from Zhang et al. [18]; copyright 2015, American Chemical Society). (B) The detection of Acidovorax avenae subsp. Citrulli (Aac) using specific monoclonal based on SPRi: The binding interaction of MAb 11E5 and Aac and PAb; SPR sensorgram illustrated the example of Aac detection after the injection of broken cells of Aac (Reproduced from Puttharugsa et al. [42]; copyright 2011, Elsevier). (C) Schematic diagram of living yeast cell observation using deep-UV SPR. Water-immersed living yeast cells autofluorescence images on the aluminum thin film with deep-UV SPR of p-polarization, without deep-UV SPR of s-polarization and conventional reflection image (Reproduced from Che et al. [43]; copyright 2022, Molecular

Diversity Preservation International). (D) Schematic illustration of the used 2D-SPR system equipped with collimator and a CCD camera with four kinds of lens for image magnification; 2D–SPR image of the sensor chip surface obtained before K+ injection and after the injection of Hanks' solution as a control experiment; 2D–SPR image of the sensor chip surface obtained after 100 mM KCl injection, which shows a robust increase of reflection intensity in PC12 cell regions (Reproduced from Shinohara et al. [44]; copyright 2013, Elsevier). (E) The working principle of Mc SPR inter. A microwell chip containing an array of microwells filled with single cells is attached to a ligand-functionalized SPR sensor surface. This combination is incubated. Product secreted by a cell diffuses via the 5 μm pore at the bottom of the microwell toward the sensor surface where it is captured (Reproduced from Abali et al. [45]; copyright 2017, Elsevier). (F) SPRi-impedance (SPRi-IMP) sensor chip and system for living cell analysis: The SPRi-IMP sensor, composed of a light source, p-polarizer, prism, objective lens, and a CMOS camera, simultaneously detects living cell-derived RI and impedance changes on a sensor chip; SPRi-IMP sensor chip is a glass slide coated with comb-shaped thin gold film in an interdigitated electrode structure with width/spacing of 20 μm (Reproduced from Yanase et al. [46]; copyright 2019, Molecular Diversity Preservation International).

3.4.2 Cell Biology Study with SPRM

SPRM are usually based on a high NA microscopic platforms, which provide high spatial resolution. Aberrations in the field of view are minimal, allowing for high-throughput analysis of multiple single-cells and for subcellular imaging.

Membrane proteins are the most popular drug targets, and play an important role in cell signaling. However, in situ measurements in membrane environments (maintaining the conformation and function of membrane proteins) remain challenging. As shown in Fig. 3.7A, label-free imaging, detection, and mass and size measurement of individual virus particles in solution could be obtained by high-resolution SPRM. The intensity of the particle image is related to the mass of the particle, and the mass detection limit is about 1 ag. This method allows the measurement of the mass of individual virus particles as well as the study of virus binding activity to ligands [47]. As shown in Fig. 3.7B, binding kinetics of ligand interactions with individual bacteria were measured based on SPRM, and statistical analysis of bacterial heterogeneity was performed. This method can be used to understand the interaction between bacteria and external ligands through label-free and real-time measurements, which has a great effect on understanding their pathogenic mechanism, antibiotic resistance, and immune evasion [48]. As shown in Fig. 3.7C, an interferometric plasmonic imaging technique was developed to track the intrinsic fluctuations of individual bacterial cells and analyze the fluctuations to determine potential energy curves and quantify the adhesion strength of single cells, providing a promising platform for studying biological and surface interactions [49].

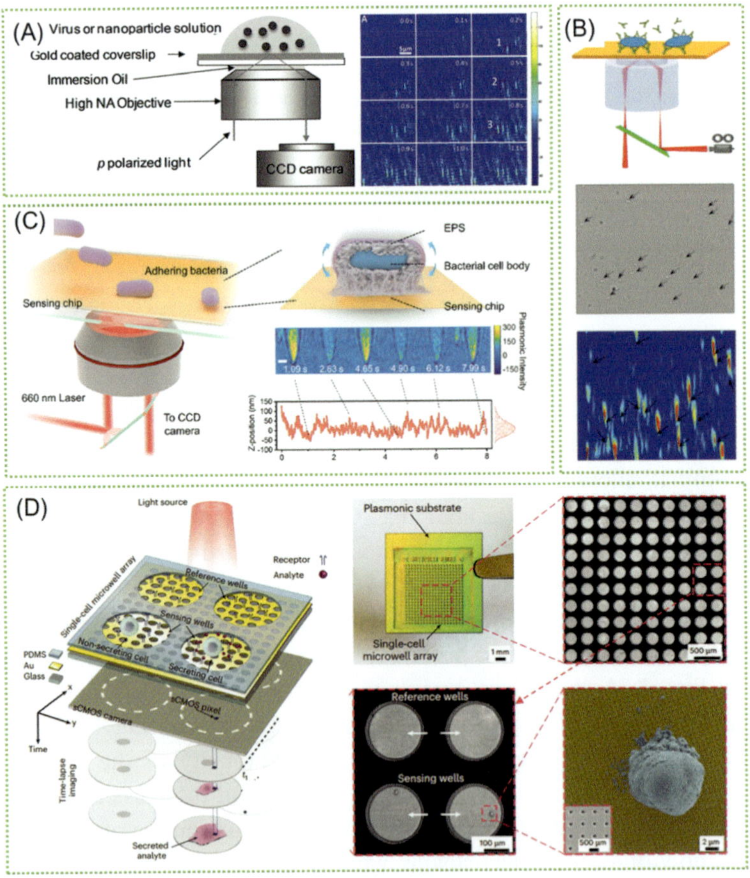

Figure 3.7 SPR microscopy for high-throughput array detection of cell biology applications. (A) Schematic of the SPRM experiment setup; SPRM images of influenza A virus on bare gold. The color map shows the relative SPR image intensity in mDeg (Reproduced from Wang et al. [47]; copyright 2010, Proceedings of the National Academy of Sciences of the United States

of America). (B) Plasmonic imaging of protein interactions with single bacterial cells: schematic of the plasmonic imaging setup using a high numerical objective and the immobilization of bacterial cells on the surface of the gold chip by covalently attached antibodies; bright-field optical image of bacteria; plasmonic image of bacteria shown as V-shaped diffraction patterns (Reproduced from Syal et al. [48]; copyright 2015, Elsevier). (C) Imaging of the vertical variation of adherent bacteria: setup of the plasma interference imaging system, schematic representation of the interface between the bacterial envelope and surface; time-resolved plasma image of a single adherent bacterium (top) and corresponding vertical distance variation (bottom) (Reproduced from Liu et al. [49]; copyright 2020,Proceedings of the National Academy of Sciences of the United States of America). (D) Microwell array-based SPRM for high-throughput detection of single-cell secretions: Schematic of the working principle of SPRM showing all four main elements: LED, PDMS micromesh, gold nanohole array substrate and sCMOS camera, Images of high-throughput microwell arrays at different scales (Reproduced from Ansaryan et al. [25]; copyright 2020, Nature).

Understanding the cellular secretion is crucial for studying cellular biological processes, investigating the evolution of dynamic diseases such as cancer, and developing new drug therapies. As shown in Fig. 3.7D, a high-throughput SPRM system based on microtiter arrays was developed that allows simultaneous real-time monitoring of hundreds of single cells, visualization, quantification, and prolonged analysis of their motility and morphology, as well as secretions. The intensity changes caused by the spectral changes of extraordinary optical transmission are directly measured by CMOS upon binding of the analytes secreted on the sensor surface. High-throughput measurements will help to study the physiological mechanisms associated with protein secretion [25].

3.5 Conclusion and Outlook

Over the past three decades, SPR technology has experienced significant growth and has been extensively utilized in various fields for biomolecular interaction analysis. Its label-free and real-time monitoring capabilities have made it a valuable tool in studying a wide range of molecular interactions. The development and integration of complementary tools and techniques have greatly enhanced the detection capabilities of SPR, particularly in high-throughput analysis and imaging assays. This chapter mainly reviews the SPR-based techniques, including cSPR, SPRi, and SPRM, and their instrumentation and application in high-throughput detection. The cSPR offers high sensitivity and is often combined with LSPR for high-throughput detection, which primarily detects average signals and lacks the analysis of heterogeneity. The imaging-based techniques like SPRi and SPRM, are often coupled with microfluidics and microarrays for high-throughput detection, providing local spatial resolution and enabling the analysis of heterogeneity in samples. However, there are still some limitations in SPR-based high-throughput detection.

First of all, the selectivity and specificity are still significant challenges in label-free SPR detection, especially for high-throughput and multiplex detection. SPR detection still relies on specific antigen-antibody or ligand-receptor reactions to enhance its specificity. It does not directly and specifically identify analytes. To improve the selectivity and specificity, proper surface modification assays, reference sensors, and more efficient detection assays are needed. As for sensitivity, although it can be enhanced through signal amplification methods and functionalization strategies, SPR is still susceptible to various sources of noise. Factors such as the light source, temperature fluctuations, environmental conditions, and mechanical noise can impact the accuracy and reliability of SPR measurements. The integration of other techniques with SPR provides new solutions to increase sensitivity for signal amplification, such as electrochemical detection, molecular modulation, and others. Furthermore, in SPR-based high thruput detection, the robustness remains a challenge, due to the limitation of both microfluidic or microarray fabrication and maturation. A more simplified strategy for high-throughput detection is needed.

In summary, while SPR has made remarkable progress and has found widespread use in biomolecular interaction and cell biology analysis, there are still limitations to be addressed. Enhancing the specificity, sensitivity, and robustness of SPR measurements, as well as developing simplified and user-friendly platforms, will be crucial for expanding its applications, particularly in high-throughput single-molecule and single-cell analysis in the biomedical field.

Acknowledgments

This work was supported by the National Natural Science Foundation of China (Grant No. 82102180) and the Fundamental Research Funds for the Central Universities 2023-KYY-515007-0003. This work was also supported by the Zhejiang Provincial Natural Science Foundation of China (Grant No. LZ24H180001).

References

1. Zhou X. L., Yang Y., Wang S., et al. Surface plasmon resonance microscopy: From single-molecule sensing to single-cell imaging. *Angew Chem Int Ed Engl*, 2020, 59(5): 1776–1785.
2. Yoo S. M., Lee S. Y. Optical biosensors for the detection of pathogenic microorganisms. *Trends Biotechnol*, 2016, 34(1): 7–25.
3. Kaur B., Kumar S., Kaushik B K. Recent advancements in optical biosensors for cancer detection. *Biosens Bioelectron*, 2022, 197: 113805.
4. Shin Y-H, Teresa Gutierrez-Wing M., Choi J-W. Review—recent progress in portable fluorescence sensors. *J Electrochem Soc*, 2021, 168(1).
5. Huang Y., Chen W., Chung J., et al. Recent progress in fluorescent probes for bacteria. *Chem Soc Rev*, 2021, 50(13): 7725–7744.
6. Yin L., Wang W., Wang S., et al. How does fluorescent labeling affect the binding kinetics of proteins with intact cells? *Biosens Bioelectron*, 2015, 66: 412–416.
7. Bockova M., Slaby J., Springer T., et al. Advances in surface plasmon resonance imaging and microscopy and their biological applications. *Annu Rev Anal Chem* (Palo Alto Calif), 2019, 12(1): 151–176.
8. Hinman S. S., McKeating K. S., Cheng Q. Surface plasmon resonance: Material and interface design for universal accessibility. *Anal Chem*, 2018, 90(1): 19–39.
9. Liu Y-N, Liu X-W. Single-cell plasmonic imaging for activity analysis. In *Surface Plasmon Resonance in Bioanalysis*, 2021. p. 107–143.
10. Gooding J. J., Gaus K. Single-molecule sensors: Challenges and opportunities for quantitative analysis. *Angew Chem Int Ed Engl*, 2016, 55(38): 11354–11366.
11. Wang W. Imaging the chemical activity of single nanoparticles with optical microscopy. *Chem Soc Rev*, 2018, 47(7): 2485–2508.

12. Prasad A., Choi J., Jia Z., et al. Nanohole array plasmonic biosensors: Emerging point-of-care applications. *Biosens Bioelectron*, 2019, 130: 185–203.
13. Liu Y., Sun L., Zhang H., et al. Microfluidics for drug development: From synthesis to evaluation. *Chem Rev*, 2021, 121(13): 7468–7529.
14. Xie X., Ma J., Wang H., et al. A self-contained and integrated microfluidic nano-detection system for the biosensing and analysis of molecular interactions. *Lab Chip*, 2022, 22(9): 1702–1713.
15. Zhang P., Ma G., Dong W., et al. Plasmonic scattering imaging of single proteins and binding kinetics. *Nat Methods*, 2020, 17(10): 1010–1017.
16. Ma G., Syu G. D., Shan X., et al. Measuring ligand binding kinetics to membrane proteins using virion nano-oscillators. *J Am Chem Soc*, 2018, 140(36): 11495–11501.
17. Yanase Y., Hiragun T., Ishii K., et al. Surface plasmon resonance for cell-based clinical diagnosis. *Sensors* (Basel), 2014, 14(3): 4948–4959.
18. Zhang F., Wang S., Yin L., et al. Quantification of epidermal growth factor receptor expression level and binding kinetics on cell surfaces by surface plasmon resonance imaging. *Anal Chem*, 2015, 87(19): 9960–9965.
19. Wang P. Phase-sensitive surface plasmon resonance sensors for highly sensitive bioanalysis. In *Surface Plasmon Resonance in Bioanalysis*, 2021. p. 55–88.
20. Su Y-W, Fang Y-M, Li T-F. Surface plasmon resonance sensing in cell biology and drug discovery. In *Surface Plasmon Resonance in Bioanalysis*, 2021. p. 1–53.
21. Chen Z., Zhang F., Li Y., et al. Bio-electron transfer modulated localized surface plasmon resonance biosensing with charge density monitoring. *Biosensors and Bioelectronics*, 2022, 201.
22. Masson J. F. Portable and field-deployed surface plasmon resonance and plasmonic sensors. *Analyst*, 2020, 145(11): 3776–3800.
23. Sikes J. C., Wonner K., Nicholson A., et al. Characterization of nanoparticles in diverse mixtures using localized surface plasmon resonance and nanoparticle tracking by dark-field microscopy with redox magnetohydrodynamics microfluidics. *ACS Phys Chem Au*, 2022, 2(4): 289–298.
24. Ma J., Du M., Wang C., et al. Advances in airborne microorganisms detection using biosensors: A critical review. *Front Environ Sci Eng*, 2021, 15(3): 47.

25. Ansaryan S., Liu Y. C., Li X., et al. High-throughput spatiotemporal monitoring of single-cell secretions via plasmonic microwell arrays. *Nat Biomed Eng*, 2023.

26. Zhang D., Lu Y., Jiang J., et al. Nanoplasmonic biosensor: Coupling electrochemistry to localized surface plasmon resonance spectroscopy on nanocup arrays. *Biosens Bioelectron*, 2015, 67: 237–242.

27. Jebelli A., Oroojalian F., Fathi F., et al. Recent advances in surface plasmon resonance biosensors for micrornas detection. *Biosensors and Bioelectronics*, 2020, 169.

28. Im H., Shao H., Park Y. I., et al. Label-free detection and molecular profiling of exosomes with a nano-plasmonic sensor. *Nat Biotechnol*, 2014, 32(5): 490–495.

29. Sonato A., Agostini M., Ruffato G., et al. A surface acoustic wave (SAW)-enhanced grating-coupling phase-interrogation surface plasmon resonance (SPR) microfluidic biosensor. *Lab Chip*, 2016, 16(7): 1224–1233.

30. Chen P., Chung M. T., McHugh W., et al. Multiplex serum cytokine immunoassay using nanoplasmonic biosensor microarrays. *ACS Nano*, 2015, 9(4): 4173–4181.

31. Zhu L., Wang K., Cui J., et al. Label-free quantitative detection of tumor-derived exosomes through surface plasmon resonance imaging. *Anal Chem*, 2014, 86(17): 8857–8864.

32. Hinman S. S., Ruiz C. J., Drakakaki G., et al. On-demand formation of supported lipid membrane arrays by trehalose-assisted vesicle delivery for SPR imaging. *ACS Appl Mater Interfaces*, 2015, 7(31): 17122–17130.

33. Wei W., Nong J., Mei Y., et al. Single-layer graphene-coated gold chip for enhanced spr imaging immunoassay. *Sensors and Actuators B: Chemical*, 2018, 273: 1548–1555.

34. Wu W., Yu X., Wu J., et al. Surface plasmon resonance imaging-based biosensor for multiplex and ultrasensitive detection of nsclc-associated exosomal mirnas using DNA programmed heterostructure of Au-on-Ag. *Biosens Bioelectron*, 2021, 175: 112835.

35. Wang W., Li M., Wei Z., et al. Bimodal imprint chips for peptide screening: Integration of high-throughput sequencing by ms and affinity analyses by surface plasmon resonance imaging. *Anal Chem*, 2014, 86(8): 3703–3707.

36. Fan Y., Duan X., Zhao M., et al. High-sensitive and multiplex biosensing assay of NSCLC-derived exosomes via different recognition sites based on SPRi array. *Biosens Bioelectron*, 2020, 154: 112066.

37. Luo Y., Yu F., Zare R. N. Microfluidic device for immunoassays based on surface plasmon resonance imaging. *Lab Chip*, 2008, 8(5): 694–700.

38. Foley K. J., Shan X., Tao N. J. Surface impedance imaging technique. *Anal Chem*, 2008, 80(13): 5146–5151.

39. MacGriff C., Wang S., Wiktor P., et al. Charge-based detection of small molecules by plasmonic-based electrochemical impedance microscopy. *Anal Chem*, 2013, 85(14): 6682–6687.

40. Jing W., Wang Y., Yang Y., et al. Time-resolved digital immunoassay for rapid and sensitive quantitation of procalcitonin with plasmonic imaging. *ACS Nano*, 2019, 13(8): 8609–8617.

41. Zeng Q., Zhou X., Yang Y., et al. Dynamic single-molecule sensing by actively tuning binding kinetics for ultrasensitive biomarker detection. *Proc Natl Acad Sci U S A*, 2022, 119(10): e2120379119.

42. Puttharugsa C., Wangkam T., Huangkamhang N., et al. Development of surface plasmon resonance imaging for detection of acidovorax avenae subsp. Citrulli (AAC) using specific monoclonal antibody. *Biosens Bioelectron*, 2011, 26(5): 2341–2346.

43. Che Lah C. N. H., Morisawa H., Kobayashi K., et al. Autofluorescence imaging of living yeast cells with deep-ultraviolet surface plasmon resonance. *Photonics*, 2022, 9(6).

44. Shinohara H., Sakai Y., Mir T. A. Real-time monitoring of intracellular signal transduction in pc12 cells by two-dimensional surface plasmon resonance imager. *Anal Biochem*, 2013, 441(2): 185–189.

45. Abali F., Stevens M., Tibbe A. G. J., et al. Isolation of single cells for protein therapeutics using microwell selection and surface plasmon resonance imaging. *Anal Biochem*, 2017, 531: 45–47.

46. Yanase Y., Yoshizaki K., Kimura K., et al. Development of spr imaging-impedance sensor for multi-parametric living cell analysis. *Sensors* (Basel), 2019, 19(9).

47. Wang S., Shan X., Patel U., et al. Label-free imaging, detection, and mass measurement of single viruses by surface plasmon resonance. *Proc Natl Acad Sci U S A*, 2010, 107(37): 16028–16032.

48. Syal K., Wang W., Shan X., et al. Plasmonic imaging of protein interactions with single bacterial cells. *Biosens Bioelectron*, 2015, 63: 131–137.

49. Liu Y. N., Lv Z. T., Lv W. L., et al. Plasmonic probing of the adhesion strength of single microbial cells. *Proc Natl Acad Sci U S A*, 2020, 117(44): 27148–27153.

Chapter 4

Hyperspectral Imaging of Optically Resonant Surfaces for Multiplexed Sensing of Biomaterials

Qinming Zhang, Shirin Parvin, Liang Dong, and Meng Lu
Department of Electrical and Computer Engineering,
Iowa State University,
Ames, IA 50011, USA
menglu@iastate.edu

The objective of this chapter is to introduce the implementation of hyperspectral imaging (HSI) technology for the detection of biomarkers using optical biosensors. These biosensors are designed to support optical resonances that are sensitive to the absorption of target biomaterials, such as cells, exosomes, proteins, and nucleic acids. Through the integration of HSI with optical biosensors, it becomes feasible to capture images of the distribution of optical resonances across both spatial and spectral dimensions. The high spatial and spectral resolutions offered by HSI microscopy can

High-Throughput Imaging Methods: Cells and Tissues
Edited by Manas Ranjan Gartia and Ram V. Devireddy

ISBN 978-981-5352-08-5 (Hardcover), 978-1-003-77417-4 (eBook)
www.jennystanford.com

DOI: 10.1201/9781003774174-4

facilitate the characterization of optical transducers, including metasurfaces, photonic crystal (PC), and plasmonic devices. This allows for label-free analysis of microarray-based assays, cell-based assays, and even single nanoparticles. To demonstrate the potential of this technology, the application of HSI to characterize a novel exosome microarray built on a PC biosensor is explained in detail.

4.1 Introduction

HSI technology is widely utilized in microscopy to capture detailed information about the spectral and spatial properties of biological samples. One of the key advantages of HSI microscopy is its ability to capture a large amount of spectral information from a single image, resulting in more accurate and detailed information. This makes HSI microscopy an ideal tool for imaging the distribution of optical resonances from nanostructured substrates that support reflection or transmission resonant modes. These substrates, such as metal surfaces, plasmonic devices, and photonic crystal (PC) sensors, can be fabricated over a large surface area. The absorption of biomolecules, nanoparticles, or cells can induce a local change in optical resonance, which can be characterized using an HSI system with high throughput and spatial resolution [1–5]. By processing and analyzing HSI microscopy data using specialized algorithms or scripts, different types of target-molecule binding can be identified based on their spectral signatures, or cell behavior can be tracked without the need for fluorescence labels.

Various HSI configurations have been demonstrated for biosensing applications to date. Selection of the HSI system depends primarily on the characteristics of the target optical resonator when used with optical biosensors. For example, Chen et al. [1] and Zhuo et al. [2] used the pushbroom imaging scheme with a PC biosensor to image single-cell behaviors and single nanoparticles. The pushbroom configuration involves an imaging spectrometer and the 2D scan of the sample. The entrance slit of the spectrometer is aligned with the excitation line, which is perpendicular to the direction of the sample scan. As the sample is scanned, the 2D detection array captures spectral images from the line. Alternatively, tunable filters, such as the liquid-crystal tunable filter, can be installed in front of the detector array to select specific wavelengths of light for imaging.

This setup allows for data acquisition without moving parts but typically provides lower spectral resolution [3]. For example, Liu et al. [4] chose the tunable-filter HSI to image the coating of materials on a surface plasmon resonance sensor. Further, Wang et al. [5] also demonstrated a snapshot HSI of exosomes on a PC sensor substrate. The snapshot imaging configuration uses a tunable light source as the excitation source and captures transmission, reflection, or scattering images using a 2D CCD or CMOS sensor. Additionally, Jahani et al. [6] integrated the metasurface sensor directly onto the CMOS detector array to enable a highly compact HSI system.

This chapter focuses on our work combining the HSI system and PC biosensor for multiplex detection of biomarkers, specifically exosomes. To begin, we will provide an overview of the microscopic HSI system and its ability to capture images of optical-resonance distribution from a PC substrate. We will then delve into the fabrication and operation of the PC substrate. Next, we will discuss the development of a label-free exosome microarray, built on the PC biosensor substrate. Finally, we will demonstrate the application of HSI microscopy for exosome microarray analysis.

4.2 Hyperspectral Imaging System

Each of the HSI configurations reviewed in the previous section has its own set of advantages. When designing an HSI system, several properties of optical sensors and instruments, including the light source, detector, and imaging components, should be considered. For optical sensors, the resonance wavelength (λr), linewidth (δλr), tuning range (Δλr), and spatial resolution (δr) are the primary factors to be considered. Based on these values, an HSI system can be designed to enable spectral and spatial resolutions better than δλr and δr, respectively. For microscopic HSI applications, the power density of the excitation light, which is defined as $W/(mm^2 \cdot nm \cdot radian)$, is critical. Although high-power LEDs can emit light with power over 10 W, they rely on arrays of LED elements, limiting the actual power density for high-resolution HSI applications. However, the lack of excitation light can be compensated for by selecting a detector or camera with high photon sensitivity and low noise. For instance, in Jones et al. [7], an electron-multiplex charge-coupled device (EMCCD) camera was used to achieve a high signal-to-noise ratio (SNR) without experiencing a long image acquisition time.

To measure the exosome microarray, we utilized the HSI configuration as shown in Fig. 4.1A. This HSI system was based on an inverted microscope (IX-81, Olympus) with minimal updates to the light source. We employed a fiber-coupled monochromator (Mini-Chrom, Optometrics Inc.) to provide the monochromatic excitation. For the light source, we chose a low-cost tungsten halogen lamp (HL-2000-HP, Ocean Optics, Inc.). The entrance and exit slits of the monochromator could be adjusted to tune the bandwidth of the light. The excitation light was collimated using a fiber tip collimator (F230SMA-850, Thorlabs, Inc.) and then linearly polarized with a polarizer (WP25L-UB, Thorlabs, Inc.) before passing through the PC substrate. The diameter of the excitation beam was about 3 mm, sufficient to cover the entire exosome microarray. The transmitted light through the PC sensor was collected using an objective lens and imaged by an EMCCD (C9100, Hamamatsu).

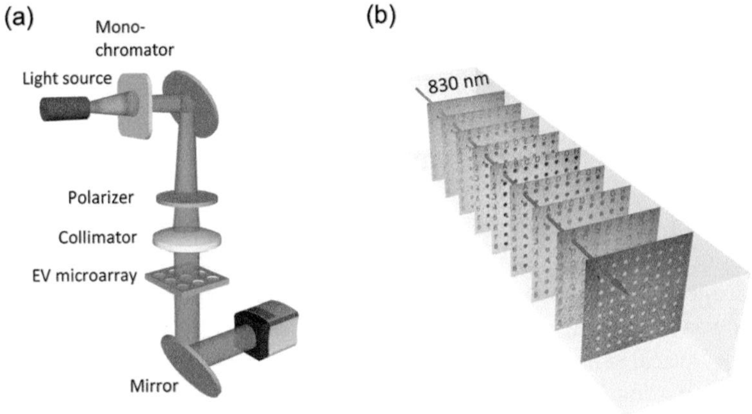

Figure 4.1 Hyperspectral imaging system for PC-based microarrays. (A) Schematic representation of the HSI detection setup that uses tunable monochromatic excitation. (B) Measured transmission images at different wavelengths (ranging from 830 nm to 870 nm), each with a spatial pixel resolution of 1.85 μm and consisting of 1000 × 1000 pixels.

To generate the HSI spectral data cube, the excitation light was scanned from 830 nm to 870 nm in 1 nm increments. The bandwidth of the monochromator output was also set to 1 nm. Figure 4.1B displays the intensity maps of monochromatic transmission captured at a specific wavelength. These images were then compiled into a three-dimensional array to form the data cube. In our experiment, the HSI data cube (x, y, λ) was constructed with dimensions of 1000 × 1000 × 80. After data acquisition, the collected data were processed to generate a transmission spectrum for each pixel in the HSI data cube. By fitting the transmission spectrum, the resonance wavelength λr(x, y) was determined for each spatial pixel (x, y). Figure 4.2A displays a transmission spectrum sampled from the HSI data cube at a given (x, y) position. The resulting label-free image in Fig. 4.2B was generated by plotting λr(x, y) to reflect the refractive index changes on the surface of the PC sensor.

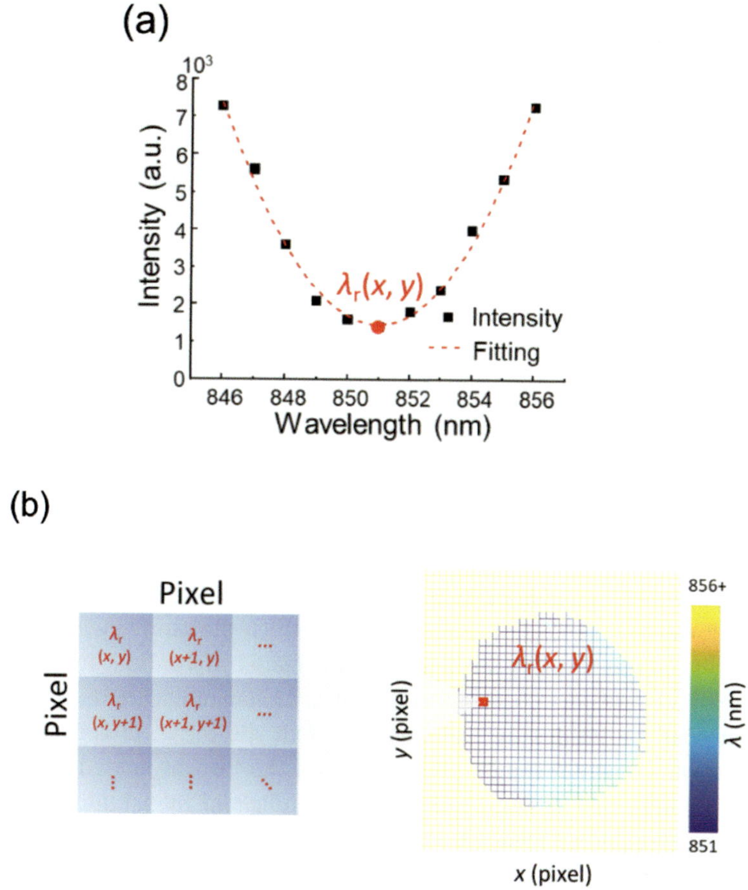

Figure 4.2 HSI data analysis. (A) Reconstructed transmission spectrum of a single pixel within the area of interest. A curve fitting algorithm determines the resonance wavelength (λr(x, y)) of this pixel. (B) Label-free image centered on one microwell.

4.2.1 A Label-Free Microarray Using Photonic Crystal Biosensor

A PC biosensor is a biosensor that utilizes a PC as its sensing element. The PC consists of periodic structures that can manipulate light in unique ways, including the creation of resonances. When a PC is exposed to a biological sample, such as blood or saliva, the presence of target molecules, such as proteins, DNA, or exosomes, can cause changes in the resonance wavelength of the sensor. These changes can be detected and measured, providing information about the presence and concentration of the target molecules.

PC biosensors have several advantages over other types of biosensors. They are highly sensitive—able to detect molecules at very low concentrations. Additionally, PCs can be manufactured using standard microfabrication techniques, making them relatively inexpensive and easy to scale up [8–10]. Importantly, sub-wavelength scale gratings on PCs can be utilized with HSI systems as an imaging biosensor to enable imaging with high spatial resolution [11]. This approach allows for label-free and non-destructive sensing, enabling real-time monitoring of biological processes and interactions at the molecular level.

The PC biosensor based on a one-dimensional (1D) grating structure was designed, fabricated, and tested. The 1D grating PC, as depicted in Fig. 4.3A, was designed to feature narrowband optical resonances that result from the guided mode resonance effect [11]. This resonance appears as a dip in the transmission spectrum of the PC substrate. The fabrication process involved utilizing a nano-replica molding method to pattern the plastic 1D grating structure, followed by the deposition of a 100 nm-thick titanium oxide (TiO_2) layer onto the grating through electron beam evaporation. To detect analytes, the PC biosensor can be functionalized with receptors, such as antibodies that are reactive with the target molecule. To immobilize the receptors, the surface of the PC was coated with a polyvinylamine (PVAm) polymer layer, followed by a bifunctional linker—glutaraldehyde (GA). Subsequently, the PVAm/GA treated sensors were immersed in phosphate-buffered saline (PBS) and assessed using the HSI system to ascertain the uniformity of the microwells' resonance wavelengths. The major steps of the fabrication and functionalization processes are summarized in

Fig. 4.3B. The 1D grating PC biosensor demonstrated high sensitivity and accuracy in detecting target molecules, making it a promising candidate for a wide range of applications.

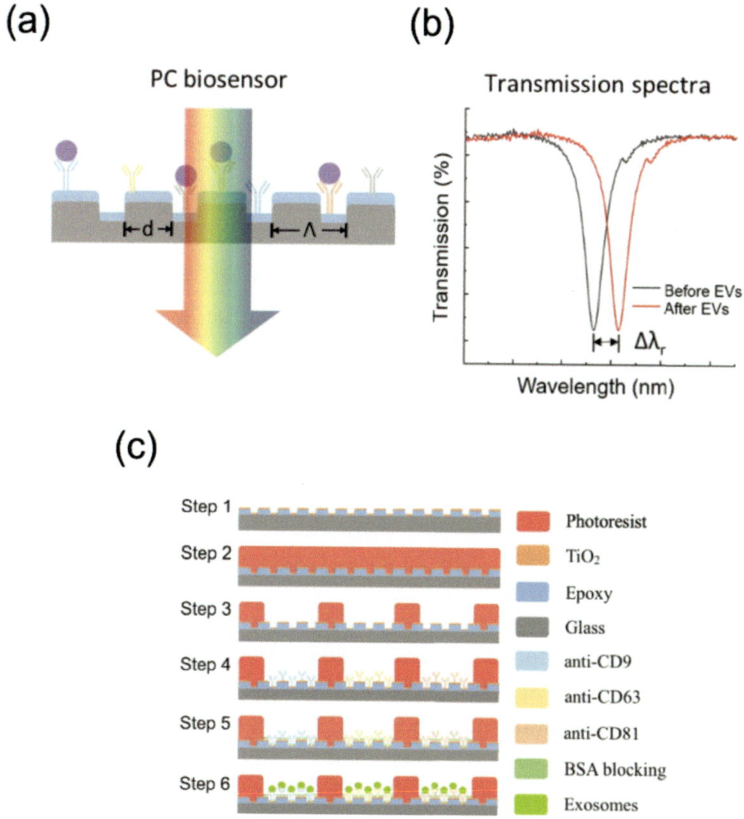

Figure 4.3 Photonic crystal (PC) biosensor used to build the exosome microarray. (A) Schematic of the PC sensor. (B) Transmission spectra measured from a PC sensor with and without the coating of EVs. (C) Flowchart for the fabrication of exosome microarray on the surface of a PC sensor.

The resonance wavelength and linewidth are key parameters of the spectral signatures obtained from PC biosensors, and have been widely utilized in the detection of various biomolecules and chemicals. When exosomes, nucleotides, proteins, or cells, which are all potential analytes, absorb onto the surface of the PC biosensor, they can induce a shift in the resonance wavelength (Δλr), as illustrated in Fig. 4.2B. This shift is directly proportional to the concentration of the analyte. By detecting and measuring the changes in the resonance wavelength of the PC biosensor, it is possible to quantify the concentration of the analyte in the sample, making this approach a highly sensitive and selective method for detecting and quantifying a wide range of biological and chemical species.

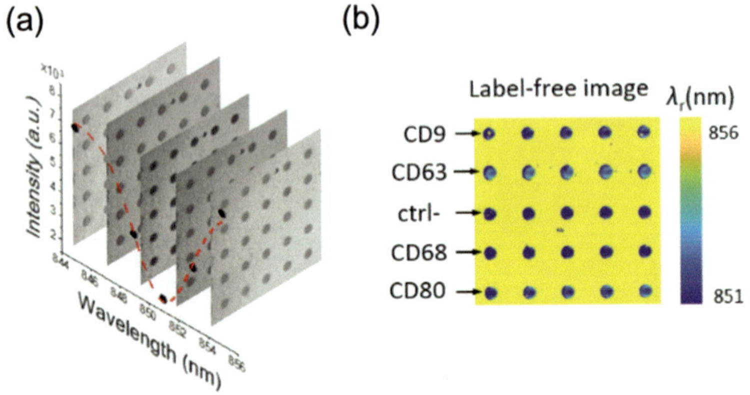

Figure 4.4 Label-free image. (A) Transmission images captured from the exosome microarray. (B) Label-free image calculated based on the HSI data cube.

To enable the detection of multiple DNAs, proteins, or exosomes, we created microarrays on the PC substrate. Initially, a layer of photoresist was patterned on the PC to create an 11 × 12 array of microwells (Fig. 4.4A). The microwell's diameter, depth, and period measured 60 μm, 25 μm, and 150 μm, respectively, with PC grating at the bottom of each microwell. For the exosome microarray, a panel of antibodies specific to exosome membrane proteins was used to functionalize the sensor. To print specific antibodies on the grating surface inside the microwells, a molecular printer was employed. The antibodies were printed in an environmental chamber maintained at 60% relative humidity. Various antibodies were printed onto the microarray, as shown in Fig. 4.4B. Each row was printed with the same material, and specific capture antibodies were used to target a range of membrane proteins. The selection of these protein markers was motivated by their variation across different macrophage phenotypes. A blank row was allocated as reference spots in every other antibody-coated row. By utilizing a microarray approach, it is possible to multiplex the detection of multiple analytes on a single substrate, which can significantly increase the throughput and reduce the assay time. This approach enables the detection and quantification of multiple biomolecules in a single experiment, making it a powerful tool for high-throughput screening and diagnostics.

The calculation of the resonance wavelength shift induced by the antibodies is illustrated in Fig. 4.4B. The λr value of each microwell was obtained by subtracting the λr value of the nearby reference microwell. This provided the average Δλr of the specific antibody, represented by the mean value of the total 11 referenced microwells in the same row. The resonance wavelength shifts caused by the blocking and EV detection steps were subsequently obtained using the λr value in the same microwell as the baseline. To visualize the exosome microarray on a PC surface and the captured exosomes on the grating, scanning electron microscopy (SEM) images were acquired and are shown in Fig. 4.5A. Additionally, transmission electron microscopy (TEM) was utilized for exosome characterization, as illustrated in the TEM image in Fig. 4.5B. By utilizing SEM and TEM imaging techniques, it was possible to visualize the captured exosomes on the grating and to confirm their presence on the sensor surface. The use of these complementary imaging techniques provides a comprehensive understanding of the binding and detection processes occurring on the PC biosensor, enabling the optimization of the sensing performance and the development of more efficient biosensing platforms.

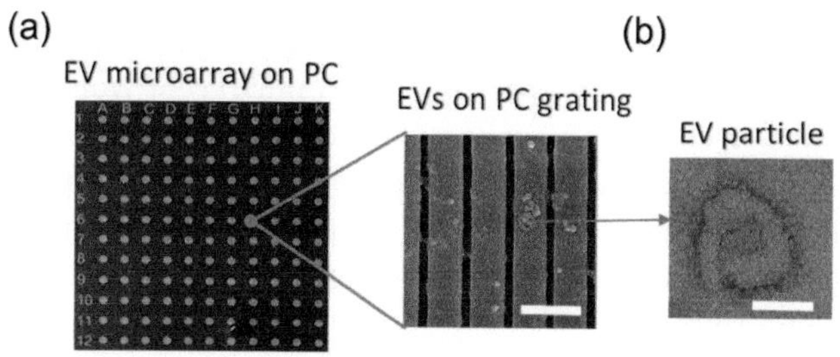

Figure 4.5 (A) SEM images showing the EV microarray (left) and the PC grating (center) with a scale bar of 800 nm). (B) TEM image shows an individual EV particle (scale bar = 80 nm).

4.3 Label-Free Exosome Microarray Assay

The PC-based HSI technology has been implemented for a variety of applications, including cell-based assays, single particle detection, and DNA and protein microarrays [2, 5, 12]. This section focuses on our recent work on the exosome microarray as a new diagnostic approach. EVs, including sub-100-nm exosomes, are nanovesicles enclosed by a membrane that are released by various cell types into their extracellular environment [13–17]. EVs play a crucial role in intercellular communication and contain a complex and varied molecular cargo, such as proteins, lipids, and microRNAs, that reflects their parental cells. The unique molecular information provided by EVs can be utilized to identify the originating cells or tissues and determine their physiological conditions, making EVs important biomarkers for disease diagnostics, prognostics, and therapeutics [18–25].

Unlike protein biomarkers, which can be affected by interfering molecules during global analysis, EVs can be isolated and enriched from clinical samples. However, quantifying EV membrane proteins with high accuracy and efficiency remains a challenge, as existing analysis methods are often time-consuming and expensive. Fluorescence-based detection methods using microbeads and magnetic particles offer multiplexing capabilities but require laborious assays [26, 27]. To address these challenges, researchers have developed low-cost, rapid, and label-free EV analysis methods, such as surface plasmonic resonance sensors and electrochemical biosensors [28–35]. Some EVs can also be distinguished from those derived from host cells or invasive parasites by their expression of specific surface antigens, which can vary significantly between different populations of EVs [36, 37]. In our recent work, we have developed an exosome microarray utilizing a PC biosensor to detect and quantify exosome membrane proteins with high sensitivity and specificity. This approach provides a rapid and low-cost method for the analysis of EVs, making it a promising tool for disease diagnosis and monitoring.

Despite significant progress in the development of EV analysis tools, a major challenge remains in distinguishing between EVs released by similar cell types or phenotypes. To address this challenge, we have demonstrated a label-free EV microarray (Fig. 4.4) capable of discriminating between EVs released by closely related cell types, such as murine macrophages. Macrophages are critical components of the immune response and are known for their plasticity, which allows them to undergo specific differentiation and develop distinct functions in response to their local tissue environments [38]. Upon activation by external stimuli, macrophages can express widely divergent phenotypes ranging from M1 macrophages, which arise in pro-inflammatory environments, to M2 macrophages, which predominate in the TH2 immune response environment [39, 40]. These patterns of macrophage activation are driven by highly dynamic transcriptional changes, resulting in differential surface expression of epitopes that are characteristic of the M1 and M2 phenotypes. These differences in cell surface markers would also be observed in the EVs released by naïve, M1, and M2 macrophages, and could be used to identify the phenotypic state of the parental cell.

Using the HSI technology, our label-free exosome microarray can effectively distinguish the EVs derived from different macrophage phenotypes. This capability provides a powerful tool for the characterization and identification of EVs, enabling the discrimination of EVs from different cell types or phenotypes. This approach offers new possibilities for the diagnosis and monitoring of diseases in which changes in the composition or concentration of EVs are indicative of specific disease states or treatment responses.

As depicted in Fig. 4.6, EVs, termed EVnaïve, EVM1, and EVM2, were generated from murine macrophages (specifically the J774A.1 cell line). These macrophages were cultured and subsequently activated to adopt one of two extreme phenotypes: M1 and M2. The activation process was carried out using lipopolysaccharide (LPS) for the M1 macrophages and interlekin-4 (IL-4) for the M2 macrophages. The concentration of the collected EVs was confirmed through a process called nanoparticle tracking analysis. Once separated, the EVs were reconstituted in a buffer for the subsequent label-free EV microarray analysis.

The EV microarray was built to measure the bindings of EV to a specific antibody printed on the PC sensor. The EV microarray consisted of an array of microwells, each with a diameter of 60 μm, patterned on the PC using photolithography (Fig. 4.7A). The SEM image in Fig. 4.7B zoomed in to a single microwell. The PC grating pattern can be seen at the bottom of the microwell. The table in Fig. 4.7C lists the printed antibodies and reference spots, along with their matching row numbers.

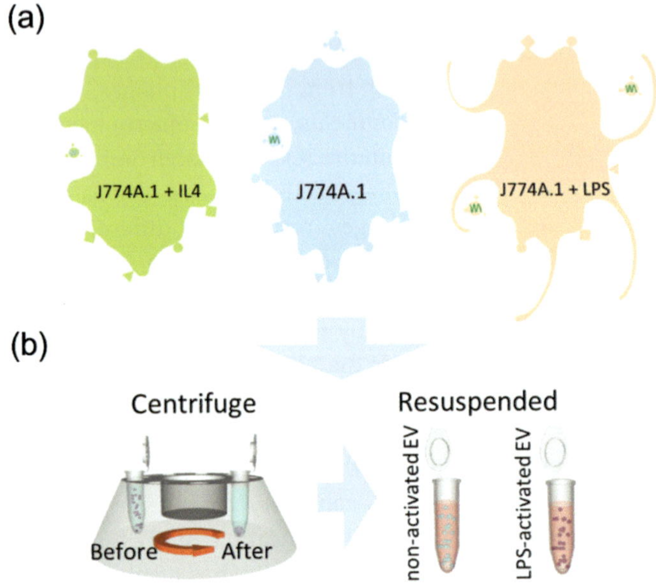

Figure 4.6 Exosomes secretion and purification. (A) The macrophage phenotypes release EVnaïve, EVM1, and EVM2. (B) The EVs are then isolated from the expended macrophage culture media using an ultracentrifuge.

To characterize the PC biosensor for EV detection, we tested the macrophage-derived EVs at four different concentrations (from 2 × 1010 EVs per mL to 2 × 1013 EVs per mL). The samples were gradually diluted in PBS by a factor of ten. The PC sensor was functionalized using the CD63 antibody. Figure 4.8 showcases the method used to ascertain the HSI image ($\Delta\lambda r$ values), resulting from the binding of the EVs to the CD63 antibody. The figure's top panel features label-free images of five microwells prior to the EVs being printed, while the bottom panel shows label-free images of the microwells after immobilization of the EVs at varying concentrations. As the EV concentration rises, the resonance wavelength also increases. Figure 4.9A illustrates the distribution profile of λr measured along a line crossing the microwells, with the black and red lines representing the λr profiles measured prior to and post the EV binding, respectively. The EV-induced Δλr corresponds to the difference, indicated as the green area, between the two lines. In Fig. 4.9B, the dose-response relationship of the EV detection is plotted by mapping Δλr as a function of the EV concentration.

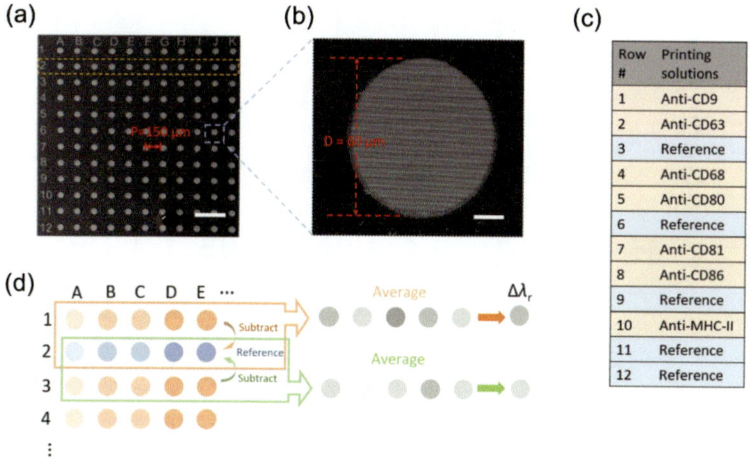

Figure 4.7 Exosome microarray of microwells patterned on the photonic crystal (PC). (A) SEM image shows the microwell with a diameter of 60 μm (scale bar = 300 μm). (B) A zoomed-in SEM image of a single microwell (scale bar = 15 μm). (C) List of the printed antibodies and reference spots, along with their matching row numbers. (D) A methodology for determining the shift in resonance wavelength following the printing of each antibody.

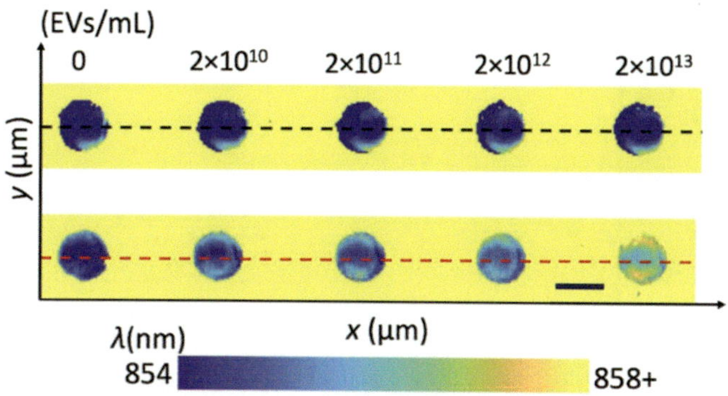

Figure 4.8 Label-free images of the EV microarray obtained by the HSI system. The top panel displays the microarray prior to the application of EVs. The bottom panel reveals the microarray post the application of EVs at varying concentrations.

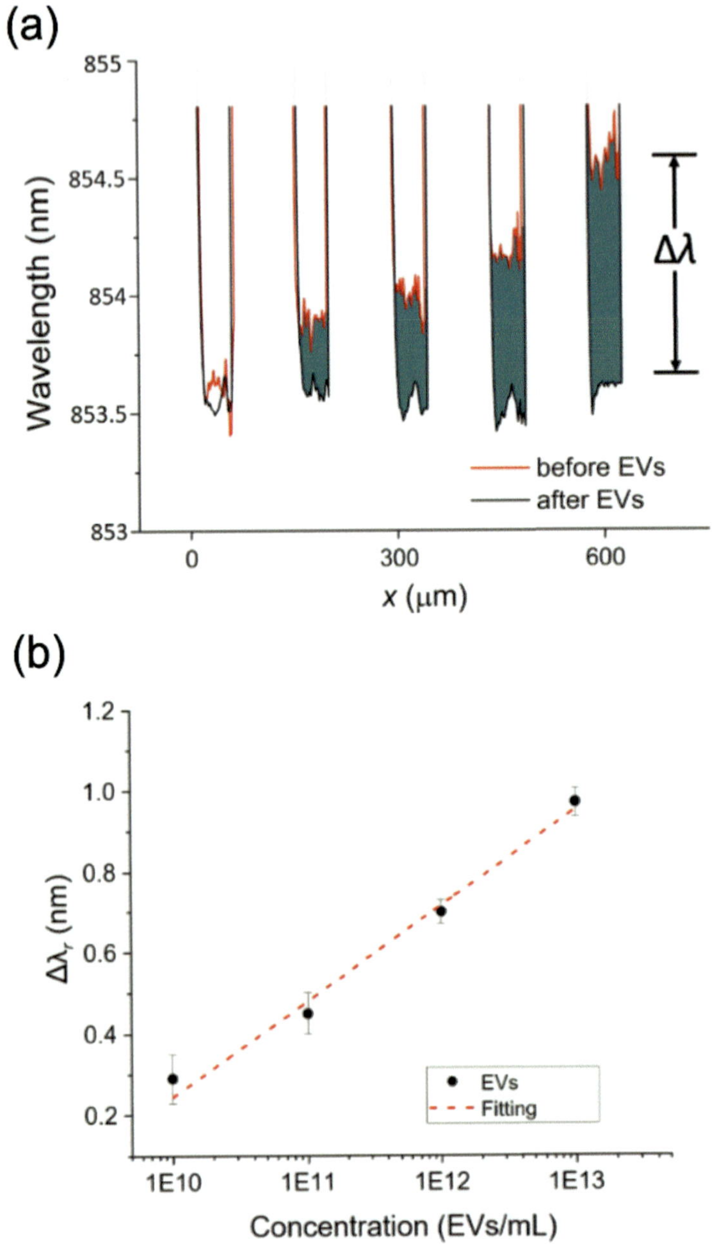

Figure 4.9 Quantification of exosomes based on the label-free image. (A) Profile plot representing the condition before and after EV treatment. (B) Dose-response curve for the detection of EVs at different concentrations.

4.4 Detection of Macrophage-Specific Exosomes

This section explains how the label-free images obtained by the HSI measurement can be utilized to distinguish EVs derived from different types of macrophages. We used the HSI technology to quantify the EV microarray after major processing steps, including coating of antibodies in the microwells, blocking using bovine serum albumin (BSA), and the absorption of EVs. The label-free images and the averaged output calculated based on the label-free images are present below.

Figure 4.10A shows the label-free image measured immediately after the printing of antibodies. The pseudo-color plot represents the resonance wavelength for the pixels inside the microwells. Since the region outside the microwell was covered by the photoresist, the high refractive index of the photoresist can cause a larger shift than that of the biomolecules. The resulting yellow color indicates the high contrast of the resonance wavelength of the photoresist by comparing with the area of the microarray. The box plot in Fig. 4.10B shows the resonance wavelength shifts for the panel of antibodies, including CD9, CD63, CD68, CD80, CD81, CD86, and MHC-II. The bars represent the mean values and the 25th and 75th percentiles and the lines are the mean values ±1.5 times the interquartile range. For each antibody, the $\Delta\lambda r$ value was calculated by subtracting the reference value from the nearby reference spot and averaging the values of all 12 spots in the row. Figure 4.10C shows the label-free microarray image measured after BSA blocking and before EV detection. The box plot in Fig. 4.10D shows the $\Delta\lambda r$ values caused by the blocking step. Here, the $\Delta\lambda r$ values were calculated by subtraction of the λr spots by the previous step value and averaging the values of all 12 spots in the row. The $\Delta\lambda r$ value of the BSA blocking step is higher if the $\Delta\lambda r$ of the antibody step is lower, and vice versa.

The label-free EV microarray assay was utilized to profile EVs derived from naïve, M1, and M2 macrophages, all at a concentration of 2 × 1013 EVs per mL. Figure 4.11A–C displays the label-free image that illustrates the binding of naive EVs, while Fig. 4.11D–F provides a comparison of Δλr for each antibody. The baseline used for Δλr calculations was the label-free image obtained post-BSA blocking step. For the naïve EVs, the highest and lowest Δλr values were recorded for CD68 (1.75 nm) and CD80 (0.32 nm), respectively. EVs from M1 macrophages exhibit the highest and lowest wavelength shifts for CD80 (1.5 nm) and CD9 (0.3 nm), respectively. EVs from M2 macrophages demonstrate the highest and lowest wavelength shifts for CD80 (1.2 nm) and MHC-II (0.1 nm), respectively.

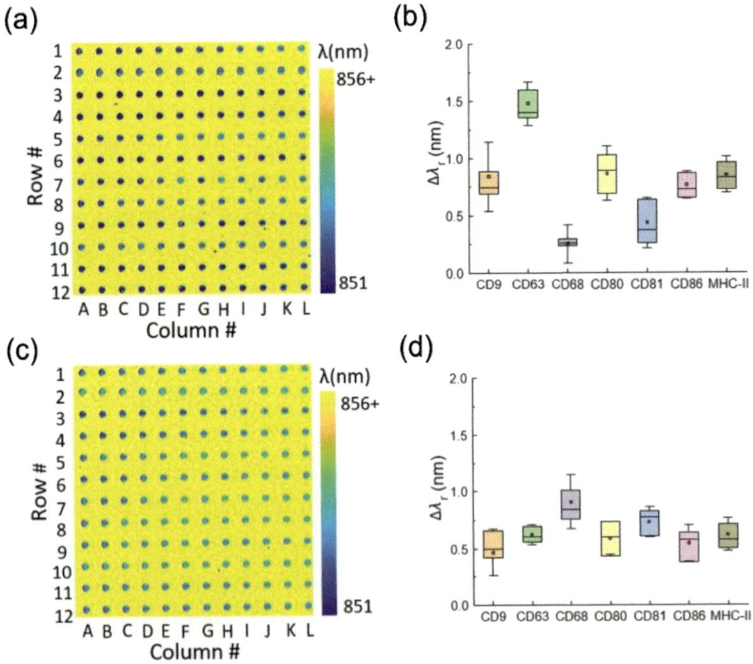

Figure 4.10 Label-free images generated by the HSI microcopy. (A) The label-free image of the EV microarray after seven antibodies have been printed is presented. Reference spots without antibodies are located at rows #3, 6, 9, 11, and 12. (B) box plot showing the average shifts in resonance wavelength for each antibody. (C) A label-free image after BSA blocking, which is used as a baseline for calculating the wavelength shifts induced by EV binding. (D) Resonance wavelength shifts, calculated by subtracting the resonance wavelengths before the BSA blocking from those measured after this process.

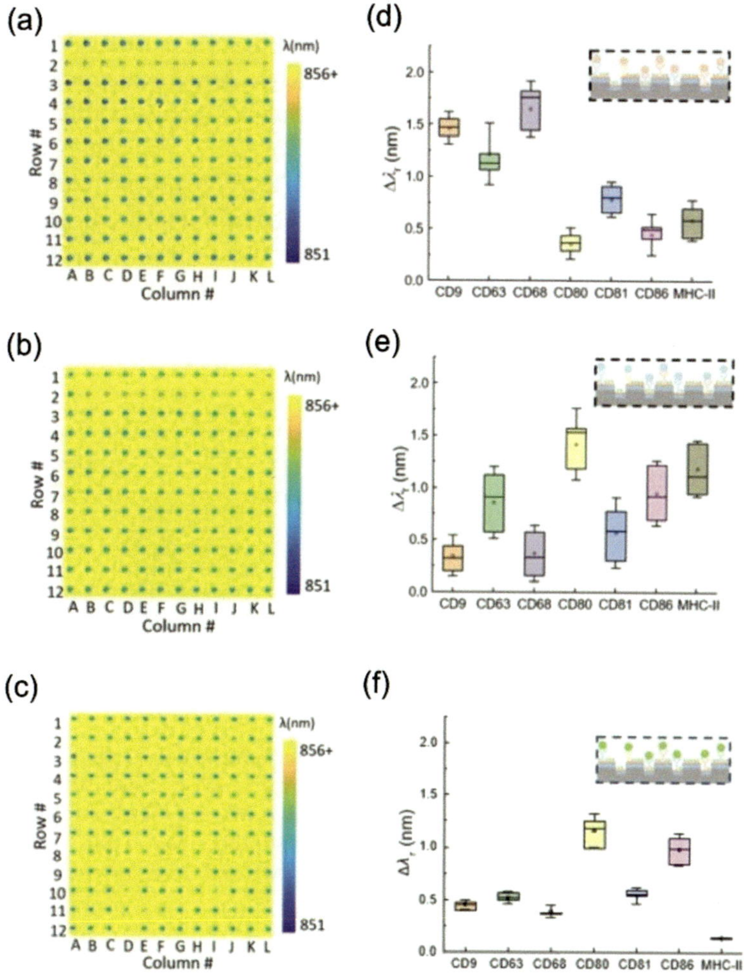

Figure 4.11 HSI analysis of EVs derived from macrophage phenotypes using the microarray. (A–C) Label-free images of the binding of naïve, M1 and M2 macrophage-derived vesicles on the EV microarray. (D–F) A box plot illustrating the average shifts in resonance wavelength caused by the binding of EVnaïve, EVM1 and EVM2 to the antibody panel.

As Fig. 4.12A highlights, the profiles of measured EV membrane protein distributions indicate that EVs from M1 macrophages carry significantly more CD80 than naïve EVs. Conversely, levels of CD68 and CD9 on EVs from M1 macrophages significantly diminished. Even though EVs from M2 macrophages also carry more CD80, the amount of MHC-II was lower. From the Δλr values for the antigens present on the EVs, it is feasible to distinguish between different macrophage phenotypes. Figure 4.12B provides a radar chart that profiles the membrane proteins of naïve EVs, EVs from M1, and EVs from M2 macrophages with the areas colored in blue, red, and green, respectively. The length of each spoke on the radar chart corresponds to the specific antibody's Δλr value, with all data points normalized to the maximum Δλr value. The chart presents distinctly different profiles of membrane proteins on EVs secreted by the three different macrophage phenotypes.

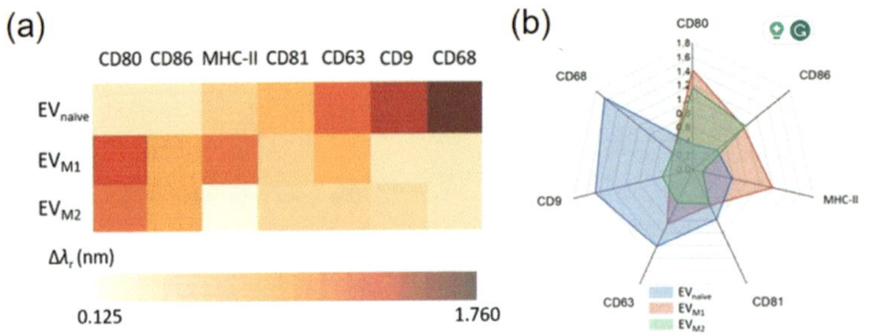

Figure 4.12 Analysis of the membrane protein composition in diverse EVs. (A) Characterization of membrane proteins on distinct EVs reveals distribution profiles of these proteins on EVs derived from naïve, M1, and M2 macrophages. (B) A radar chart depicting the wavelength shifts observed in the binding interactions between antibodies and EVnaïve, EVM1, and EVM2, respectively.

4.5 Conclusion

The combination of HSI and optical resonators has greatly expanded the range of applications for optical biosensors, transforming them from simple analytical tools into powerful biomedical research instruments. This chapter provides a review of the methods for combining HSI setups, optical resonant biosensors, and data analysis. HSI technology can be integrated with various optical biosensors that enable continuous monitoring of changes in biomaterial absorption at microscopic scales. As an illustration, this chapter focuses on the use of EV microarrays patterned on PC substrates as a potential tool for disease diagnosis. The HSI configuration used to map the distribution of optical resonance from EV microarrays is discussed in detail. The outcome of HSI acquisition, defined as a label-free image, directly correlates with the concentration of target EVs in clinical samples. Moreover, analysis of a panel of EV membrane proteins enables the determination of EV host cells. In addition to microarrays, the optical sensor-based HSI approach can also be applied to image cell behaviors, 2D materials, and nanoparticles such as viruses and bacteriophages, without the use of labeling materials.

Future research directions in the field of HSI for biosensors and diagnosis include several aspects. Firstly, the development of a low-cost HSI microscopy system with a higher degree of integration and a smaller footprint is highly desirable. Secondly, there is a need for a wider wavelength range with sufficient spectral resolution, which can be achieved using tunable laser sources or innovations in tunable optical filters. Lastly, improving SNR using a high-intensity light source or a scientific CMOS camera is also essential.

References

1. Chen W., Long K. D., Yu H., Tan Y., et al. Enhanced live cell imaging via photonic crystal enhanced fluorescence microscopy. *Analyst.* 2014;139(22):5954–63.
2. Zhuo Y., Hu H., Chen W., Lu M., et al. Single nanoparticle detection using Photonic Crystal Enhanced Microscopy. *Analyst.* 2014;139(5):1007–15. doi:10.1039/c3an02295a.
3. Wang X., Zhang Y., Ma X., Xu T., et al. Compressive spectral imaging system based on liquid crystal tunable filter. *Opt Express.* 2018;26(19):25226. doi:10.1364/oe.26.025226
4. Liu Z., Wu J., Cai C., Yang B., et al. Flexible hyperspectral surface plasmon resonance microscopy. *Nat Commun.* 2022;13(1). doi:10.1038/s41467-022-34196-7.
5. Wang Y., Zhang Q., Yuan W., Wang Y., et al. Hyperspectral imaging-based exosome microarray for rapid molecular profiling of extracellular vesicles. *Lab Chip.* 2021;21(1):196–204. doi:10.1039/d0lc01006e.
6. Jahani Y., Arvelo E. R., Yesilkoy F., Koshelev K., et al. Imaging-based spectrometer-less optofluidic biosensors based on dielectric metasurfaces for detecting extracellular vesicles. *Nat Commun.* 2021;12(1). doi:10.1038/s41467-021-23257-y.

7. Jones H. D., Haaland D. M., Sinclair M. B., Melgaard D. K., et al. Weighting hyperspectral image data for improved multivariate curve resolution results. *J Chemom.* 2008;22(9):482–90. doi:10.1002/cem.1170.
8. Cunningham B. T., Li P., Schulz S., Lin B., et al. Label-free assays on the bind system. *SLAS Discov.* 2004;9(6):481–90. doi:10.1177/1087057104267604.
9. Zhang Q., Loghry H. J., Qian J., Kimber M. J., et al. Towards nanovesicle-based disease diagnostics: A rapid single-step exosome assay within one hour through in situ immunomagnetic extraction and Nanophotonic label-free detection. *Lab Chip.* 2021;21(18):3541–9. doi:10.1039/d1lc00446h.
10. Zhang Q., Rawal G., Qian J., Ibrahim H., et al. An integrated magneto-opto-fluidic biosensor for rapid on-chip assay of respiratory viruses of livestock. *Lab Chip.* 2022;22(17):3236–44. doi:10.1039/d2lc00406b.
11. Wang S. S., Moharam M. G., Magnusson R., Bagby J. S. Guided-mode resonances in planar dielectric-layer diffraction gratings. *J Opt Soc Am A.* 1990;7(8):1470. doi:10.1364/josaa.7.001470.
12. Chen W., Long K. D., Lu M., Chaudhery V., et al. Photonic crystal enhanced microscopy for imaging of live cell adhesion. *Analyst.* 2013;138(20):5886. doi:10.1039/c3an01541f.
13. Meldolesi J. Exosomes and ectosomes in intercellular communication. *Curr Biol.* 2018;28(8). doi:10.1016/j.cub.2018.01.059.
14. Tkach M., Théry C. Communication by extracellular vesicles: Where we are and where we need to go. *Cell.* 2016;164(6):1226–32. doi:10.1016/j.cell.2016.01.043.
15. Johnstone R. M. Exosomes biological significance: A concise review. *Blood Cells Mol Dis.* 2006;36(2):315–21. doi:10.1016/j.bcmd.2005.12.001.
16. Zhang J., Li S., Li L., Li M., et al. Exosome and exosomal MicroRNA: Trafficking, sorting, and function. *Genomics, Proteomics Bioinformatics.* 2015;13(1):17–24. doi:10.1016/j.gpb.2015.02.001.
17. Azmi A. S., Bao B., Sarkar F. H. Exosomes in cancer development, metastasis, and drug resistance: A comprehensive review. *Cancer Metastasis Rev.* 2013;32(3–4):623–42. doi:10.1007/s10555-013-9441-9.
18. Schorey J. S., Bhatnagar S. Exosome function: From tumor immunology to pathogen biology. *Traffic.* 2008;9(6):871–81. doi:10.1111/j.1600-0854.2008.00734.x.

19. De Toro J., Herschlik L., Waldner C., Mongini C. Emerging roles of exosomes in normal and pathological conditions: New insights for diagnosis and therapeutic applications. *Front Immunol.* 2015;6. doi:10.3389/fimmu.2015.00203.

20. Simpson R. J., Lim J. W., Moritz R. L., Mathivanan S. Exosomes: Proteomic insights and diagnostic potential. *Expert Rev Proteomics.* 2009;6(3):267–83. doi:10.1586/epr.09.17.

21. Properzi F., Logozzi M., Fais S. Exosomes: The future of biomarkers in medicine. *Biomark Medi.* 2013;7(5):769–78. doi:10.2217/bmm.13.63.

22. Barile L., Vassalli G. Exosomes: Therapy delivery tools and biomarkers of diseases. *Pharmacol Ther.* 2017;174:63–78. doi:10.1016/j.pharmthera.2017.02.020.

23. Masyuk A. I., Masyuk T. V., LaRusso N. F. Exosomes in the pathogenesis, diagnostics and therapeutics of liver diseases. *J Hepatol.* 2013;59(3):621–5. doi:10.1016/j.jhep.2013.03.028.

24. Qian J., Zhang Q., Lu M. Integration of on-chip lysis and paper-based sensor for rapid detection of viral and exosomal RNAS. *Biosens Bioelectron.* 2023;226:115114. doi:10.1016/j.bios.2023.115114.

25. Shao H., Chung J., Balaj L., Charest A., et al. Protein typing of circulating microvesicles allows real-time monitoring of glioblastoma therapy. *Nat Med.* 2012;18(12):1835–40. doi:10.1038/nm.2994.

26. Qian J., Zhang Q., Liu M., Wang Y., et al. A portable system for isothermal amplification and detection of exosomal micrornas. *Biosens Bioelectron.* 2022;196:113707. doi:10.1016/j.bios.2021.113707.

27. Hartjes T., Mytnyk S., Jenster G., van Steijn V., et al. Extracellular vesicle quantification and characterization: Common methods and emerging approaches. *Bioengineering.* 2019;6(1):7. doi:10.3390/bioengineering6010007.

28. Im H., Shao H., Park Y. I., Peterson V. M., et al. Label-free detection and molecular profiling of exosomes with a nano-plasmonic sensor. *Nat Biotechnol.* 2014;32(5):490–5. doi:10.1038/nbt.2886.

29. Goda T., Masuno K., Nishida J., Kosaka N., et al. A label-free electrical detection of exosomal micrornas using Microelectrode Array. *Chem Commun.* 2012;48(98):11942. doi:10.1039/c2cc36111f.

30. Zhu L., Wang K., Cui J., Liu H., et al. Label-free quantitative detection of tumor-derived exosomes through surface plasmon resonance imaging. *Anal Chem.* 2014;86(17):8857–64. doi:10.1021/ac5023056.

31. Zubiri I., Posada-Ayala M., Sanz-Maroto A., Calvo E., et al. Diabetic nephropathy induces changes in the proteome of human urinary

exosomes as revealed by label-free comparative analysis. *J Proteomics.* 2014;96:92–102. doi:10.1016/j.jprot.2013.10.037.

32. Su J. Label-free single exosome detection using frequency-locked microtoroid optical resonators. *ACS Photonics.* 2015;2(9):1241–5. doi:10.1021/acsphotonics.5b00142.
33. Wang S., Zhang L., Wan S., Cansiz S., et al. Aptasensor with expanded nucleotide using DNA nanotetrahedra for electrochemical detection of cancerous exosomes. *ACS Nano.* 2017;11(4):3943–9. doi:10.1021/acsnano.7b00373.
34. Tian F., Liu C., Lin L., Chen Q., et al. Microfluidic analysis of circulating tumor cells and tumor-derived extracellular vesicles. *TrAC Trends Anal Chem.* 2019;117:128–45. doi:10.1016/j.trac.2019.05.013.
35. He M., Zeng Y. Microfluidic exosome analysis toward liquid biopsy for cancer. *SLAS Technol.* 2016;21(4):599–608. doi:10.1177/2211068216651035.
36. Wang Y., Yuan W., Kimber M., Lu M., et al. Rapid differentiation of host and parasitic exosome vesicles using microfluidic photonic crystal biosensor. *ACS Sens.* 2018;3(9):1616–21. doi:10.1021/acssensors.8b00360.
37. Liu C., Zhao J., Tian F., Cai L., et al. Low-cost thermophoretic profiling of extracellular-vesicle surface proteins for the early detection and classification of cancers. *Nat Biomed Eng.* 2019;3(3):183–93. doi:10.1038/s41551-018-0343-6.
38. Mosser D. M., Edwards J. P. Exploring the full spectrum of macrophage activation. *Nat Rev Immunol.* 2008;8(12):958–69. doi:10.1038/nri2448.
39. Murray P. J. Macrophage polarization. *Annu Rev Physiol.* 2017;79(1):541–66. doi:10.1146/annurev-physiol-022516-034339.
40. Lawrence T., Natoli G. Transcriptional regulation of macrophage polarization: Enabling diversity with identity. *Nat Rev Immunol.* 2011;11(11):750–61. doi:10.1038/nri3088.

Chapter 5

Single-Molecule Junctions for Electrical Detection of Biomolecules

Ian Bunker[a] and Kun Wang[b,c]
[a]*Department of Mechanical Engineering, Mississippi State University, Mississippi State, MS 39762, USA*
[b]*Department of Physics, University of Miami, Coral Gables, FL 33146, USA*
[c]*Department of Chemistry, University of Miami, Coral Gables, FL 33146, USA*
kunwang@miami.edu

5.1 Introduction

Biotechnology is an interdisciplinary field of study concerned with the application of biological systems and organisms to develop new technologies crucial to agriculture, healthcare, environmental science, forensic analysis, and industry. At its core, biotechnology is inherently a molecular science, one that strives to manipulate and understand the basic components of biological molecules and living organisms. Nanotechnology, in contrast, is a field of science and engineering concerned with the study of nanoscale materials and the

High-Throughput Imaging Methods: Cells and Tissues
Edited by Manas Ranjan Gartia and Ram V. Devireddy

ISBN 978-981-5352-08-5 (Hardcover), 978-1-003-77417-4 (eBook)
www.jennystanford.com

DOI: 10.1201/9781003774174-5

development of suitable techniques to characterize their properties. One of the leading challenges in nanotechnology was to explore the possibility of using single molecules as functional electronic components in nanodevices, also known as single-molecule electronics (SME). Since its inception by Ratner and Aviram in their influential 1974 paper [1], SME has primarily focused on probing and understanding the electrical properties of single-molecule structures. As a result of this endeavor, researchers have developed experimental platforms and methods allowing for the precise measurement of charge transport through individual molecules. To date, these single-molecule experimental techniques adopted in SME have made it possible to electrically monitor crucial chemical and biological processes with single-molecule precision, representing the ultimate analytical limit of molecular science. It is appropriate to consider the study and manipulation of biological systems at the nanoscale, spanning both fields, as nanobiotechnology. Due to the inherent complexity of measuring phenomena at the molecular level, single-molecule research into individual biomolecules is still in its infancy. Despite this, nanobiotechnology and its transformative tools have opened previously unattainable avenues of research, reshaping our understanding of fundamental biomolecules. It is the aim of this chapter to shed light on these recent discoveries, the experimental platforms they employed, and their direct impact on our fundamental understanding of critical biomolecular systems.

Before engaging directly with research frontiers, a brief history and description of experimental platforms is necessary. In 1982, Gerd Binnig and Heinrich Rohrer, two scientists working at the International Business Machines Corporation (IBM) Zurich laboratory, created a novel type of microscope, the Scanning Tunneling Microscope (STM) for which in 1986 they were awarded the Nobel Prize in Physics [2]. Employing an atomically sharp conductive tip, the STM forms measurable quantum tunneling currents between the surface of a sample and the tip. This allowed for the creation of highly detailed images of the atomic structure of conductive materials from the electron densities of individual atoms. The same year of his joint Nobel Prize, Gerd Binnig was involved in another phenomenal development, the invention of the

Atomic Force Microscope (AFM) with Christoph Gerber and Calvin Quate [3]. Unlike the STM, the AFM utilizes spring compression of a cantilevered tip to measure the height of nanoscale structures. In 1993, Murrell et al. [4] modified the AFM to perform conductivity measurements, creating the conductive atomic force microscopy (CAFM). Since their invention, these scanning probe microscopy techniques, including STM and AFM, have played a central role in characterizing the physical properties of a wide range of microscopic and nanoscopic materials and structures. In what follows, we will describe how the capability of these techniques were further expanded to enable precise and versatile single-molecule electrical detection under a variety of environmental conditions.

In 2003, researchers at Arizona State University, Bingqian Xu and Nongjian Tao, expanded the functionality of conventional STM beyond imaging to enable direct measurement of electrical current through a single molecule [5]. Today, their technique is known as the STM-break junction or STM-BJ technique. In contrast to the imaging mode where a gap between the STM tip and sample is maintained by sustaining a preset tunneling current, the STM-BJ technique harnesses the precise movement of STM tip to "fish" or "trap" a single molecule from the sample surface. The STM-BJ working principle is illustrated in Fig. 5.1. Specifically, a sharp STM metal tip is brought in contact with a metal substrate coated with a monolayer of target molecules. The target molecules are modified with chemical linkers (i.e., anchor groups) at the two terminals to covalently bind with the metal substrate and STM tip. Once contact is made between the tip and surface, the STM tip slowly retracts away from the substrate. A metal-molecule-metal junction is created momentarily during the tip retraction process, which could be detected by monitoring the current flowing (or electrical conductance) between the tip and substrate. By repeating the same process hundreds of times, one can collect many conductance traces and construct a conductance histogram which indicates the most probable conductance of the measured single molecule. This technique has now become the most widely adopted method in studying charge transport through single molecules and has been extensively employed in many of the studies presented in the following sections.

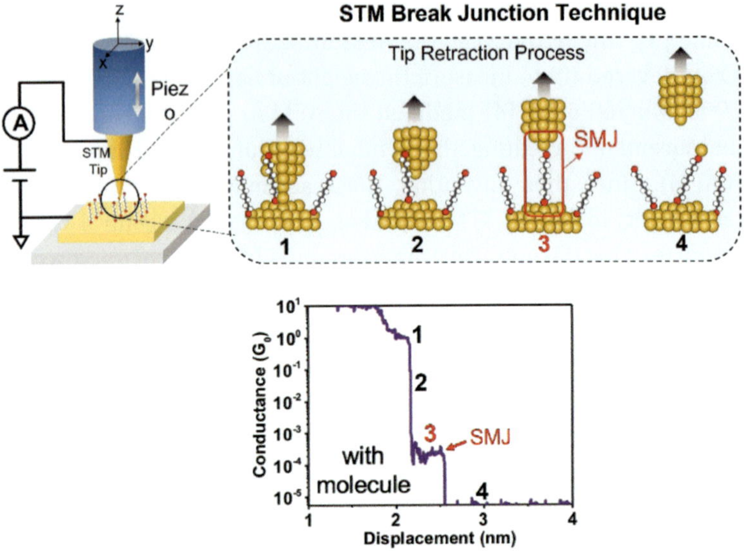

Figure 5.1 The STM-BJ working principle. Left: the STM-BJ experimental setup. Middle: Schematic illustration of STM tip retraction from the sample surface, during which a single-molecule junction is created (as highlighted in the red box). Note that the red dots on each molecule shown in the middle panel are chemical linkers which could covalently connect the molecule with the metal electrodes. Right: A typical conductance vs. tip displacement trace collected during one cycle of the tip retraction. Note that the red arrow points at a conductance plateau due to the formation of a single-molecule junction.

In addition to STM-BJ technique, several other approaches adopting similar break junction methods have also been developed to connect an individual molecule between two conductive electrodes (Fig. 5.2). These techniques include CAFM Break Junction (CAFMBJ) [4], Mechanically Controllable Break Junction (MCBJ) [6, 7], Graphene Molecule Graphene (GMG) [8], and Electromigration Junction (EMJ) [9]. For more detailed information about these techniques, we refer readers to relevant review articles [10–12]. These technical methods share mostly the same working principles: a molecule is momentarily or stably sandwiched between two microscopic electrodes (Fig. 5.2A: source (S) and drain (D) electrode), forming an SMJ. The electrical, conformational (or structural geometry), and chemical properties of an individual molecule or a cluster of molecules can then be investigated by monitoring the electrical current flowing across the entire SMJ. Collectively, these aforementioned techniques allow researchers to manipulate materials at the single-molecule scale, providing invaluable insights unattainable at the macroscale. In what follows, we focus our attention on elaborating how these SMJ techniques have enabled researchers to probe and understand the physical and chemical properties of a variety of biomolecules.

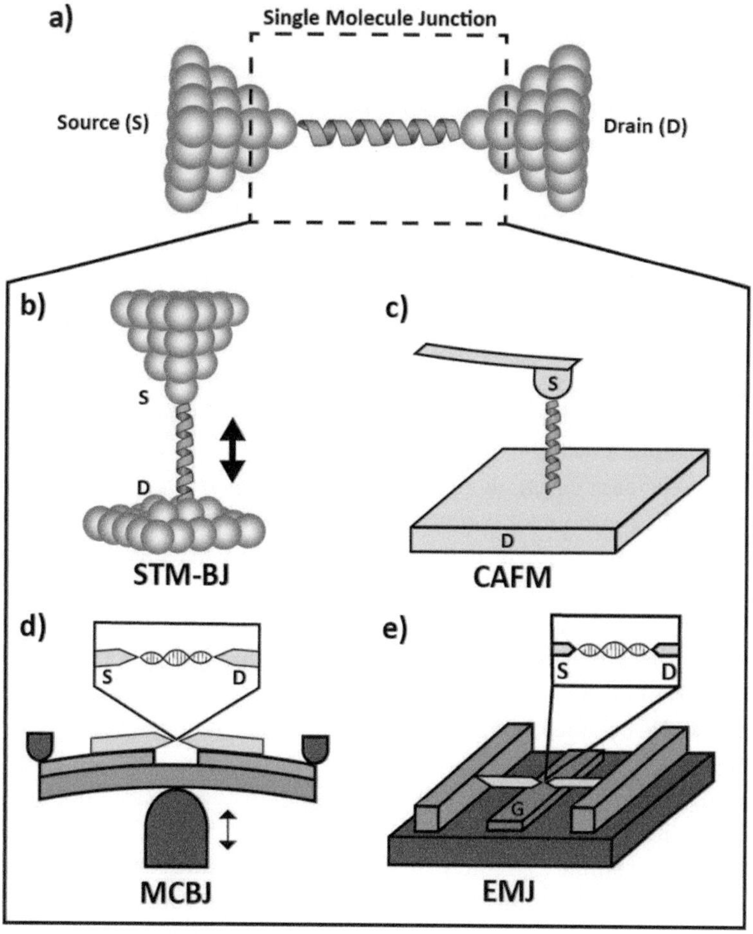

Figure 5.2 Single-Molecule Junction (SMJ) Techniques. (A) Schematic of a single-molecule junction where an individual molecule is sandwiched between two microscopic electrodes (i.e., the source (S) and drain (D) electrodes). (B–E) Schematics of four representative STM techniques, including STM-BJ, CAFM, MCBJ, and EBJ techniques, respectively.

5.2 Electrical Conductance of Single DNA Molecules

We start our discussion with the measurement of the electrical conductance of deoxyribonucleic acid (DNA) using the above-mentioned SMJ techniques. Known as the repository of genetic information, DNA molecules have garnered tremendous attention in nanoelectronics in recent years because their structural stability, diversity, and programmability render them promising building blocks for molecular-scale electronics [13]. Eley and Spivey's observation [14] of DNA and RNA conductivity in 1962 provided crucial early experimental evidence that would guide future research into the possible incorporation of DNA into electronics. However, understanding the electrical properties of single DNA molecules has come a long way. Debate ensued [15–17] as difficulties in experimental design and conditions resulted in numerous seemingly contradictory observations in early measurements. This issue was primarily due to a lack of suitable experimental tools and poor control of the environmental conditions. Thanks to the development of SMJ-based techniques, electron transport in single DNAs (or other biomolecules) can now be systematically investigated under well-defined experimental conditions, such as wet vs. dry. Many recent studies have implied that several important factors, including environmental conditions, molecular conformation, sequence, and length, could significantly influence the electronic properties of DNA. It is now widely recognized that DNA and RNA, under certain conditions, can facilitate efficient charge transport. For instance, the stacked π orbitals between neighboring DNA base pairs (bps) make DNA a good medium for electron transmission. In addition, DNA's remarkable self-assembly ability, as manifested in the emerging DNA origami technologies [18–20], opens the opportunity to fabricate complex circuitry at the molecular scale, circumventing the costly and sophisticated processes involved in conventional nanolithography approaches. Equally important is that probing electron transport through individual biomolecules, including DNA, offers new opportunity to address questions associated with many fundamental biological processes, such as oxidative damage of DNA [21, 22]. In this section, we provide a brief overview of recent advances in probing charge transport through single DNA molecules using SMJ-based methods.

5.2.1 Impact of Environmental Condition and Molecular Conformation

Biological activity and molecular structure are significantly influenced by the environmental conditions surrounding biomolecules. In fundamental biology, the folding of proteins (and many other biological structures) is demonstrated by the numerous interactions of the molecular structure with both itself and its environment [23]. For example, DNA is commonly recognized by its helical structure, but could also exist in multiple fundamentally distinct forms. These forms are the single-stranded DNA (ssDNA), which can interact together to form helical double-stranded DNA (dsDNA). The preference of ssDNA to form dsDNA arises from the complimentary hydrogen bonding between base pairs facilitated by an aqueous environment the DNA is contained within [24]. The first conductance experiments reported on DNA involved dsDNA rather than ssDNA, which at the time was believed to be electrically insulating but this was later proved invalid due to the lack of an appropriate characterization method. In a biological environment, such as the physiological condition in the human body, several conformations of dsDNA have been observed. These include the duplex right-handed B-DNA, left-handed A-DNA [25], and zig-zagged Z-DNA [26]. While these general forms of dsDNA are discussed, numerous conformations are also possible—for example, even chemically rare triple-stranded (triplex) H-DNA [27] and quadruple-stranded (quadruplex) G-DNA [28], and more complex two- and three-dimensional nanostructures made from DNA origami technology [29, 30]. It is the versatility of DNA to adopt a seemingly endless array of sequences and conformations at the nanometer length scale that has made it an attractive candidate for nanotechnology.

In an early work in 2000, Tran et al. [31] explored the effect of buffer condition on lambda-phage DNA (λ-DNA) extracted from *E. Coli* bacteria. Conductance measurements were performed on DNA in water and a purified dry state, respectively. The λ-DNA within the water solution exhibited an electrical conductance one order of magnitude higher than that measured in the dry state. The higher conductance was attributed to the effect of water molecules surrounding the λ-DNA molecule, stabilizing the structure through a combination of dipole-dipole and electrostatic interactions with the phosphate-sugar backbone. This outlook is supported by a 2006 theoretical model by Brovchenko et al. [32] that modeled water surrounding B-DNA (CGCGAATTCGCG) as a network of stabilizing bonds, facilitating the formation of the stable double helix B-DNA. In addition, the formation of spanning water networks results in an acceleration of long-range ion transport, thereby enhancing the conductance. However, these early experimental investigations suffered from poor control over the molecule-electrode contact, which led to discrepancies among experimental data collected from different research labs. The key to robustly quantify electrical conductance of a biomolecule is to wire the target molecule to two probing electrodes in a reliable and repeatable manner. This task remained challenging until the invention of SMJ techniques, especially the STM-BJ technique. Since the benchmark work [5] by Xu and Tao in 2003, the field has been utilizing SMJ as a main workhorse for single-molecule electrical measurements.

In what follows, we continue the discussion with the electrical conductance of DNA in different conformations, but focus more on experimental studies employing SMJ techniques. The less-observed Z-DNA is biologically relevant, particularly within gene expression [33, 34]. Understanding the mechanisms involved when B-DNA transitions to Z-DNA further elucidates its biological role. Environmental conditions, such as the presence of highly concentrated cations, generally from salts, can induce a conformation change of B-DNA to Z-DNA [35], permitting laboratory study of this unique form. An atomistic model of this transition by Moradi et al. [36] in 2012 examined the repeating CG (also GC) nucleotide sequence formed within left-handed Z-DNA. Their results indicated multiple competing mechanisms for the formation of Z-DNA and the apparent difficulty with traditional experimental observation. Employing the STM-BJ method, Wang et al. [37] probed the B to Z conformational transition of a poly $d(GC)_4$ DNA duplex with the increase of Mg^{2+} ions in a buffer solution. Their experimental results are shown in Fig. 5.3. By monitoring the conductance change in real time and in situ, they observed a dramatic conductance decrease by almost two orders of magnitude (from $1 \times 10^{-3}G_0$ to $3 \times 10^{-5}G_0$, where G_0 is the quantum conductance of an atomic point contact) when the DNA transformed from the right-handed B- to the left-handed Z-form (Fig. 5.3B and 5.3C). Interestingly, the observed conductance reduction is caused by structural change-induced breaking of π-π orbital stacking between neighboring bps, leading to the disruption of the main charge transport pathway. As shown in Fig. 5.3D, the GC(CG) bps in B-DNA is well stacked vertically along the same axis, forming a stabilized π-π stacking interactions as well as a shorter path for charge carriers to travel through the structure. This aligned structure can conduct current more readily compared to the Z-DNA form (Fig. 5.3E), which has a twisted structure that forms a gap on the interior, leading to poor charge transport along the helix.

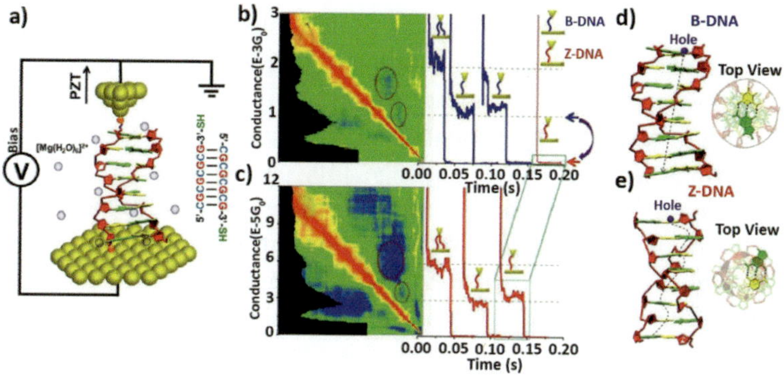

Figure 5.3 (A) Experimental schematic of the STM-BJ measurement for poly d(GC)4 DNA surrounded by $[Mg(H_2O)_6]^{2+}$ ions. (B) The DNA conductance measurement results. In 1 M Mg^{2+} solution, two sets of conductance values (1×10^{-3} G_0 in blue and 1×10^{-5} G_0 in red) were observed for dsDNA. The high conductance state is for the B-conformation, and the low conductance state is due to the Z-conformation of the DNA. (C) and (D) show the structural side view and top view of B- and Z-DNA, respectively. The dashed lines demonstrate the intra-strand path for charge carriers (i.e., hole in this case) migration.

Similarly, one could ask, how would the electrical conductance change if a dsDNA changes its conformation from B- to A-form? This question was addressed by Artes et al. [38] in a subsequent study. Their STM-BJ measurement results of a GC-rich DNA sequence (5′-CCCGCGCGCCC-3′) showed that the conductance of a dsDNA increases by approximately one order of magnitude during a transition from B-form to A-form, contradicting an earlier theoretical prediction [39] that the A-form is less conductive due to poor base alignment. Further theoretical calculations of the DNA's electronic structures for the two conformations revealed that the highest occupied molecular orbital (HOMO) of the DNA is 20% more delocalized in the A-form than in the B-form, therefore leading to a higher conductance of the A-form DNA. These recent studies employing the STM-BJ technique help to rationalize the large dispersion of DNA conductivity found in early literature, and, more importantly, highlight the strong relationship between molecular structure and environment.

5.2.2 Sequence and Length Dependance

Elucidating the impact of sequence and molecular length on DNA conductance is of fundamental importance for the future design and development of nanoelectronics and biosensing technologies. However, varied experimental platforms and environmental conditions have posed a challenge [40, 41]. Delaney et al. [42] observed in 2003 that alignment of guanine nucleotides facilitates electron movement through DNA via oxidation. Guanine contains aromatic rings which, when fully (or partially) overlapping with another aromatic ring (such as another Guanine), are believed to form a stabilizing electron π-π stacking interaction. Considered an important effect, the quantum mechanical nature of π-π stacking itself is still being realized [43, 44]. While our focus on DNA conductance has been on dsDNA, these base pair interactions can be found within ssDNA. Cohen et al. in 2005 [45] employed CAFM to simultaneously measure applied force and conductance of dsDNA chemically bonded by propyl thiol linkers to gold nanoparticles (GNPs) on a metal substrate. Measurements were performed on (5'-CAT TAA TGC TAT GCA GAA AAT CTT AG-3'-C_3H_6-SH) dsDNA and its complementary ssDNA. The dsDNA form was measured to be conductive, while the ssDNA was found to be insulating, a view that has since expanded.

Compared to dsDNA, ssDNA conductance is highly specific to the sequence involved. Illustrating this concept, Zalinge et al. [46] performed the hover-mode STM (also known as STM I(t)) measurements on ss and ds oligonucleotides. Similar to other results, dsDNA exhibited higher conductance when compared to individual ssDNA of the same base pair length. In 2004 Xu et al. [47] further explored this relationship employing STM-BJ conductance measurements on a series of short dsDNA oligonucleotides. Thiol (SH) linkers were added at the 3' ends of the two complementary strands of the dsDNA to covalently bind to the gold electrodes (Figure 5.4A). Their measurements were performed within buffer solution to preserve the dsDNA conformation. Charge transport was found to occur predominantly through stacked base pair interactions, rather than the DNA skeleton. To test the effect of conductance on length, the researchers tested a variety of base pair lengths. Interestingly, they found that DNA's conductance depends on both sequence and length. Specifically, as shown in Fig. 5.4D, the conductance was observed to be inversely proportional to the length for alternating $(GC)_n$ sequences. When inserting $(A{:}T)_m$ into GC-rich regions, DNA conductance decayed exponentially with the length of A:T base pairs, with a decay constant of 0.43 $Å^{-1}$ (Fig. 5.4C). This observation immediately implies that electron conduction through dsDNA could be governed by different physical mechanisms for different sequences. The results reported in Xu's work were later confirmed by Li et al. [48] in even longer dsDNAs. In Li's work, they reported that STM-BJ measurements for an $ACGC(AT)_mGCGT$/ $ACGC(AT)_mAGCGT$ dsDNA sequence. In this sequence, for $m = 0$, 1, and 2, the resistance (inverse of conductance) increases more rapidly with length, agreeing well with the exponential function observed in Xu's work. For $m > 2$, the resistance only weakly depends on length, suggesting a clear transition of charge transport mechanisms from the coherent tunneling to the hopping regime. To summarize these experimental observations, it has been widely accepted that the charge transport through dsDNA is dominated by coherent tunneling when the G:C bps are separated by A:T bps, and the conductance has an exponential dependence on the length of A:T bps (i.e., the tunneling barrier). If the number of A:T bps is increased, incoherent hopping (linear dependence of conductance on length) would govern the transport. However, for alternating (GC) n sequences (also known as alternating GC, where n is the number of repeating base pair unit), the conductance of DNA is usually governed by thermally activated hopping, where the G bases serve as the hopping sites regardless of the molecular length. It is evident that the interplay of sequence and length has a deterministic impact on the electrical properties of DNA molecules.

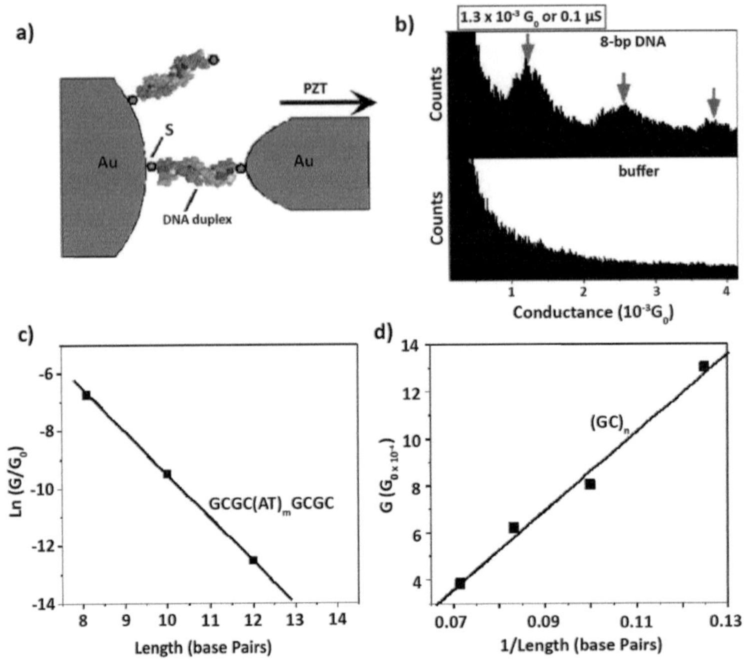

Figure 5.4 Sequence and length dependent conductance of single dsDNA molecule. (A) Schematic of a metal-dsDNA-metal junction. (B) Conductance histograms of an 8-bp dsDNA (top) and pure buffer solution without molecule (bottom), respectively. (C) ln(G/G0) vs. molecular length plot for dsDNA with ACGC(AT)mGCGT sequences. (D) Conductance (G) vs. 1/length plot for dsDNA with (GC)n sequences. Reprinted with permission from Xu et al. [47]. Copyright 2004 American Chemical Society.

In contrast, more recently, the widely adopted hopping model for understanding charge transport in G-C rich dsDNA was found to be not always valid when the DNA sequence contains stacked G instead of alternating GC. Using STM-BJ technique, Xiang et al. [49] measured the resistance of dsDNA with alternating GC and stacked G of different lengths. Their experimental results are illustrated in Fig. 5.5. They found that the resistance of DNA with alternating GC increased linearly with the length, which agreed with previous observation. Surprisingly, the resistance of DNA with stacked G showed a unique pattern of oscillating up and down as the length of the DNA increases. With the assist of theoretical simulation, they attributed such oscillating resistance-length relationship in stacked G sequences to a new transport mechanism called intermediate tunneling-hopping regime, that is, the combination of partially coherent (organized) tunneling and partially hopping (unorganized). Further calculations [50] revealed that the HOMOs in the stacked G are delocalized over several G bases, supporting the intermediate coherent tunneling and incoherent hopping mechanism. Coherence tunneling dominates the conductance for odd-n sequences, whereas incoherent hopping dominates even-n sequences. These findings indicate how sequence can be exploited to alter the balance between coherent and incoherent transport in DNA molecules.

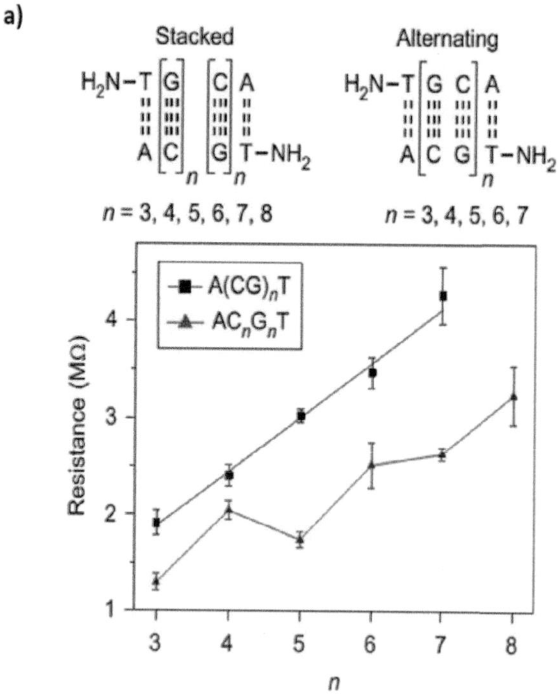

Figure 5.5 STM-BJ single-molecule conductance measurements of dsDNA with alternating G (A(CG)$_n$T) and stacked G (AC$_n$G$_n$T) sequences. The number of repeated sequences, n, is plotted against resistance. Reprinted with permission from Xiang et al. [49]. Copyright 2015 Nature Publishing Group.

5.3 Electronic Diode and Switch Made of Single DNA

An important step toward incorporating molecular components into electronic devices is to search for molecular candidates that could imitate the behavior of traditional silicon-based semiconductors, such as diodes and/or switches. Diodes are a fundamental building block in today's microelectronic devices that conduct electrical current predominantly in a single direction. Thanks to the remarkable structural flexibility of DNA, the first DNA-based single-molecule diode was reported by Guo et al. in 2016 [51]. In their work, they harnessed the strategy of intermolecular interaction, that is, site-specific intercalation of conjugated molecules into a native DNA duplex, to construct a DNA diode with asymmetric electron transmission. As shown in Fig. 5.6, they experimentally intercalated two coralyne molecules into a native duplex DNA molecule (5′-CGCGAAACGCG-3′). The intercalation effect significantly modulated the electronic structure of the native DNA. Their electrical measurements of the DNA-coralyne complex using the STM-BJ method showed clear asymmetric diode-like I–V characteristics with a high rectification ratio of around 15 at 1.1 V (Fig. 5.6C and 5.6D). Further theoretical simulation of such a DNA–coralyne complex system revealed that the diode behavior was induced by the bias-dependent coupling of the molecular HOMO-1 level to the electrode. This study paved the path toward designing functional DNA-based computing nanodevices in the future.

The unique feature of a semiconductor logic device is that the device conductance can be effectively switched "on" and "off" through a third-gate electrode. Following a similar design principle, researchers have also explored the possibility of creating a DNA molecular switch through a gating effect. While adding a physical gate electrode to an SMJ device poses a significant challenge in device fabrication, a new strategy taking advantage of electrochemical (EC) effect in an ionic environment has been developed in recent years [52, 53]. More importantly, the promising EC gating method can be integrated into an STM-BJ setup to achieve efficient gate coupling that manipulates the energy alignment and the molecular redox processes. For example, Xiang et al. [54] reported their conductance

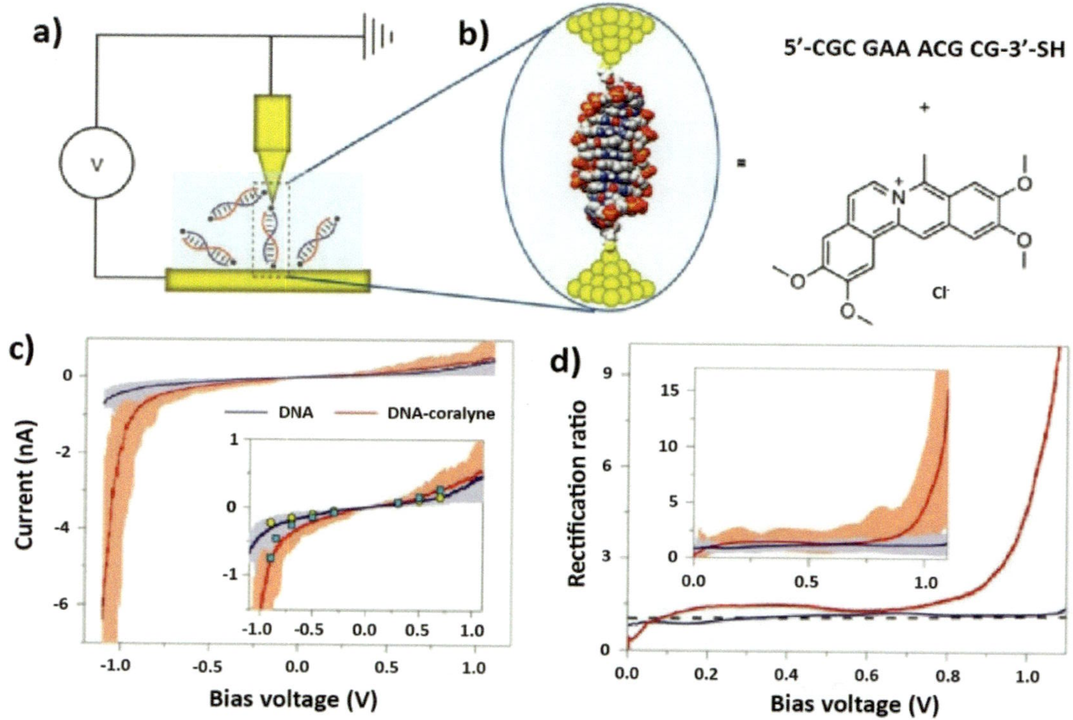

Figure 5.6 Molecular diode made of dsDNA. (A) Schematic of the STM-BJ setup; (B) Schematic of a dsDNA-coralyne complex molecular junction; (C, D) show the current-voltage (I–V) curves and rectification ratio curves of native DNA (blue) and DNA-coralyne complex (red). Note that strong current rectification with a rectification ratio of 15 was observed for the DNA-coralyne complex molecular junction. Reprinted with permission from Guo et al. [51]. Copyright 2016 Nature Publishing Group.

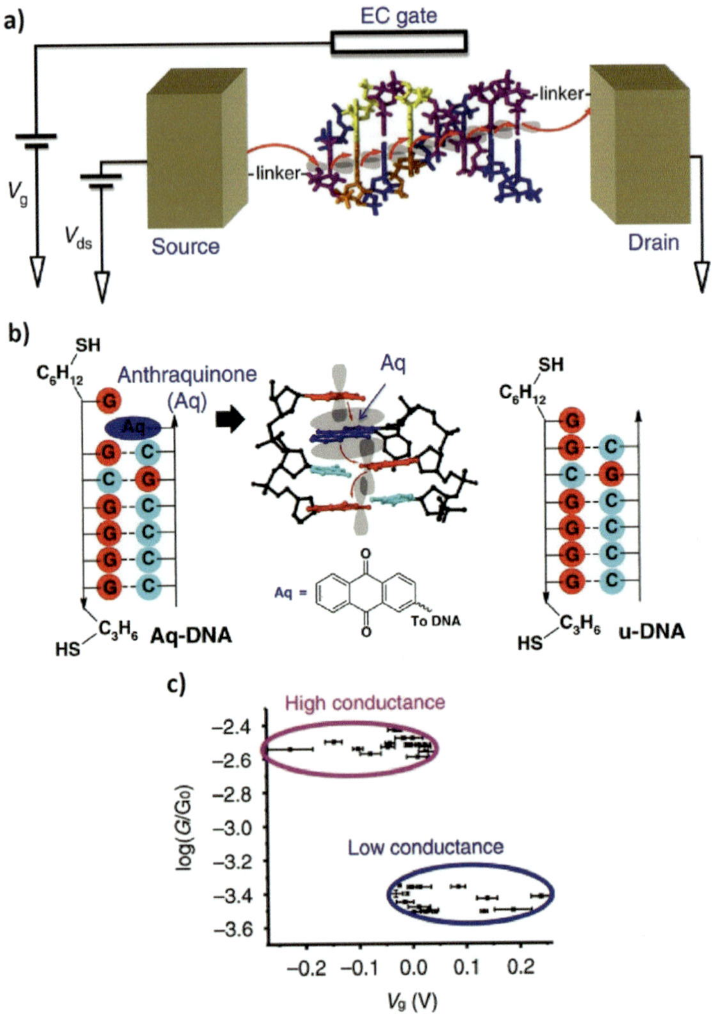

Figure 5.7 DNA conductance switching via electrochemical gating. (A) Experiment schematic of an EC-gated STM-BJ setup where the source and drain electrodes are the STM tip and substrate, and EC gate is a silver electrode inserted in the solution. (B) Structure of a redox modified DNA (Aq-DNA), where a base was replaced with an anthraquinone (Aq) moiety (highlighted in blue) at the 3′-end of a DNA strand. (C) Conductance values measured under different gate voltages showing two discrete conductance states, high (magenta circle) and low (blue circle) conductance states. Reprinted with permission from Xiang et al. [54]. Copyright 2017 Nature Publishing Group.

measurements of a modified DNA duplex molecule where one base of the dsDNA was replaced with a redox group for optimal EC control. By applying an EC gate voltage to the molecule, they demonstrated that switching between the oxidized and reduced states of the redox group can turn on and off the electrical conductance of the DNA (Fig. 5.7). Their theoretical calculation further suggested that the conductance switching is induced by the variation of the molecular energy alignment associated with the redox state switching. This work implies that one can introduce an active control to DNA conductance via redox reaction. The implementation of this strategy can be extended to more advanced DNA nanostructures to create active building blocks for DNA-based nanodevices.

5.4 SMJ for Biosensing and Sequencing

Applications have also been explored utilizing these above-mentioned SMJ techniques to create biosensors capable of measuring a specific analyte, such as an enzyme, antibody, nucleic acid, or other interacting biomolecule [55]. This is primarily because SMJ-based techniques open unique opportunity in electrical detection of single molecules in aqueous solution, which allows for the ultrasensitive biosensing and biosequencing. For instance, in a custom-designed biomolecular SMJ, the conductance variation specific to changes in molecular length [42], conformation [38], and base-pairing could be characterized with high precision [56, 57]. Note that this approach is fundamentally different from the well-known nanopore sequencing technologies, which mainly rely on measuring the change of ionic current when a base of the nucleic acid passes through the nanopore [58, 59]. SMJ-based single-molecule biosensing technique also helps to overcome major drawbacks associated with conventional approaches, such as time-consuming cell-culturing and the necessary enzymatic amplification processes [60, 61]. While the SMJ-based approaches can detect known sequences rapidly and accurately, they face challenges to identify unknown or new sequences, which is the major limitation of these techniques.

Here we present a recent experimental breakthrough in the biosensing application of SMJ. The bacteria *E. coli* is a widely studied and prolific bacteria found within the environment, untreated water sources, and intestinal systems. While often harmless, *E. coli* comprises multiple contagious strains notable for producing Shiga Toxin (Stx), causing minor to severe illness in humans. Commonly used methods to identify the sequence of DNA:RNA involve careful and time-consuming cell culturing before determination by polymerase chain reactions (PCRs) or nucleic acid. Recently, remarkable progress has been made by Li et al. [62] to adapt the STM-BJ method to single biomolecular sequencing applications, especially via combining with powerful machine learning assisted data analysis. In their work, STM-BJ-based single-molecule electrical detection was applied to particular *E. coli* strains. As shown in Fig. 5.8, single-base mismatches can be detected and identified with high specificity. Subtle variation in an individual base can dramatically alter the electrical conductance of the SMJ. The perfect match wild-type for the pathogenic *E. coli* O157:H7 yields a conductance of $2.5 \times 10^{-4}G_0$. When an adenine close to the 3′ in the RNA fragment was changed to guanine, the SMJ conductance rapidly reduced to only 20% of the wild-type conductance. This change enables direct detection of other pathogenic *E. coli* strains. When these substitutions are made between the *E. coli* variants, the geometric alignment of the RNA helix is altered. Consistent with examples discussed earlier, this causes the molecules to adopt conformations that change the effectiveness of the π-π base stacking interactions, previously mentioned with guanine. Furthermore, conductance is affected by the altering of band gaps and ionization potentials when the bases are substituted. It was also found that these conductance signatures can be identified even when interfering species are present. This brings such a detection procedure into practical relevance as a powerful tool genotyping, without the time-consuming and chemically altering effects of traditional methods. By performing thousands of these measurements with the high-specificity present, in a relatively short time span, the relevance to many other biological applications is possible.

We note that these recent advancements in single-molecule biosensing are not limited to nanoengineering applications but can also catalyze new opportunities spanning across various fields, including biophysics, biochemistry, biotechnology, and medicinal chemistry, where understanding biomolecular interactions is crucial for unveiling the molecular basis of numerous diseases. While the

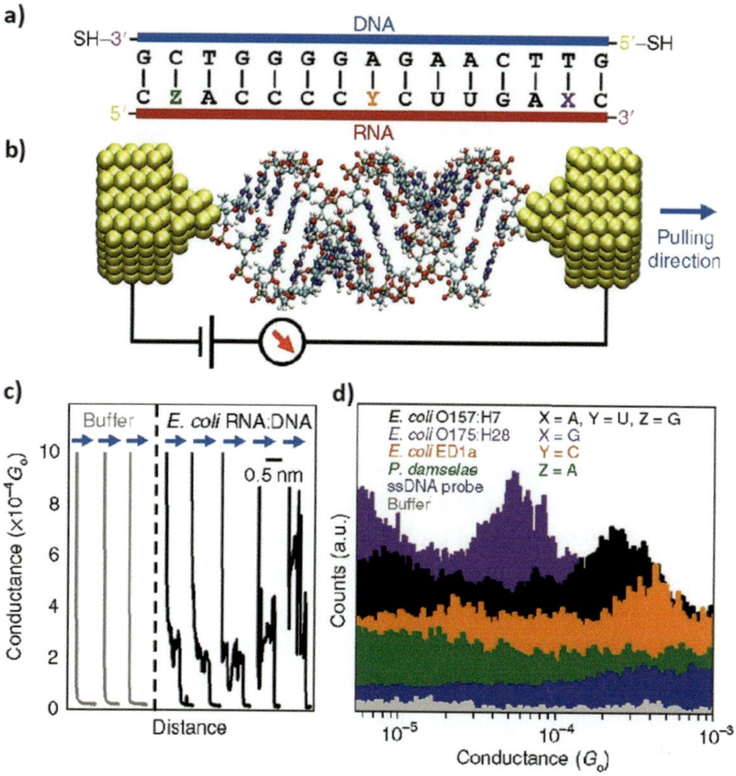

Figure 5.8 Single-molecule electrical sensing on biologically relevant oligonucleotide from *E. coli* strains. (A) 15 bp RNA:DNA sequences studied. The blue line represents the DNA probe with thiol linkers, and the red line represents the targeted RNA sequences. (B) Experimental schematic of the RNA:DNA molecule connected between two gold electrodes in an STM-BJ setup. (C) Example conductance vs. distance traces obtained from the O157:H7 hybrids. The black curves (with conductance plateaus) are obtained when a molecule binds between the electrodes; and the grey curves were collected in a buffer solution without molecule as a control. (D) Conductance histograms for four different RNA:DNA hybrids and two control experiments measured for the ssDNA probe and blank buffer. Reprinted with permission from Li et al. [62]. Copyright 2018 Nature Publishing Group.

commonly adopted sequencing and sensing techniques based on optical detection of fluorescence tagged molecules have proven to be powerful for deciphering these interactions, comparing their results with the bulk behavior of cellular interactions has remained a challenge for scientists. To overcome this, SMJ methods can help study the basics of biophysical interactions at the single-molecule level via quantifying electron transport between biomolecules and in biomolecular complexes. In the future, a well-designed version of the SMJ-based approach could also validate the concept of various single biomolecular sensors for critical interactions that may profoundly impact human health and the environment, such as the interaction between antigen and antibody, RNA-induced silencing complex, and even cancer detection.

5.5 Summary and Outlook

Remarkable progress in the field of single-molecule electronics (SME) over the past two decades has catalyzed a vibrant development of powerful experimental toolboxes capable of measuring electrical properties of individual molecules. When applied to biomolecules, these novel experimental approaches open new perspectives for single-molecule biophysics, biochemistry, and biomedicine. In this chapter, we present recent advances in employing SMJ-based techniques for direct and label-free electrical detection and sensing of individual biomolecules. These SMJ approaches have enabled researchers to delve further into the structure–property relationships of a broad range of biologically relevant materials, including DNA, RNA, and proteins. As we strive to understand an increasingly complex biosystem, governed by chemical and physical interactions, the necessity to employ techniques capable of providing precise individual analysis is paramount [63–67]. We envision that when equipped with ultrafast measurement schemes and machine learning-assisted data processing, the potential of SMJ methods could be fully realized to uncover the essential dynamics of single-molecule chemical/biological processes with high spatial and temporal resolution, which is otherwise difficult to access. Insights from this research will be of fundamental significance in elucidating the intrinsic mechanisms relevant to existing challenges in today's health care and life sciences, such as early and accurate molecular diagnoses [62, 68, 69] and high-efficiency genetic testing [60, 70].

Acknowledgments

Kun Wang gratefully acknowledges the financial support from U.S. Department of Energy, Basic Energy Sciences (Award No. DE-SC0024924).

References

1. Aviram, A. and Ratner, M. A. (1974). Molecular rectifiers, *Chem Phys Lett*, **29**, pp. 277–283.
2. Binnig, G., Rohrer, H., Gerber, Ch., and Weibel, E. (1982). Tunneling through a controllable vacuum gap, *Appl Phys Lett*, **40**, pp. 178–180.
3. Binnig, G., Quate, C. F., and Gerber, Ch. (1986). Atomic force microscope, *Phys Rev Lett*, **56**, pp. 930–934.
4. Murrell, M. P., Welland, M. E., O'Shea, S. J., Wong, T. M. H., et al. (1993). Spatially resolved electrical measurements of SiO gate oxides using atomic force microscopy, *Appl Phys Lett*, **62**, pp. 786–788.
5. Bingqian, X. and Nongjian, J. T. (2003). Measurement of single-molecule resistance by repeated formation of molecular junctions, *Science*, **301**, pp. 1221–1223.
6. Muller, C. J., van Ruitenbeek, J. M., and de Jongh, L. J. (1992). Conductance and supercurrent discontinuities in atomic-scale metallic constrictions of variable width, *Phys Rev Lett*, **69**, pp. 140–143.
7. Reed, M. A., Zhou, C., Muller, C. J., Burgin, T. P., et al. (1997). Conductance of a molecular junction, *Science*, **278**, pp. 252–254.
8. Prins, F., Barreiro, A., Ruitenberg, J. W., Seldenthuis, J. S., et al. (2011). Room-temperature gating of molecular junctions using few-layer graphene nanogap electrodes, *Nano Lett.*, **11**, pp. 4607–4611.
9. Park, H., Lim, A. K. L., Alivisatos, A. P., Park, J., et al. (1999). Fabrication of metallic electrodes with nanometer separation by electromigration, *Appl Phys Lett*, **75**, pp. 301–303.
10. Xie, X., Li, P., Xu, Y., Zhou, L., et al. (2022). Single-molecule junction: A reliable platform for monitoring molecular physical and chemical processes, *ACS Nano*, **16**, pp. 3476–3505.
11. Wang, K. and Xu, B. (2017). Modulation and control of charge transport through single-molecule junctions, *Top Curr Chem*, **375**, p. 17.
12. Xiang, D., Wang, X., Jia, C., Lee, T., and Guo, X. (2016). Molecular-scale electronics: from concept to function, *Chem Rev*, **116**, pp. 4318–4440.

13. Wang, K. (2018). DNA-based single-molecule electronics: From concept to function, *J Funct Biomater*, **9**, p. 8.
14. Eley, D. D. and Spivey, D. I. (1962). Semiconductivity of organic substances. Part 9—Nucleic acid in the dry state, *Trans Faraday Soc*, **58**, pp. 411–415.
15. Fink, H.-W. and Schönenberger, C. (1999). Electrical conduction through DNA molecules, *Nature*, **398**, pp. 407–410.
16. Kasumov, A. Yu., Kociak, M., Guéron, S., Reulet, B., et al. (2001). Proximity-induced superconductivity in DNA, *Science*, **291**, pp. 280–282.
17. Porath, D., Bezryadin, A., de Vries, S., and Dekker, C. (2000). Direct measurement of electrical transport through DNA molecules, *Nature*, **403**, pp. 635–638.
18. Nangreave, J., Han, D., Liu, Y., and Yan, H. (2010). DNA origami: A history and current perspective, *Curr Opin Chem Biol*, **14**, pp. 608–615.
19. Dey, S., Fan, C., Gothelf, K. V., Li, J., et al. (2021). DNA origami, *Nature Reviews Methods Primers*, **1**.
20. Hong, F., Zhang, F., Liu, Y., and Yan, H. (2017). DNA Origami: Scaffolds for creating higher order structures, *Chem Rev*, **117**, pp. 12584–12640.
21. Saito, I., Nakamura, T., Nakatani, K., Yoshioka, Y., et al. (1998). Mapping of the hot spots for DNA damage by one-electron oxidation: Efficacy of GG doublets and GGG triplets as a trap in long-range hole migration, *J Am Chem Soc*, **120**, pp. 12686–12687.
22. Hall, D. B., Holmlin, R. E., and Barton, J. K. (1996). Oxidative DNA damage through long-range electron transfer, *Nature*, **382**, pp. 731–735.
23. Crothers, D. M. (1986). Principles of nucleic acid structure. Von W. Saenger. Springer-Verlag, Berlin 1984. XX, 556 S., geb. DM 79.00. - ISBN 3-540-90761-0xs, *Angewandte Chemie*, **98**, pp. 762–762.
24. Travers, A. and Muskhelishvili, G. (2015). DNA structure and function, *FEBS Journal*, **282**, pp. 2279–2295.
25. Arnott, S. and Hukins, D. W. L. (1972). Optimized parameters for A-DNA and B-DNA, *Biochem Biophys Res Commun*, **47**, pp. 1504–1509.
26. Rich, A. and Zhang, S. (2003). Z-DNA: The long road to biological function, *Nat Rev Genet*, **4**, pp. 566–572.
27. Htun, H. and Dahlberg, J. E. (1989). Topology and formation of triple-stranded H-DNA, *Science*, **243**, pp. 1571–1576.

28. Di Antonio, M., Ponjavic, A., Radzevičius, A., Ranasinghe, R. T., et al. (2020). Single-molecule visualization of DNA G-quadruplex formation in live cells, *Nat Chem*, **12**, pp. 832–837.

29. Liu, W., Duan, H., Zhang, D., Zhang, X., et al. (2021). Concepts and application of DNA origami and DNA self-assembly: A systematic review, *Appl Bionics Biomech*, **2021**, pp. 1–15.

30. Majikes, J. M. and Liddle, J. A. (2021). DNA origami design: A how-to tutorial, *J Res Natl Inst Stand Technol*, **126**, p. 126001.

31. Tran, P., Alavi, B., and Gruner, G. (2000). Charge transport along the L-DNA double helix, *Phys Rev. Lett.*, **85**, p. 1564.

32. Brovchenko, I., Krukau, A., Oleinikova, A., and Mazur, A. K. (2006). Water percolation governs polymorphic transitions and conductivity of DNA, *Phys Rev Lett*, **97**.

33. Herbert, A., Karapetyan, S., Poptsova, M., Vasquez, K. M., et al. (2021). A, B and Z: The structure, function and genetics of Z-DNA and Z-RNA, *Int J Mol Sci*, **22**, p. 7686.

34. Herbert, A. (2019). Z-DNA and Z-RNA in human disease, *Commun Biol*, **2**, p. 7.

35. Guéron, M., Demaret, J.-Ph., and Filoche, M. (2000). A unified theory of the B-Z transition of DNA in high and low concentrations of multivalent ions, *Biophys J*, **78**, pp. 1070–1083.

36. Moradi, M., Babin, V., Roland, C., and Sagui, C. (2013). Reaction path ensemble of the B–Z-DNA transition: A comprehensive atomistic study, *Nucleic Acids Res*, **41**, pp. 33–43.

37. Wang, K., Hamill, J. M., Wang, B., Guo, C., et al. (2014). Structure determined charge transport in single DNA molecule break junctions, *Chem Sci*, **5**, pp. 3425–3431.

38. Artés, J. M., Li, Y., Qi, J., Anantram, M. P., et al. (2015). Conformational gating of DNA conductance, *Nat Commun*, **6**.

39. Endres, R. G., Cox, D. L., and Singh, R. R. P. (2004). Colloquium: The quest for high-conductance DNA, *Rev Mod Phys*, **76**, pp. 195–214.

40. Lincoln, P., Tuite, E., and Nordén, B. (1992). Short-Circuiting the molecular wire: Cooperative binding of Δ-[Ru(Phen) 2 Dppz]$^{2+}$ and Δ-[Rh(Phi) 2 Bipy]$^{3+}$ to DNA, *Molecules*, **24**, p 769.

41. Kelley, S. O., Holmlin, R. E., Stemp, E. D. A., and Barton, J. K. (1997). Photoinduced electron transfer in ethidium-modified DNA duplexes: Dependence on distance and base stacking, *J Am Chem Soc*, **119**, pp. 9861–9870.

42. Delaney, S. and Barton, J. K. (2003). Long-range DNA charge transport, *J Org Chem*, **68**, pp. 6475–6483.

43. Sinnokrot, M. O. and Sherrill, C. D. (2006). High-accuracy quantum mechanical studies of π–π interactions in benzene dimers, *J Phys Chem A*, **110**, pp. 10656–10668.

44. Parrish, R. M. and Sherrill, C. D. (2014). Quantum-mechanical evaluation of π–π versus substituent–π interactions in π stacking: Direct evidence for the Wheeler–Houk picture, *J Am Chem Soc*, **136**, pp. 17386–17389.

45. Cohen, H., Nogues, C., Naaman, R., and Porath, D. (2005). Direct measurement of electrical transport through single DNA molecules of complex sequence. *Proc Natl Acad Sci*, **102**, pp. 11589–11593.

46. Van Zalinge, H., Schiffrin, D. J., Bates, A. D., Haiss, W., et al. (2006). Single-molecule conductance measurements of single- and double-stranded DNA oligonucleotides, *ChemPhysChem*, **7**, pp. 94–98.

47. Xu, B., Zhang, P., Li, X., and Tao, N. (2004). Direct conductance measurement of single DNA molecules in aqueous solution, *Nano Lett*, **4**, pp. 1105–1108.

48. Li, Y., Xiang, L., Palma, J. L., Asai, Y., et al. (2016). Thermoelectric effect and its dependence on molecular length and sequence in single DNA molecules, *Nat Commun*, **7**, p. 11294.

49. Xiang, L., Palma, J. L., Bruot, C., Mujica, V., et al. (2015). Intermediate tunnelling-hopping regime in DNA charge transport, *Nat Chem*, **7**, pp. 221–226.

50. Liu, C., Xiang, L., Zhang, Y., Zhang, P., et al. (2016). Engineering nanometre-scale coherence in soft matter, *Nat Chem*, **8**, pp. 941–945.

51. Guo, C., Wang, K., Zerah-Harush, E., Hamill, J., et al. (2016). Molecular rectifier composed of DNA with high rectification ratio enabled by intercalation, *Nat Chem*, **8**, pp. 484–490.

52. Baghernejad, M., Manrique, D. Z., Li, C., Pope, T., et al. (2014). Highly-effective gating of single-molecule junctions: An electrochemical approach, *Chem Commun*, **50**, pp. 15975–15978.

53. Huang, C., Rudnev, A. V., Hong, W., and Wandlowski, T. (2015). Break junction under electrochemical gating: Testbed for single-molecule electronics, *Chem Soc Rev*, **44**, pp. 889–901.

54. Xiang, L., Palma, J. L., Li, Y., Mujica, V., et al. (2017). Gate-controlled conductance switching in DNA, *Nat Commun*, **8**, p. 14471.

55. Chen, H., Jia, C., Zhu, X., Yang, C., et al. (2022). Reactions in single-molecule junctions, *Nat Rev Mater*, **8**, pp. 165–185.

56. Li, Y., Artés, J. M., Qi, J., Morelan, I. A., et al. (2016). Comparing charge transport in oligonucleotides: RNA:DNA hybrids and DNA duplexes, *J Phys Chem Lett*, **7**, pp. 1888–1894.

57. Hihath, J., Xu, B., Zhang, P., and Tao, N. (2005). Study of single-nucleotide polymorphisms by means of electrical conductance measurements, *Proc Natl Acad Sci USA*, **102**, pp. 16979–16983.

58. Kono, N. and Arakawa, K. (2019). Nanopore sequencing: Review of potential applications in functional genomics, *Dev Growth Differ*, **61**, pp. 316–326.

59. Wang, Y., Zhao, Y., Bollas, A., Wang, Y., et al. (2021). Nanopore sequencing technology, bioinformatics and applications, *Nat Biotechnol*, **39**, pp. 1348–1365.

60. Wang, Y., Alangari, M., Hihath, J., Das, A. K., et al. (2021). A machine learning approach for accurate and real-time DNA sequence identification, *BMC Genomics*, **22**, p. 525.

61. Bartlett, J. M. S. and Stirling, D. (2003). *PCR Protoc*, Humana Press, New Jersey.

62. Li, Y., Artés, J. M., Demir, B., Gokce, S., et al. (2018). Detection and identification of genetic material via single-molecule conductance, *Nat Nanotechnol*, **13**, pp. 1167–1173.

63. Mallajosyula, S. S. and Pati, S. K. (2010). Toward DNA conductivity: A theoretical perspective, *J Phys Chem Lett*, **1**, pp. 1881–1894.

64. Wang, K. (2018). DNA-based single-molecule electronics: From concept to function, *J Funct Biomater*, **9**.

65. Genereux, J. C. and Barton, J. K. (2010). Mechanisms for DNA charge transport, *Chem Rev*, **110**, pp. 1642–1662.

66. Travers, A. and Muskhelishvili, G. (2015). DNA structure and function, *FEBS J*, **282**, pp. 2279–2295.

67. Taniguchi, M. and Kawai, T. (2006). DNA electronics, *Physica E Low Dimens Syst Nanostruct*, **33**, pp. 1–12.

68. Mishra A. K. (2019). Application of nanotechnology in diagnosis, drug dissolution, drug discovery, and drug carrier, *Nanobiotechnol Bioformul*, Springer Nature, pp. 449–475.

69. Pattiya Arachchillage, K. G. G., Chandra, S., Williams, A., Pham, J., et al. (2023). Electrical detection of RNA cancer biomarkers at the single-molecule level, *Sci Rep*, **13**, p. 12428.

70. Xu, M., Fujita, D., and Hanagata, N., (2009). Perspectives and challenges of emerging single-molecule sequencing technologies, *Small*, **5**, pp. 2638–2649.

Chapter 6

Tissue Engineering Approach: 3D Printing

Mahima Gupta and Ram V. Devireddy
Department of Mechanical Engineering, Louisiana State University, Baton Rouge, LA 70803, USA
rdevir1@lsu.edu

Over the past decades, significant advancement has been achieved in making cellularized extracellular scaffolds to regenerate diverse tissues including bone, cartilage, heart, blood vessels, nerve, liver, and numerous other types of tissue. However, the majority of conventional scaffolds exhibit intricate construction and do not facilitate sufficient cell migration to establish adequate cell–cell adhesion, nutrient diffusion, cell–cell communication, and cell–extracellular matrix (ECM) interactions, which are all critically important for tissue-level functions. 3D printing and tissue engineering offer a promising alternative, allowing for precise deposition of cells and biomaterials to create functional tissue constructs by improving scaffold design and cell integration. Additionally, high throughput and label-free imaging technologies such as hyperspectral imaging (HSI) play

High-Throughput Imaging Methods: Cells and Tissues
Edited by Manas Ranjan Gartia and Ram V. Devireddy

ISBN 978-981-5352-08-5 (Hardcover), 978-1-003-77417-4 (eBook)
www.jennystanford.com

DOI: 10.1201/9781003774174-6

a crucial role in monitoring scaffold integrity, cellular responses, and other biochemical changes in engineered tissues. To address certain limitations associated with traditional scaffold-based tissue engineering, this article explores the development of cell-sheet engineering as an alternative approach in tissue engineering, and also the integration of 3D printing with tissue engineering, and highlights the role of HSI in advancing regenerative medicine.

6.1 Introduction

The proliferation and differentiation of stem cells into structured three-dimensional tissues often requires the introduction of isolated cell suspensions into biodegradable scaffolds [1]. This process is typically performed by seeding cells at a high density within bioreactors, either static or dynamic, to facilitate cell–extracellular matrix (ECM) and cell–cell communications [2–10]. Obviously, an optimal distribution and seeding of cells (which is most often unknown a priori) within the scaffolds structure is critical for regulating tissue formation and generating 3D cell culture systems [11]. This process is further hindered by the lack of vascular channels, which in turn causes the seeded cells to solely rely on diffusion for solute-nutrient entrance and waste-product exit [12]. Therefore, most scaffold-based cell seeding techniques are limited by the inability to obtain nutrients, and consequently, cells closest to the exterior are better serviced than cells that are in the interior or in the nutrient depletion region [13–18].

This article presents cell sheet engineering as an alternative method of tissue engineering that can help engineers to get around some of the drawbacks of conventional scaffold-based tissue fabrication methods [19–29]. Without the need of proteolytic enzymes, cultivated cells were successfully extracted whole sheets using temperature-responsive poly (N-isopropyl acrylamide) (PNIPAAm) coated Tissue Culture Polystyrene (TCPS) dishes by simple temperature changes (Fig. 6.1). Essentially, cell sheet engineering removes the need for biodegradable scaffolds. Several studies have shown that thermoresponsive hydrogel coated on TCPS dishes can be used for harvesting living cell sheets [20, 30]. Aqueous mixtures containing specific salts were poured over TCPS dishes at room temperature to create the hydrogel, which was then gelled at 37°C for cell culture [30].

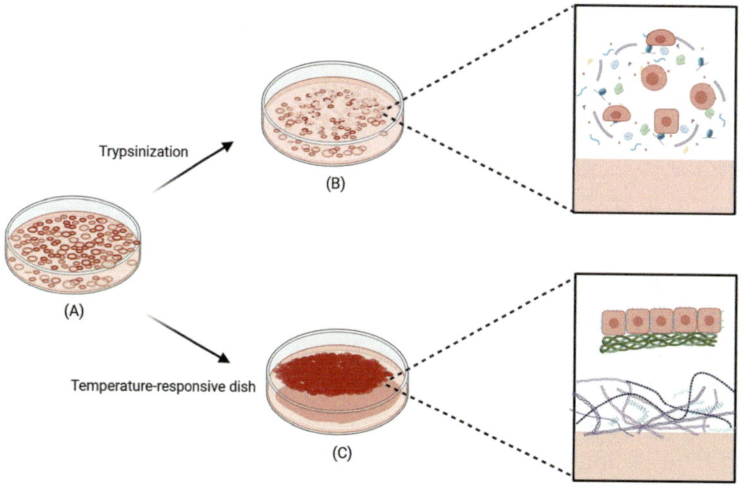

Figure 6.1 Temperature-responsive cell sheet fabrication vs. traditional trypsinization methods.

Gels were created at physiologically relevant temperatures using methylcellulose (MC), which has a viscosity of 4000 cP at a 2% solution. Nevertheless, the process outlined was intricate and did not always result in stable hydrogels, since the MC formulation was shown to be excessively viscous for effective manipulation [30]. Additionally, the high viscosity of MC formulation makes it challenging to apply MC hydrogel uniformly to the tissue culture plate, which leads to an uneven coating and an unstable hydrogel system. Considering that the main requirement for cell sheet engineering applications is that the MC hydrogel systems produced be continuously stable as the duration of culture grows, this is a significant drawback. Chen et al. [30] and Thirumala et al. [31] created a modified MC–collagen hydrogel method to create cell sheets, specifically designed to work with adipose tissue-derived stromal/stem cells (ASCs) and for the creation of multidimensional cell sheets, in order to address the drawbacks of the previous MC-based cell sheet engineering technology [30, 31]. This process has been modified and enhanced by Forghani et al. [32] and by Forghani and Devireddy [33]. After creating cell sheets with MC gel and PNIPAAm, Forghani et al. [32]

conducted parallel investigations for 21 days to directly compare the development of ASC sheets, cell morphology, viability, proliferation, and differentiation capacity [32]. Additionally, alkaline phosphatase (ALP) expression and mineralization were used to colorimetrically evaluate osteogenic differentiation, and RT-PCR was used to evaluate osteocalcin gene expression [32]. Tissue reestablishing is one of the several uses for single or multilayer cell sheets [34–44].

6.2 Adipose Tissue-Derived Adult Stem Cells

Stem cells are essential building blocks of multicellular organisms and offer a helpful tool for understanding cellular and developmental processes. It has been postulated and proven that the unique ability of stem cells to differentiate along multiple lineages and pathways is a very powerful tool in fabricating specialized 3D tissue structures. Some examples of these include elucidating embryonic development as well as developing novel organ regeneration and transplantation strategies. The stem cell research field has, in the past few years, generated several seminal studies in culturing early human embryonic stem cells, and human embryonic germ cells, as well as manipulating their differentiation in vitro. Additional evidence suggests that adult stem cells also have a higher degree of developmental potential and plasticity than previously mentioned [45–47]. This alleviates some of the concerns regarding the legal and moral concerns with the use of embryonic stem cells. An additional concern is the tendency of embryonic stem cells to form malignant teratomas after injection into animal models. These drawbacks have refocused the attention on adult stem cell models.

Over the past two decades, adult stem cells have been extracted from many organs. For example, the central nervous system yields neuronal stem cells that can differentiate and produce not only neuronal cells but also other tissues, including blood and muscle. Stem cells from the liver can be transformed, in vitro, into insulin-secreting pancreatic cells. Bone marrow has also been a major source of adult stem cells. Bone marrow biopsies are invasive and painful procedures that can extract volumes of tissue measured in milliliters. However, adult stem cells similar to the bone marrow mesenchymal stem cells are plentiful in human fat available via liposuction harvesting, which, in contrast, routinely delivers liter-sized volumes of tissue. More importantly, the adipose tissue-derived progenitor cells exhibit multi-lineage potential, express biochemical markers, and feature in vitro behaviors consistent with bone, cartilage, epithelial, fat, liver, muscle, and nerve cells [48]. A study conducted by Hicok et al. [49], and confirmed by Justesen et al. [50], showed that these cells derived from human fat are capable of forming osteoid in vivo once implanted subcutaneously into mice. Furthermore, human adipose-derived stem cells (ASCs) have been used successfully to repair a damaged region in a skull. The potential benefits of these ASCs can only be fully realized for tissue engineering strategies and cell-based therapies, if and when an efficient method of delivery and harvesting of these cells in vivo without losing their multipotentiality can be developed.

6.3 Fabrication of Adult Stem Cell Sheets

Extracts of subcutaneous adipose tissue need to be acquired from several donors through liposuction. In all of the results reported in this article, the tissues were obtained from LaCell LLC (New Orleans, LA) with informed consent under a clinical protocol approved by either the Pennington Biomedical Research Center Institutional Review Board (Baton Rouge, LA) or Western Institutional Review Board (Puyallup, WA). The ASC isolation procedure has been previously described [8, 45, 48, 50–56]. Briefly, liposuction aspirates from subcutaneous adipose tissue are obtained from male and female subjects undergoing elective procedures in local plastic surgical offices. Tissues are washed three to four times with phosphate-buffered saline (PBS) and suspended in an equal volume of PBS supplemented with 1% bovine serum and 0.1% collagenase type I prewarmed to 37°C. The tissue is then placed in a shaking water bath at 37°C with continuous agitation for 60 min and centrifuged for 5 min at 300–500 × g at room temperature. The supernatant, containing mature adipocytes, is aspirated. The pellet is identified as the stromal vascular fraction (SVF) and suspended/plated immediately in T225 flasks in stromal medium (DMEM/F 12 Ham's, 10% fetal bovine serum (FBS) (Hyclone, Logan, UT), 100 U penicillin/100 μg streptomycin/0.25 μg Fungizone) at a density of 0.156 ml of tissue digest/cm^2 of surface area for expansion and culture. The primary cell culture's initial passage is referred to as "Passage 0" and denoted as P0. The cells are removed from the dish using 5 mL trypsin followed by incubation at 37°C for 7–10 min and plated at a density of 5,000 cells per cm^2 ("Passage 1") in T125 flasks for expansion and allowed to grow to 80% confluency. Passage 2 cells were used for cell viability testing and osteogenesis [32].

6.3.1 Nunc™ Dishes with UpCell™ Surface

One method is to plate the ASCs on a thermoresponsive 3.5 cm culture dish (Nunc™ Dishes with UpCell™ Surface, Thermo Scientific). Briefly, 1 mL of pre-warmed stromal medium (37°C) was added to the culture dish at a cell density of 1 × 10^6 cells per mL. The culture dish is then placed on a hot plate (37°C) for 30 min for cell attachment. After this 30-min incubation, an additional 1 mL of pre-warmed stromal medium is added. The cells are then incubated at 37°C in 5% CO^2 for a further 24 h. The culturing dish needs to be taken out of the incubator after 24 h, and the cell sheet detaches spontaneously after incubation at room temperature for 15–20 min (Fig. 6.2A).

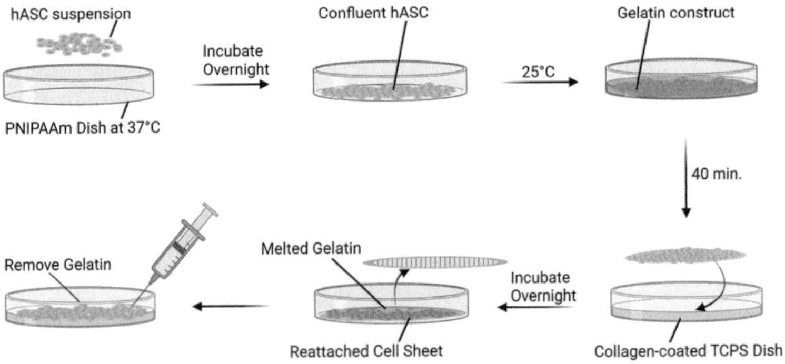

Figure 6.2(A) Schematic illustrations of cell sheet fabrication, detachment, transfer, and reattachment methods using PNIPAAm. ASCS cell sheet formed after overnight incubation using PNIPAAm. Redrawn with permission from Forghani et al. (2017).

6.3.2 Cell Sheets Using MC Hydrogel-Coated Dishes

Fabrication of cell sheets using MC hydrogel-coated dishes can also be performed as described by Thirumala et al. [31], based on prior work by Sarkar [57], Sarkar and Walker [58], Ford [59], Hirose et al. [60], Tate et al. [61], Li et al. [62], Zheng et al. [63], and Tang et al. [64]. Briefly, 14% aqueous MC (viscosity 15 cp) solutions are prepared by dissolving commercially available MC powder in heated water with the addition of PBS at room temperature. After being thoroughly shaken and mixed, the solutions are refrigerated overnight until homogeneous MC solutions are obtained.

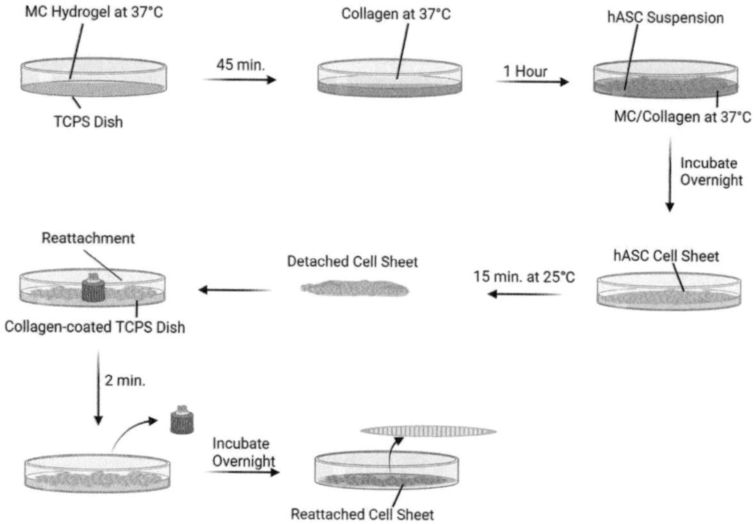

Figure 6.2(B) Schematic illustrations of cell sheet fabrication, detachment, transfer, and reattachment methods using MC. MC hydrogel was fabricated in a TCPS dish followed by coating with collagen. An intact cell sheet was formed after overnight incubation. The cell sheet was transferred to a collagen-coated TCPS dish. A metal mesh disc and weight are placed on top to facilitate reattachment. Redrawn with permission from Forghani et al. (2017) [32].

Approximately 500 μL of the 14% MC solution is then poured into the center of each TCPS (3.5 cm) dish and placed on ice. A thin, transparent layer of the solution needs to be evenly distributed on the TCPS dish. Subsequently, the TCPS dish is incubated at 37°C for 45 min to form an opaque gelled layer. To improve cell attachment, 1 mL of aqueous collagen type I (3 mg/ml, Rat tail collagen Type I, Sigma-Aldrich), needs to be evenly spread over the gelled MC at 37°C. Approximately, 1.0×10^6 ASCs are then seeded evenly on the MC-coated TCP culture system on a hot plate (37°C), and a cell sheet will be formed after culturing for 24 h at 37°C, 5% CO_2. When placed at room temperature (20°C), the cell sheet detaches gradually from the MC surface (Fig. 6.2B).

6.4 In Vitro Degradation, Swelling, and Osmotic Stability of MC Gels

Thirumala et al. (2013) assessed the initial weight of the gelled MC hydrogel layer and denoted it as "MC weight" [31]. Subsequently, to investigate the degradation and swelling behavior of the gelled MC hydrogel layer, approximately 2 mL of cell culture media (DMEM-F12 and 10% FBS) were added to the TCPS dishes. The weight of the gelled MC hydrogel layer immediately (~10 s) after the addition of cell incubation media was also measured and denoted as "Wet MC weight @ t = 0". The TCPS dishes with the 2 mL of media were then incubated at 37°C for several days, subsequently, the cell culture media was removed from the TCPS dishes at various time (t) points (1, 4, 7, and 10 days) of incubation and the weight of the MC hydrogel was recorded and denoted as "Wet MC weight @ t". The MC hydrogel layers were then dried overnight in a convective heater (VWR Scientific, Bridgeport, NJ, USA, 1602 HAFO series) at 70°C, and the dried MC weight was recorded and denoted as "Dried MC weight @ t". The % of MC swelling was quantified, as a function of time (t) as a ratio of "Wet MC weight @ t" to "Wet MC weight @ t = 0" while the percentage of degradation was quantified as the ratio of the difference between "Dried MC weight @ t = 0" and "Dried MC weight @ t" to the "Dried MC weight @ t = 0". The results are shown in Fig. 6.3.

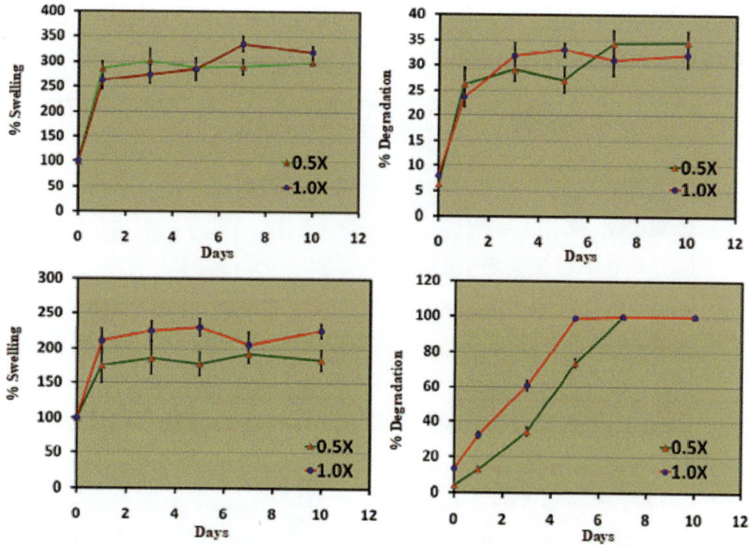

Figure 6.3 Percent swelling and percent degradation values of MC (mol. wt. = 15,000) system as determined in our laboratory—top two figures. Percent swelling and percent degradation values of hydrogel system developed by Chen et al. [30]—bottom two figures. Note the different values on the y-axis between the top and bottom figures.

6.5 Thermal Reversibility of Attachment

Upon confluence, a continuous monolayer ASC sheet was formed on the surface of the MC hydrogel system (Fig. 6.4A). When the confluent cell sheet was removed from the incubator and exposed to room temperature (~30°C), it controllably detached from the surface of the thermoresponsive hydrogel (without the use of any detachment enzymes, like trypsin). The ASC cell sheet started detaching from its edge at ~1 min after being exposed to room temperature. Detachment of the entire cell sheet was completed within 2 to 3 mins (Fig. 6.4B). Additional control experiments with just a TCPS dish coated with collagen did not result in the formation and spontaneous detachment of an ASC sheet (data not shown). All the proposed experiments will be conducted using detached ASC sheets (Fig. 6.4C). As these cells were never exposed to trypsin and other detachment enzymes, it is to be expected that the binding integrins are not compromised in the cell culture/growth process.

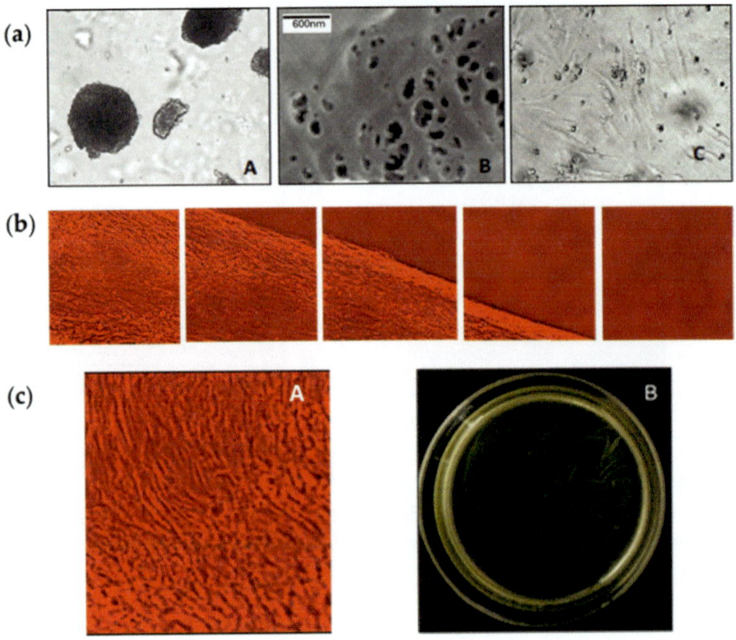

Figure 6.4 (A) ASCs grown on MC hydrogel system (B) SEM image of the collagen network formed onto the MC coated TCPS dish (C) ASCs grown on MC hydrogel system coated with collagen. (B) A series of photomicrographs of the detaching ASC cell sheet with time (20×). The cell sheet detached rapidly (~1 min) after it was transferred to room temperature (~30°C) from the incubator (~37°C). (C) Image of a detached ASC sheet obtained with the MC-collagen hydrogel system exhibiting the characteristics of a monolayer at ~32°C. (B) Floating cell sheet detached from plate.

6.6 Analysis of Cell Sheets Using Histology

Following detachment, the ASC cell sheets were processed and stained as described previously. Briefly, the cell sheets were rinsed with Dulbecco's phosphate-buffered saline (DPBS), fixed in 10% formalin for 24 h, and embedded in paraffin wax for sectioning. To evaluate the composition of the cell sheets, 10 μm sections were cut and stained with H&E and Masson's Trichrome (American MasterTech, Lodi, CA, Item No. KTTRBPT), according to the manufacturer's protocols and imaged under bright-field illumination with an Olympus BX46 microscope at 10× magnification (Fig. 6.5). Collagen appears pink with H&E and blue with Masson's Trichrome stains. Both H&E pink and Masson Trichrome blue stains confirmed the existence of a considerable amount of collagen in addition to ASCs on MC cell sheets, while PNIPAAm sheets are primarily comprised of ASCs.

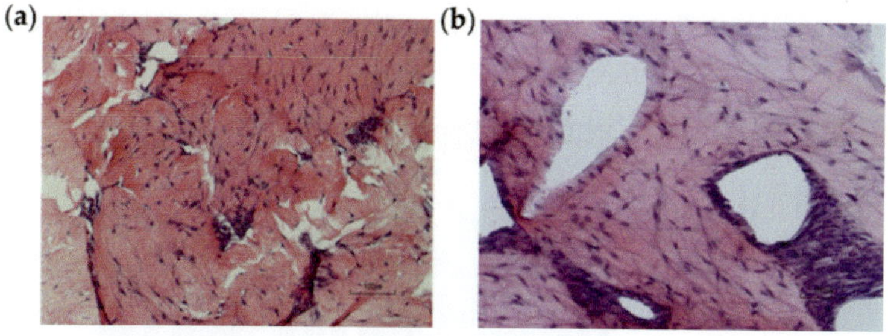

Figure 6.5 (A) Histology analysis of cell sheets fabricated with MC and (B) PNIPAAm using H&E and trichrome blue staining.

6.7 Cell Viability in Cell Sheets

Cell viability was analyzed over 21 days of culture in the stromal medium. PicoGreen® total DNA quantification was used to determine the number of ASCs in cell sheets on days 1, 7, 14, and 21. DNA content analysis indicates that ASCs in MC cell sheets proliferated until reaching confluence by day 14 and proliferation decreased after day 14, while ASCs in PNIPAAm cell sheets continued to proliferate until day 21 [32].

There was a significant difference $^{*}p < 0.05$ in cell number between day 1 and day 14 in MC sheets, but no significant difference between day 1 and day 21. Conversely, statistical analysis of PNIPAAm cell sheets revealed a significant difference between days 1 and 14 as well as days 1 and 21, with $^{*}p < 0.05$ and $^{**}p < 0.01$, respectively. ASC viability was also visualized with LIVE/DEAD® staining, representative images of cell sheets at day 1, 7, 14, and 21 with live cells in green and dead cells in red. The LIVE/DEAD® staining indicates that ASCs in both MC and PNIPAAm cell sheets were viable and proliferated until reaching confluence. Cells retained a healthy spindle-shaped morphology in cell sheets (Fig. 6.6). Images from

LIVE/DEAD® staining support the data generated from DNA content analysis that indicate the ASCs in MC cell sheets proliferated until reaching confluence by day 14 and the proliferation stopped at that day. ASCs in PNIPAAm cell sheets appear to continue to proliferate until day 21, in agreement with DNA content analysis. Analysis of LIVE/DEAD® staining also shows ASC migration/reorganization over time. It was observed that ASCs in both MC and PNIPAAm cell sheets migrated to achieve a regular distribution of cells throughout the cell sheet and on the culture dish.

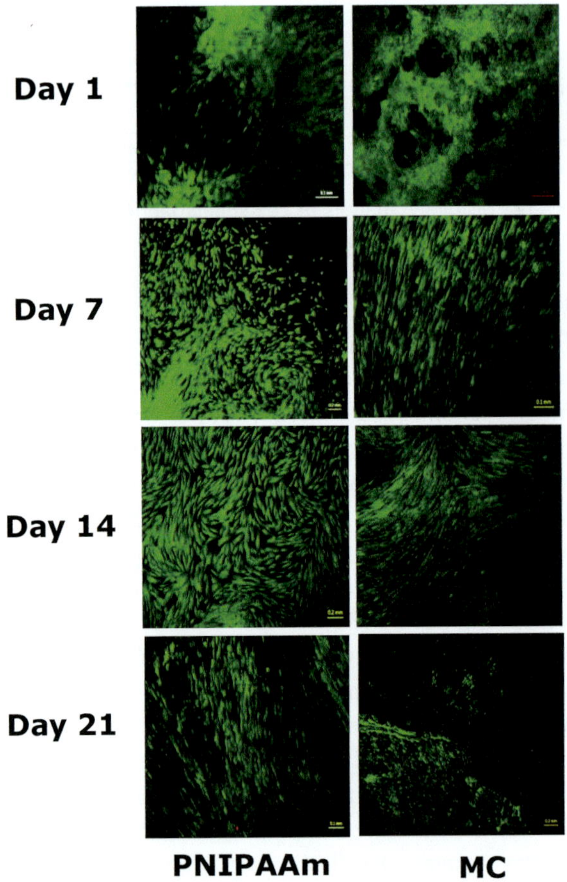

Figure 6.6 LIVE/DEAD® staining assessment over 21-day culture period. LIVE/DEAD® assay stains live cells green and dead cells red. Representative overlay images of MC and PNIPAAm cell sheets on days 1, 7, 14, and 21.

6.8 Confocal Microscopy

MC and PNIPAAm cell sheets were stained to show F-actin distribution in ASCs one day after formation. Representative images are shown in (Fig. 6.7). MC cell sheets composed of a collagen layer with ASCs, had more depth in a three-dimensional structure compared to the more two-dimensional PNIPAAm cell sheets composed only of ASCs and their ECM. F-actin in MC cell sheets, imaged as a z-stack, appeared to be less organized as cells migrated into the three-dimensional structure of the collagen gel compared to ASCs in PNIPAAm cell sheets which had a highly aligned F-actin structure.

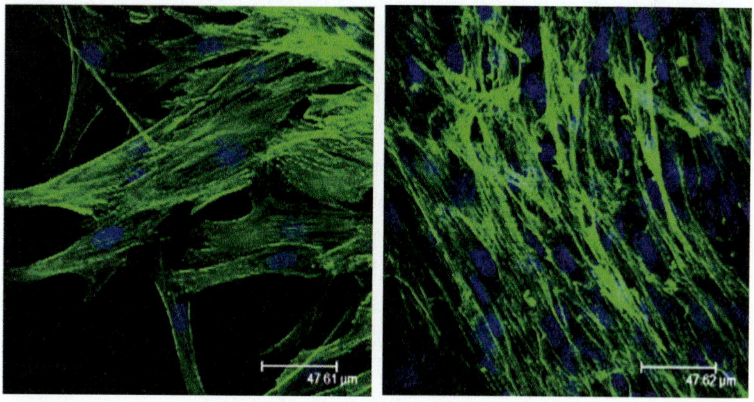

Figure 6.7 Confocal images of ASCs in cell sheets one day after reattachment. Immunofluorescent staining shows cell nuclei (blue) and F-actin (green) in a representative z-stack of the MC cell sheet (left image) and a representative image of the PNIPAAm cell sheet (right image).

6.9 Osteogenic Potential of ASCs Embedded in Cell Sheets

Extensive literature has demonstrated that ASCs exhibit differentiation potential and an immunophenotype based on flow cytometric analysis resembling that of bone marrow-derived mesenchymal stem cells [8, 32, 55, 56, 65–73]. Dr. Gimble, in collaboration with Dr. Guilak at Duke University, developed conditions that promote ASC differentiation along the chondrocyte and osteoblast pathways

[65, 74–76]. When suspended in calcium alginate and incubated in the presence of ascorbate, dexamethasone, and TGF-β, human ASCs exhibit an induction in chondrogenic markers, including collagen types II and VI and proteoglycans. When cultured in the presence of 1,25 dihydroxy vitamin D3, dexamethasone, ascorbate, and β-glycerophosphate, human ASCs secrete osteocalcin and mineralize their ECM, hallmarks characteristic of osteoblast function. In vivo, human ASCs combined with a hydroxyapatite biomaterial synthesize osteoid matrix when implanted subcutaneously into immunodeficient mice [50].

The ability of ASCs to mineralize ECM was quantified using Alizarin Red S staining. ASCs from three to five donors were plated at 50,000 cells per well in 24-well plates in triplicate and incubated for 24 h in a maintenance medium. Cells were then fed either maintenance medium or osteoblast medium composed of DMEM/F-12 Ham's with 10% FBS, 10 mM β-glycerophosphate, 50 μg/ml sodium ascorbate2-phosphate, 100 U penicillin/100 μg streptomycin/0.25 μg Fungizone. Cultures were fed with fresh maintenance medium or osteogenic induction medium every 3–4 days for a period of up to 3 weeks. Cells were fixed in neutral buffered formalin for 1 h and then stained for ALP, using fast blue BB naphthol AS-MX as substrate as previously described [50, 67, 69–71]. ALP levels were quantitated by measuring trypan blue dye levels in individual wells, using the Bio-Quant Image Analysis program (BioQuant Image Analysis, Nashville, TN). To determine the potential to mineralize ECM, ASCs were plated in 24-well plates at 20,000 cells per well and incubated for 24 h in a maintenance medium. Cells were fed either a maintenance medium or osteoblast medium for 21 days. Cells were fixed in 70% ice-cold ethanol and stained with 40 mM alizarin red staining on cultures 21 days after the induction of osteogenesis. The percentage area staining positive was then quantified by scanning the plate and then analyzing with the MetaVue Software image analysis program. Alternatively, or in combination, we can measure the ALP enzyme activity in cultures at 7 days under osteogenic conditions and normalize the degree of substrate conversion to the number of cells based on DNA content. Both approaches, in combination, will provide a measure of osteoblast function quantitatively, as presented recently [32].

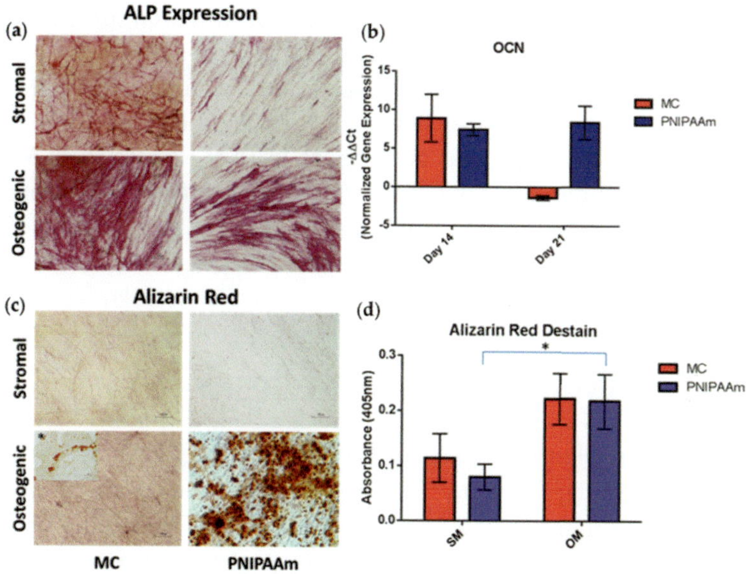

Figure 6.8 Markers of osteogenesis of ASCs in MC and PNIPAAm cell sheets. (A) ALP expression of ASCs on day 7 cultured in stromal compared to osteogenic conditions. (B) Upregulation of OCN in cell sheets at days 14 and 21 cultured in osteogenic conditions compared to stromal conditions. Error bars represent the standard error of the mean. (C) Calcium deposition in osteogenic cell sheets at day 21 stained by Alizarin Red (D) Destained cell sheets read at an absorbance of 405 nm to semi-quantitatively analyze Alizarin Red. Reprinted with permission from Forghani et al. (2017) [32].

To demonstrate the osteogenic potential of ASCs cultured in sheets using MC and PNIPAAm methods, cell sheets were cultured in an osteogenic medium for 21 days and compared to ASC cell sheet controls cultured in a stromal medium. Osteogenesis results are summarized in (Fig. 6.8). Upregulation of ALP expression was observed in osteogenic cell sheets at day 7. At day 21, cell sheets were stained with Alizarin Red to show calcium deposition. Moderate mineralization was observed in MC cell sheets, compared to the more substantial mineralization of PNIPAAm cell sheets. Cell sheets were destained to allow semi-quantitative analysis of Alizarin Red results. In general, an increase in calcium deposition was observed in osteogenic samples relative to those cultured in stromal medium for cell sheets prepared by both methods. However, the difference was only statistically significant for PNIPAAm sheets. Due to higher background staining in stromal controls, the increase in Alizarin Red was not statistically significant. For further validation of osteogenesis, OCN expression was also quantified via RT-PCR at days 14 and 21. There was an increase in OCN expression of MC cell sheets cultured in osteogenic conditions compared to stromal conditions at day 14, with a $-\Delta\Delta$Ct value of 8.895. OCN expression of PNIPAAm cell sheets was upregulated at days 14 and 21—with an increased $-\Delta\Delta$Ct value from 7.435 at day 14 to 8.395 at day 21.

6.10 Polyethylene Glycol Hydrogels

A recent example of polyethylene glycol (PEG) hydrogels, incorporated with adipose-derived ECM fabricated via a thiolene platform and Michael addition reaction, is shown in (Fig. 6.9). PEG could be a biocompatible hydrogel system (as opposed to methylcellulose) that could support stem cell growth and adipose differentiation for potential in vivo fat regeneration applications. The materials were obtained as follows: Four arm PEG-Acrylate (MW 10000, 5 g) from Laysan Bio Inc. (AL, USA); Dialysis sacks (MWCO 3500) from Fisher Scientific (PA, USA); Ethoxylated trimethylolpropane tri-3-mercaptopropionate (ETTMP, Thiocure 1300) from Bruno Bock (Marschacht, Germany); Pierce BCA Assay Kit, Quant-iT PicoGreen dsDNA Assay Kit, LIVE/DEAD Viability/Cytotoxicity Kit, Dulbecco's Modified Eagle's Medium/F12 (DMEM-F12), FBS, Rat Tail Collagen I Hydrogel, Goat Anti-Rabbit IgG (H+L) Secondary Anti-body, DyLight 650 anti-perilipin-2/ADFP polyclonal rabbit primary antibody from Thermo Fisher Scientific (PA, USA). Other chemical reagents were from Sigma Aldrich (PA, USA). All chemicals were reagent grade and used as received.

ECM-PEG hydrogels were prepared according to the protocol previously published. Specifically, poly(ethylene glycol)-based hydrogels were formed using precursors containing multiple thiol and acrylate functional groups capable of reacting in an aqueous solvent to form a network structure [78]. Aqueous stock solutions of the two precursors, PEG-400-DA, and ETTMP, were prepared by dissolving the polymers in PBS at 4°C for 24 h. To fabricate the hydrogels, appropriate volumes of the two stocks solutions, resulting in an equimolar stoichiometric ratio of thiol and acrylate functional groups, were combined, yielding a final total polymer concentration of 25 wt.% in a Falcon tube at 4°C. Following a 5-min incubation period at 4°C to increase the dissolution of the precursors in the mixture, molds of a desired size were filled with the solution and transferred to an incubator at 37°C to complete gelation. Gelation occurred within 5 min of incubation. However, samples produced for structural characterization and mechanical testing were maintained in the 37°C environment for 20 min to promote the complete conversion of functional groups. The gelation protocol described above was applied to all hydrogels prepared in PBS. For cell culture studies, DMEM was incorporated as the solvent phase in the hydrogel by incubating the cured hydrogel in the media for a 24-h period, prior to cell seeding. For MPSS encapsulation and release studies, weighed samples of the steroid were incorporated into the PEG-400-DA solution for a final concentration of 0.5, 1, or 2 mg/mL in the hydrogels. Functionalization of the hydrogel with an oligomycin peptide was performed by dissolving the peptide in the PEG-400-DA in solution, prior to mixing with the ETTMP stock solution. The oligomycin was reacted into the hydrogel via its thiol-containing cysteine end residues. Solutions were sterilized by syringe filtration through a 0.2 mm nylon filter. The concentrations of oligopeptide used were 5, 50, and 500 mg/mL.

Hydrogel reagents were reacted together using Michael addition (Fig. 6.9). Adipose ECM proteins in PBS were preincubated for 30 min with PEG-acrylate. ECM additives beyond 1% w/v% of total volume did not disperse homogeneously, leading to an unstable hydrogel. During preincubation, a solution of ETTMP, and 1 m NaOH was made. After the initial reaction of ECM proteins with PEG-acrylate in PBS, the solution containing PEG-acrylate and ECM was then mixed with the solution containing ETTMP and NaOH, resulting in polymerization. Thirteen milli-grams of PEG, 2.03 μL ETTMP, 1.35 μL NaOH, and ECM at various concentrations (0%, 0.01%, 0.1%, and 1%) were added in a 135 μL hydrogel system. For the cell encapsulation study, ASCs were mixed with PEG-ECM-PBS solution first and encapsulated in hydrogel after adding ETTMP-NaOH solution.

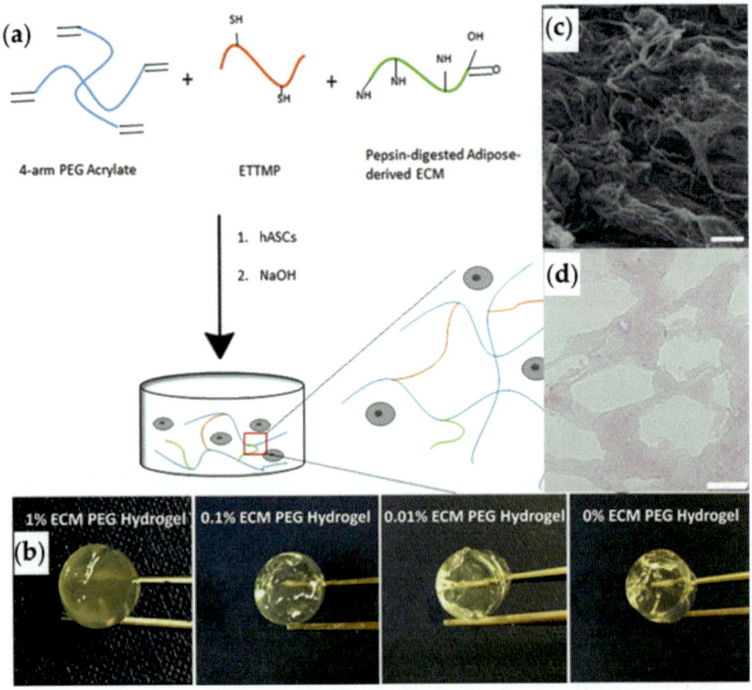

Figure 6.9 Extraction of adipose-derived ECM and fabrication of ECM-PEG hydrogel: (A) Schematic illustration of ECM-PEG hydrogel via a Michael addition reaction. (B) Representative photographs of hydrogel with different ECM concentrations. Macroscopically, the color and transparency of hydrogel differ among groups, especially the 1% ECM-PEG hydrogel group. (C) Porous structures of ECM under scanning electrical microscope. Scale bar: 10 μm. (D) Hematoxylin and eosin staining of ECM-PEG hydrogel, which shows the porous network under the microscope. Scale bar: 50 μm. Reprinted with permission from Li et al. (2018) [77].

6.11 Cell Viability and Proliferation in ECM-PEG Hydrogels

The viability of ASCs encapsulated in ECM-PEG hydrogel evaluated at days 1, 14, 21, and 28 by Live/Dead Staining is shown in Fig. 6.10. Round-shaped ASCs were uniformly distributed in all the groups. ASCs maintained viability and proliferated for up to 28 days. One percent ECM-PEG group had the highest cell proliferation rate, which was demonstrated via cell counting ($p < 0.05$) (Fig. 6.11). A positive correlation between ECM concentrations and cell proliferation rate was also observed. ECM is highly porous, hydrophilic, and can be remodeled by cells. Therefore, it promotes cell migration and proliferation. Most of the encapsulated ASCs appear to survive and proliferate in ECM-PEG hydrogel. Cryo SEM analysis of the cell-laden hydrogel suggests that rounded cells are present in the interconnected pore structure in agreement with fluorescent microscopy analysis [77]. When cells are encapsulated in ECM-PEG hydrogel, the interconnectivity may promote cell viability and proliferation by providing better nutrient and waste transport. The round morphologies of ASCs in the interior were in accordance with the low stiffness of hydrogels, which promote rounded cell morphologies and imply potential application in adipose tissue regeneration.

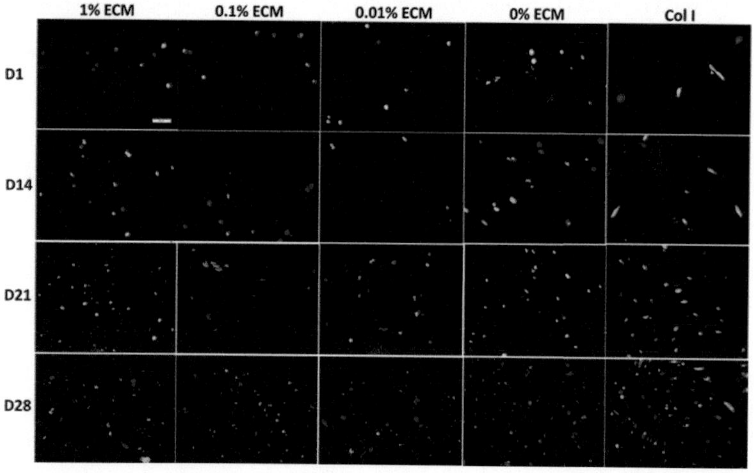

Figure 6.10 Viability of ASCs encapsulated in PEG hydrogel with different ECM concentrations at different time points. Representative live/dead fluorescence images of hASCs within different ECM-PEG hydrogels (1%, 0.1%, 0.01%, and 0 %) after 1, 14, 21, and 28 days of culture, cells stained with green fluorescence are alive and red fluorescence indicates dead cells. Nucleus were stained by DAPI (blue). Scale bar: 100 μm. Reprinted with permission from Li et al. (2018) [77].

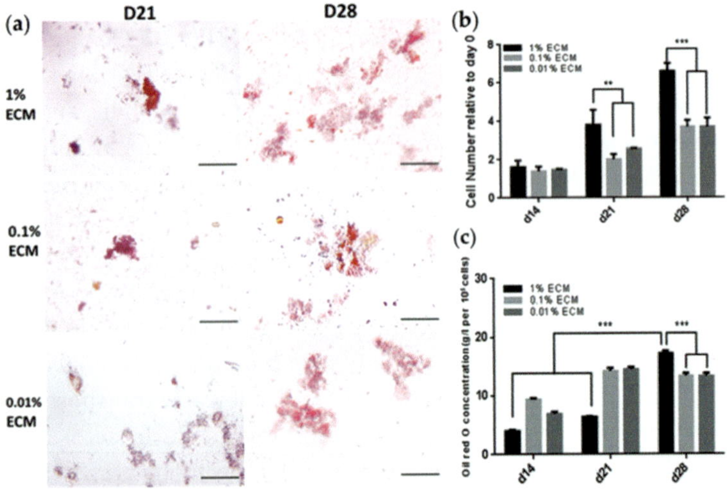

Figure 6.11 Proliferation rate and adipogenic differentiation of hASCs encapsulated in ECM-PEG hydrogels: (A) Oil Red O staining of hASCs encapsulated in the ECM-PEG hydrogel at Day 14 and then cultured for another 2 weeks in adipogenic medium. (B) Cell counting using ImageJ for Live/Dead staining at days 14, 21, and 28. (C) Quantitative analysis of Oil Red O staining. ** indicates $p < 0.01$ and *** indicates $p < 0.001$. Scale bar: 100 μm. Reprinted with permission from Li et al. (2018) [77].

6.12 In Vivo Proliferation

Four ECM-PEG hydrogel scaffolds were implanted subcutaneously in the back of each Bal/C mouse for all treatment groups for 2 weeks [77]. Observed behavior and weight gain were normal (3.2 ± 0.6 g) for all mice after surgery. Histological analysis (H&E staining) showed that a large number of host cells infiltrated and migrated homogeneously into the hydrogel for up to 2 weeks after subcutaneous implantation (Fig. 6.12). No evidence of inflammation was found in the implant site, and few inflammatory cells were found in H&E staining sections. In vivo, adipogenesis was evaluated by Oil Red O staining. The results showed that the Oil Red O positive area was more apparent in the 1% ECM group than the others, indicating that the addition of ECM in the PEG hydrogel system can enhance adipogenesis in vivo.

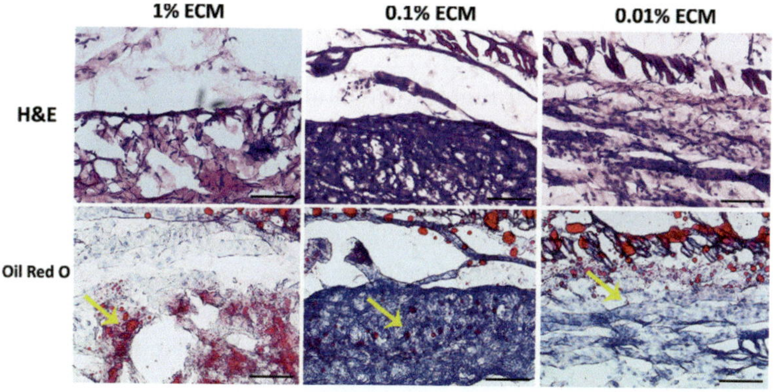

Figure 6.12 In vivo study on adipose regeneration of ECM-PEG hydrogels at day 14 after implantation. more lipids were found in the 1% ECM group. The yellow arrow indicates lipids. Scale bar: 100 MM. Reprinted with permission from Li et al. (2018) [78].

6.13 3D Printing in Tissue Engineering

3D printing has revolutionized tissue engineering in the fabrication of biomimetic tissue structures with superior cell viability, vascularization, and mechanical properties [79]. Most of the conventional scaffold-based approaches have not been able to truly mimic the anisotropic geometrical architecture of the native tissues. This leads to some problems concerning cell infiltration and nutrient diffusion. Three-dimensional bioprinting can overcome those challenges associated with conventional techniques by depositing biomaterials, growth factors, and living cells in a layer-by-layer manner with the help of computer-aided design models [80, 81]. The various bioprinting techniques, such as extrusion-based, inkjet, and laser-assisted bioprinting, manufacture personalized tissue constructs to meet specific patient needs [82]. The nature of bioinks used for printing provides critical support for cell adhesion, proliferation, and differentiation, comprising natural and synthetic polymers, hydrogels, and decellularized ECM [83]. Vascularized bioprinting, 4D bioprinting, and organ-on-a-chip systems are some of the recent related developments [84]. All these new technologies have extended the prospects for the application of 3D-printed tissues to regenerative medicine, drug screening, and transplantation. Some major advantages that 3D printing offers over tissue engineering allow the control of scaffold architecture by porosity, mechanical properties, and degradation rates [85, 86]. All the techniques adopted so far, such as electrospinning or freeze-drying-normally, come up with irregular pore structures; therefore, 3D printing has an opening through which custom-designed microenvironments suitable for various types of cells or tissues with given functions can be constructed [87]. Although for maintaining proper oxygen and diffusion of the nutrients, 3D-printed scaffold can be fabricated to include a vascular network, therefore maintaining the proper integration in bone, liver, and cardiac tissues to survive and function in the long term. The development of technology that helps in the simultaneous deposition of cells, small biomolecules, growth factors, and biomaterials is one of the major advancements in 3D bioprinting. Examples include osteochondral grafts, for which stiff, mineralized bone and softer, cartilaginous layers are fabricated by using a combination of hydrogels, calcium phosphate, and collagen-based bioinks [88]. Similarly, 3D printing of muscle tissues does allow the alignment of muscle fibers and incorporation of nerve conduits, hence making their functional tissue regeneration possible [89].

Furthermore, bioprinting enables the incorporation of multiple cell types that are consistent and proportionate with natural tissues, essential for the heterogeneous composition in functional tissue regeneration [90]. Microfluidic bioprinting has been developed by assembling vascularized constructs that can ensure good oxygen and nutrient supply to the printed tissues. Multi-material printing enables a combination of soft and rigid biomaterials, enhancing the durability and mechanical strength of tissues [91]. In some ways, the use of stem cells, iPS cells, and cells derived from the patient in 3D bioprinting increases biocompatibility and immune acceptance of engineered tissues. Machine learning and AI-assisted bioprinting have been explored to optimize printability, the composition of bioink, and the biofabrication of tissue maturation [92]. Moreover, although there are still challenges with regulatory approval and translational efforts, advances in the field of bioprinting technologies, bio inks, and strategies for tissue maturation are being made to functional and transplantable tissues for personalized medicine. Although significant advances have been made, still scalability, standardization, and long-term functionality remain critical challenges. Future bioprinting techniques will be developed with the integration of artificial intelligence, real-time monitoring, and dynamic biomaterials in continued advances toward the next generation of functional and clinically transplantable tissues.

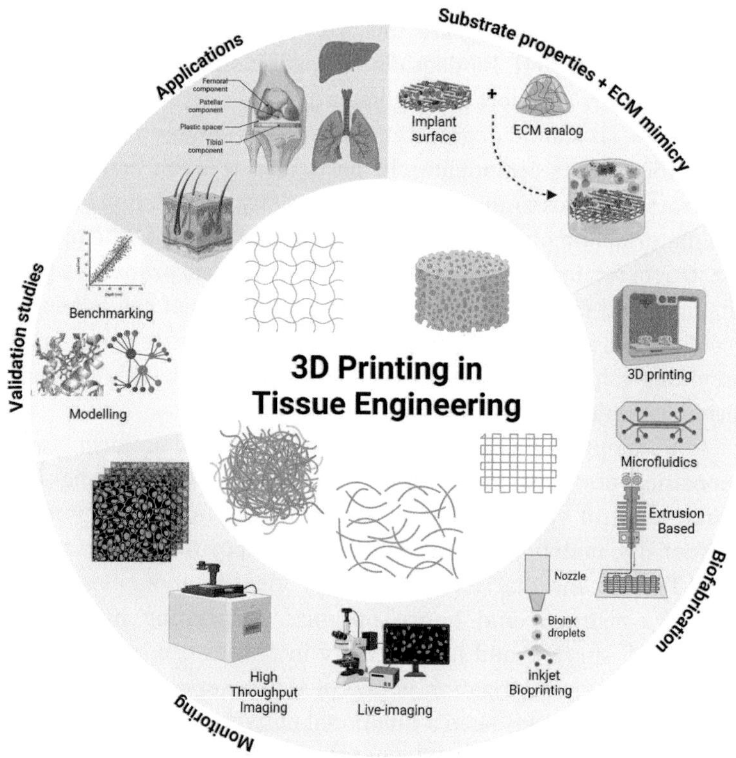

Figure 6.13 An overview of various types of 3D scaffolds, fabrication methods together with their applications in tissue engineering.

6.14 Hyperspectral Imaging in Tissue Engineering and 3D Bioprinting

Hyperspectral imaging (HSI) is advancing as an important technology in tissue engineering and 3D bioprinting to provide label-free, high throughput, noninvasive analysis of engineered tissues and biomaterials. Most of the conventional imaging techniques are based on fluorescent labeling or destructive assays that can harm cellular behavior and are not convenient for long-term experiments. On the other hand, HSI allows real-time information of integrity of scaffold, cell viability, and biochemical composition and is therefore well suited for the testing of tissue constructs at different stages of development [93]. The other most important applications of HSI in tissue engineering are the characterizations of 3D-printed scaffolds system [94]. Biodegradable scaffolds are used to provide initial structural support for engineered tissue constructs. However, such materials should degrade at a rate that is similar to new tissue formation and its remodeling. In particular, HSI can make use of spectral signatures to analyze scaffold porosity, degradation kinetics, and material homogeneity of the scaffold to ensure that the printed constructs retain their mechanical and biochemical properties [95]. Similarly, HSI can be employed in the monitoring of cell adhesion, proliferation, and differentiation on a variety of biomaterial surfaces, providing valuable insights into cellular interactions with variously formulated bioinks [96].

In 3D bioprinting, complete cell distribution and uniform bioink deposition are hard to achieve. HSI enables real-time, label-free quality control via detecting spectral changes that correspond to cell density, metabolic activity, and the composition of the bioink [97]. This means that researchers can identify anomalies in the printed constructs and hence optimize bioprinting parameters so that cell survival and homogeneity in tissues can be enhanced. Besides that, vascularization is one of the major challenges in 3D bioprinting, as tissues need a functional network of blood vessels to maintain cellular activity. HSI can be used to monitor oxygenation levels and perfusion efficiency as real-time feedback on vascular formation in engineered tissues [98]. Beyond scaffold monitoring and vascularization, HSI also plays an important role in tracking tissue maturation and differentiation. In that respect, most of the tissue-engineered constructs need in vitro culturing for longer periods to drive cells through a differentiation process to more specialized tissue types. HSI supports a noninvasive tracking of such cellular differentiation processes through the capturing of spectral fingerprints distinguishing undifferentiated stem cells from progenitor cells into fully differentiated tissue-specific cells [99].

This is of high benefit in personalized regenerative medicine, where patient-derived stem cells fabricate customized grafts and implants.

The early detection of abnormalities within tissues—whether in the form of inflammation, fibrosis, or necrosis—is important to ensure the viability of engineered tissues before implantation. Thus, HSI detects biochemical changes due to cellular stress and disease progression, demonstrating its power for preclinical validation of tissue constructs. Integrating HSI with different bioreactors and microfluidic cell culture platforms to create automation in monitoring tissue-engineered construction helps in better reliable and scalable tissue-engineered products [100]. With the advancement of AI and machine learning, AI-powered HSI analysis enhanced the applications of tissue engineering and 3D bioprinting. These automated image-processing algorithms will analyze hyperspectral data in high-throughput analyses for subtle changes in tissue composition, scaffold degradation, and cell behavior not visible in traditional imaging techniques [101]. This integration of AI and HSI will lead to the optimization of workflows in bioprinting, allowing the making of real-time adjustments in print parameters, bioink formulations, and culture conditions.

In the future, compact HSI devices may be used for site-specific assessment of engineered tissues and printed grafts prior to implantation in point-of-care applications. Integration of HSI into clinical and translational research will be important for bridging the gap between laboratory constructed tissues and clinical-grade implants. With high-resolution, real-time, and noninvasive imaging capabilities, HSI is well placed to become an important technology in the advancement of tissue engineering, bioprinting, and regenerative medicine.

6.15 Clinical Problem

Musculoskeletal defects restrict the lives of millions of Americans and require orthopedic surgeries to accelerate or improve bone repair [102, 103]. It is clear that autologous bone does not accommodate every clinical situation, and there is a need for bone-grafting materials and stem cells in orthopedic surgery [104]. Osteoinductive growth factors such as bone morphogenetic proteins (BMPs) are now commercially available (Infuse™ from Sofamor/Medtronic), but they require the presence of osteogenic cells within the spinal fusion site to promote new bone formation. The development of an osteogenic stem cell has the potential to improve the outcome when combined with alternative spinal fusion materials.

Soft tissue defects represent a second opportunity for the use of ASCs in regenerative medicine. The global plastic surgery industry was valued at $67.3 billion in 2021 ($27 billion in the US) and is expected to reach $203 billion in 2031. This comes to a total of almost 15.6 million procedures over the year in the United States. Of those procedures, almost 3.4 million were surgical procedures dealing with soft tissue fillers, in combination with liposuction and abdominoplasty, which had close to 300,000 procedures. Many of these surgeries require graft tissue, ideally from the patients themselves to alleviate transplant rejection and immunological rejection response. Additionally, the American Cancer Society reports that over 200,000 women will be diagnosed with new cases of breast cancer each year. A high percentage of these women will undergo partial or complete mastectomy in the course of their treatment. Plastic and reconstructive surgeons are seeking novel methods to repair the resulting soft tissue defects in these patients. We postulate that adipogenic ASCs, in combination with cell sheet-engineered scaffolds, may prove to be a cell-based therapeutic tool for oncological surgeons. All of these procedures have the potential of needing fresh adipose tissue on hand for further procedures or to consider the preservation of tissue with the idea of harvesting ASCs. As an aside, recent studies suggest that ASC cryopreservation was associated with a reduced adipogenic capacity in vitro and osteogenic differentiation in vitro and in vivo [105, 106], thus, necessitating further studies in optimizing the freeze-storage process for the tissue constructs as well as in the surrounding media [107]. Given the two-dimensional nature of the cell sheet engineered materials, the complexity of freezing process optimization is reduced when compared to the complexity associated with three-dimensional large-scale traditional bio-scaffold-based tissue constructs.

6.16 Closing Remarks

Several techniques are currently available for cell sheet fabrication. The two techniques demonstrated in this chapter allow users to transfer and manipulate the cell sheets in vitro while maintaining sheet morphology, ECM structure, and cell–cell interactions largely intact. Cell sheets fabricated by using MC or PNIPAAm methods, maintained cell viability for up to 21 days. It is clear that ASC cell sheets made by either MC or PNIPAAm have the potential as a research tool to examine complex cell–cell interactions and for potential clinical use in tissue engineering applications. Cell sheets made by the PNIPAAm method demonstrated greater ossification compared with the MC method, indicating the PNIPAAm fabrication approach may be better suited for application in bone tissue engineering research. Future studies will explore the potential of cell sheets fabricated by both methods for different applications in tissue engineering [108–110].

Author Contributions: Conceptualization, M.G. and R.D.; formal Analysis, M.G.; writing—original draft preparation, M.G.; writing review and editing, R.D.; supervision, R.D.; funding acquisition, R.D. All authors have read and agreed to the published version of the manuscript. In addition, sections of this chapter have been published earlier in other book chapters [31 and 33] and are related to specific methods and assays. These are reproduced verbatim to preserve their accuracy and veracity.

Funding: Research reported in this publication was partly supported by the Louisiana Board of Regents, the Department of Mechanical Engineering at the Louisiana State University (LSU) as well as the NSF EPSCoR LA-SiGMA project under award #EPS-1003897. Additional funding from National Institutes of Health (NIH) R01 (DE024790) through the National Institute of Craniofacial and Dental Research as well as the Institute of Diabetes and Digestive and Kidney Diseases of the NIH under award number R21DK91852 and by the Institute of General Medical Sciences of the NIH under award number R15GM141653.

Acknowledgments

The authors thank Dr. Elizabeth Clubb and Dr. James Wade at the Pennington Biomedical Research Center (PBRC) for supplying the liposuction aspirates and their many patients for consenting to participate in this protocol. Acknowledgments are also due to Drs. Gimble and Hayes, long-time colleagues and friends, and without whose help this work would never have been initiated. And finally, the acknowledgments are also due to Dr. S. Thirumala and Dr. A. Forghani for performing most of the experiments reported in this article. This work was supported in part by funding from the Louisiana Board of Regents, the Department of Mechanical Engineering at the Louisiana State University (LSU) as well as the NSF EPSCoR LA-SiGMA project under award #EPS-1003897. Additional funding from National Institutes of Health (NIH) R01 (DE024790) through the National Institute of Craniofacial and Dental Research as well as the Institute of Diabetes and Digestive and Kidney Diseases of the NIH under award number R21DK91852 and by the Institute of General Medical Sciences of the NIH under award number R15GM141653.

References

1. Yang, J., Yamato, M., Kohno, C., Nishimoto, A., et al. Cell sheet engineering: Recreating tissues without biodegradable scaffolds. *Biomaterials* **2005**, *26*, 6415–6422, doi:10.1016/j.biomaterials.2005.04.061.
2. Hubbell, J. A., Massia, S. P., Desai, N. P., Drumheller, P. D. Endothelial cell-selective materials for tissue engineering in the vascular graft via a new receptor. *Biotechnology* (N Y) **1991**, *9*, 568–572, doi:10.1038/nbt0691-568.
3. Sittinger, M., Bujia, J., Rotter, N., Reitzel, D., et al. Tissue engineering and autologous transplant formation: Practical approaches with resorbable biomaterials and new cell culture techniques. *Biomaterials* **1996**, *17*, 237–242, doi:10.1016/0142-9612(96)85561-x.
4. Bauer, T. W., Muschler, G. F. Bone graft materials. An overview of the basic science. *Clin Orthop Relat Res* **2000**, 10–27.
5. Griffith, L. G., Naughton, G. Tissue engineering–current challenges and expanding opportunities. *Science* **2002**, *295*, 1009–1014, doi:10.1126/science.1069210.

6. Griffith, L. G. Emerging design principles in biomaterials and scaffolds for tissue engineering. *Ann N Y Acad Sci* **2002**, *961*, 83–95, doi:10.1111/j.1749-6632.2002.tb03056.x.
7. Anil Kumar, P. R., Varma, H. K., Kumary, T. V. Cell patch seeding and functional analysis of cellularized scaffolds for tissue engineering. *Biomed Mater* **2007**, *2*, 48–54, doi:10.1088/1748-6041/2/1/008.
8. Chinnasami, H., Gimble, J. M., Devireddy, R. Structure–property relation of porous poly (l-lactic acid) scaffolds fabricated using organic solvent mixtures and controlled cooling rates and its bio-compatibility with human adipose stem cells. *J Bioacti Compat Polym* **2018**, *33*, 397–415.
9. Muschler, G. F., Nakamoto, C., Griffith, L. G. Engineering principles of clinical cell-based tissue engineering. *J Bone Joint Surg Am* **2004**, *86*, 1541–1558, doi:10.2106/00004623-200407000-00029.
10. Sikavitsas, V. I., Bancroft, G. N., Mikos, A. G. Formation of three-dimensional cell/polymer constructs for bone tissue engineering in a spinner flask and a rotating wall vessel bioreactor. *J Biomed Mater Res* **2002**, *62*, 136–148, doi:10.1002/jbm.10150.
11. Kumar, P. R. A., Sreenivasan, K., Kumary, T. V. Alternate method for grafting thermoresponsive polymer for transferring in vitro cell sheet structures. *J Appl Polym Sci* **2007**, *105*, 2245–2251, https://doi.org/10.1002/app.26221.
12. Avgoustiniatos, E. S., Colton, C. K. Effect of external oxygen mass transfer resistances on viability of immunoisolated tissue. *Ann N Y Acad Sci* **1997**, *831*, 145–167, doi:10.1111/j.1749-6632.1997.tb52192.x.
13. Hutmacher, D. W. Scaffolds in tissue engineering bone and cartilage. *Biomaterials* **2000**, *21*, 2529–2543, doi:10.1016/s0142-9612(00)00121-6.
14. Martin, I., Wendt, D., Heberer, M. The role of bioreactors in tissue engineering. *Trends Biotechnol* **2004**, *22*, 80–86, doi:10.1016/j.tibtech.2003.12.001.
15. Janssen, F. W., Oostra, J., Oorschot, A., van Blitterswijk, C. A. A perfusion bioreactor system capable of producing clinically relevant volumes of tissue-engineered bone: in vivo bone formation showing proof of concept. *Biomaterials* **2006**, *27*, 315–323, doi:10.1016/j.biomaterials.2005.07.044.
16. Khong, Y. M., Zhang, J., Zhou, S., Cheung, C., et al. Novel intra-tissue perfusion system for culturing thick liver tissue. *Tissue Eng* **2007**, *13*, 2345–2356, doi:10.1089/ten.2007.0040.

17. Dvir, T., Benishti, N., Shachar, M., Cohen, S. A novel perfusion bioreactor providing a homogenous milieu for tissue regeneration. *Tissue Eng* **2006**, *12*, 2843–2852, doi:10.1089/ten.2006.12.2843.

18. Sistino, J. J. Bioreactors for tissue engineering—a new role for perfusionists? *J Extra Corpor Technol* **2003**, *35*, 200–202.

19. Yamato, M., Utsumi, M., Kushida, A., Konno, C., et al. Thermo-responsive culture dishes allow the intact harvest of multilayered keratinocyte sheets without dispase by reducing temperature. *Tissue Eng* **2001**, *7*, 473–480, doi:10.1089/107632701152436517.

20. Yamato, M., Okano, T. Cell sheet engineering. *Materials Today* **2004**, *7*, 42–47, doi:https://doi.org/10.1016/S1369-7021(04)00234-2.

21. Shimizu, T., Sekine, H., Yang, J., Isoi, Y., et al. Polysurgery of cell sheet grafts overcomes diffusion limits to produce thick, vascularized myocardial tissues. *Faseb J* **2006**, *20*, 708–710, doi:10.1096/fj.05-4715fje.

22. Akiyama, Y., Kushida, A., Yamato, M., Kikuchi, A., et al. Surface characterization of poly(N-isopropylacrylamide) grafted tissue culture polystyrene by electron beam irradiation, using atomic force microscopy, and X-ray photoelectron spectroscopy. *J Nanosci Nanotechnol* **2007**, *7*, 796–802, doi:10.1166/jnn.2007.509.

23. da Silva, R. M., Mano, J. F., Reis, R. L. Smart thermoresponsive coatings and surfaces for tissue engineering: Switching cell-material boundaries. *Trends Biotechnol* **2007**, *25*, 577–583, doi:10.1016/j.tibtech.2007.08.014.

24. Hayashida, Y., Nishida, K., Yamato, M., Yang, J., et al. Transplantation of tissue-engineered epithelial cell sheets after excimer laser photoablation reduces postoperative corneal haze. *Invest Ophthalmol Vis Sci* **2006**, *47*, 552–557, doi:10.1167/iovs.05-0995.

25. Kim, J. H., Ko, S. Y., Lee, J. H., Kim, D. H., et al. Evaluation of the periodontal regenerative properties of patterned human periodontal ligament stem cell sheets. *J Periodontal Implant Sci* **2017**, *47*, 402–415, doi:10.5051/jpis.2017.47.6.402.

26. Kobayashi, J., Okano, T. Fabrication of a thermoresponsive cell culture dish: A key technology for cell sheet tissue engineering. *Sci Technol Adv Mater* **2010**, *11*, 014111, doi:10.1088/1468-6996/11/1/014111.

27. Kobayashi, J., Yamato, M., Okano, T. On-off affinity binding modulation on thermoresponsive polymer-grafted surfaces for capture and release of proteins and cells. *J Biomater Sci Polym Ed* **2017**, *28*, 939–957, doi:10.1080/09205063.2017.1295508.

28. Matsuda, N., Shimizu, T., Yamato, M., Okano, T. Tissue engineering based on cell sheet technology. *Adv Mat* **2007**, *19*, 3089–3099, doi:https://doi.org/10.1002/adma.200701978.

29. Tang, Z., Okano, T. Recent development of temperature-responsive surfaces and their application for cell sheet engineering. *Regen Biomater* **2014**, *1*, 91–102, doi:10.1093/rb/rbu011.

30. Chen, C. H., Tsai, C. C., Chen, W., Mi, F. L., et al. Novel living cell sheet harvest system composed of thermoreversible methylcellulose hydrogels. *Biomacromolecules* **2006**, *7*, 736–743, doi:10.1021/bm0506400.

31. Thirumala, S., Gimble, J. M., Devireddy, R. V. Methylcellulose based thermally reversible hydrogel system for tissue engineering applications. *Cells* **2013**, *2*, 460–475, doi:10.3390/cells2030460.

32. Forghani, A., Kriegh, L., Hogan, K., Chen, C., et al. Fabrication and characterization of cell sheets using methylcellulose and PNIPAAm thermoresponsive polymers: A comparison Study. *J Biomed Mater Res A* **2017**, *105*, 1346–1354, doi:10.1002/jbm.a.36014.

33. Forghani, A., Devireddy, R. Methylcellulose based thermally reversible hydrogels. *Methods Mol Biol* **2018**, *1773*, 41–51, doi:10.1007/978-1-4939-7799-4_5.

34. Akizuki, T., Oda, S., Komaki, M., Tsuchioka, H., et al. Application of periodontal ligament cell sheet for periodontal regeneration: A pilot study in beagle dogs. *J Periodontal Res* **2005**, *40*, 245–251, doi:10.1111/j.1600-0765.2005.00799.x.

35. Fortier, G. M., Gauvin, R., Proulx, M., Vallée, M., et al. Dynamic culture induces a cell type-dependent response impacting on the thickness of engineered connective tissues. *J Tissue Eng Regen Med* **2013**, *7*, 292–301, doi:10.1002/term.522.

36. Harimoto, M., Yamato, M., Hirose, M., Takahashi, C., et al. Novel approach for achieving double-layered cell sheets co-culture: Overlaying endothelial cell sheets onto monolayer hepatocytes utilizing temperature-responsive culture dishes. *J Biomed Mater Res* **2002**, *62*, 464–470, https://doi.org/10.1002/jbm.10228.

37. Ide, T., Nishida, K., Yamato, M., Sumide, T., et al. Structural characterization of bioengineered human corneal endothelial cell sheets fabricated on temperature-responsive culture dishes. *Biomaterials* **2006**, *27*, 607–614, https://doi.org/10.1016/j.biomaterials.2005.06.005.

38. Labbé, B., Marceau-Fortier, G., Fradette, J. Cell sheet technology for tissue engineering: The self-assembly approach using adipose-derived

stromal cells. *Methods Mol Biol* **2011**, *702*, 429–441, doi:10.1007/978-1-61737-960-4_31.

39. Larouche, D., Paquet, C., Fradette, J., Carrier, P., et al. Regeneration of skin and cornea by tissue engineering. *Methods Mol Biol* **2009**, *482*, 233–256, doi:10.1007/978-1-59745-060-7_15.

40. Mota, C., Puppi, D., Chiellini, F., Chiellini, E. Additive manufacturing techniques for the production of tissue engineering constructs. *J Tissue Eng Regen Med* **2015**, *9*, 174–190, doi:10.1002/term.1635.

41. Nishida, K., Yamato, M., Hayashida, Y., Watanabe, K., et al. Functional bioengineered corneal epithelial sheet grafts from corneal stem cells expanded ex vivo on a temperature-responsive cell culture surface. *Transplantation* **2004**, *77*, 379–385, doi:10.1097/01.Tp.0000110320.45678.30.

42. Obokata, H., Yamato, M., Yang, J., Nishida, K., et al. Subcutaneous transplantation of autologous oral mucosal epithelial cell sheets fabricated on temperature-responsive culture dishes. *J Biomed Mater Res A* **2008**, *86*, 1088–1096, doi:10.1002/jbm.a.31659.

43. Shimizu, T., Yamato, M., Kikuchi, A., Okano, T. Cell sheet engineering for myocardial tissue reconstruction. *Biomaterials* **2003**, *24*, 2309–2316, doi:10.1016/s0142-9612(03)00110-8.

44. Vermette, M., Trottier, V., Ménard, V., Saint-Pierre, L., et al. Production of a new tissue-engineered adipose substitute from human adipose-derived stromal cells. *Biomaterials* **2007**, *28*, 2850–2860, doi:10.1016/j.biomaterials.2007.02.030.

45. Gimble, J., Guilak, F. Adipose-derived adult stem cells: Isolation, characterization, and differentiation potential. *Cytotherapy* **2003**, *5*, 362–369, doi:10.1080/14653240310003026.

46. Gimble, J. M., Nuttall, M. E. Bone and fat: old questions, new insights. *Endocrine* **2004**, *23*, 183–188, doi:10.1385/endo:23:2-3:183.

47. Pittenger, M. F., Mackay, A. M., Beck, S. C., Jaiswal, R. K., et al . Multilineage potential of adult human mesenchymal stem cells. *Science* **1999**, *284*, 143–147, doi:10.1126/science.284.5411.143.

48. Gimble, J. M., Katz, A. J., Bunnell, B. A. Adipose-derived stem cells for regenerative medicine. *Circ Res* **2007**, *100*, 1249–1260, doi:10.1161/01.Res.0000265074.83288.09.

49. Hicok, K. C., Du Laney, T. V., Zhou, Y. S., Halvorsen, Y. D., et al. Human adipose-derived adult stem cells produce osteoid in vivo. *Tissue Eng* **2004**, *10*, 371–380, doi:10.1089/107632704323061735.

50. Justesen, J., Pedersen, S. B., Stenderup, K., Kassem, M. Subcutaneous adipocytes can differentiate into bone-forming cells in vitro and in vivo. *Tissue Eng* **2004**, *10*, 381–391, doi:10.1089/107632704323061744.

51. Aust, L., Devlin, B., Foster, S. J., Halvorsen, Y. D., et al. Yield of human adipose-derived adult stem cells from liposuction aspirates. *Cytotherapy* **2004**, *6*, 7–14, doi:10.1080/14653240310004539.

52. Mitchell, J. B., McIntosh, K., Zvonic, S., Garrett, S., et al. Immunophenotype of human adipose-derived cells: temporal changes in stromal-associated and stem cell-associated markers. *Stem Cells* **2006**, *24*, 376–385, doi:10.1634/stemcells.2005-0234.

53. Thirumala, S., Gimble, J. M., Devireddy, R. V. Transport phenomena during freezing of adipose tissue derived adult stem cells. *Biotechnol Bioeng* **2005**, *92*, 372–383, doi:10.1002/bit.20615.

54. Thirumala, S., Zvonic, S., Floyd, E., Gimble, J. M., et al. Effect of various freezing parameters on the immediate post-thaw membrane integrity of adipose tissue derived adult stem cells. *Biotechnol Prog* **2005**, *21*, 1511–1524, doi:10.1021/bp050007q.

55. Zuk, P. A., Zhu, M., Mizuno, H., Huang, J., et al. Multilineage cells from human adipose tissue: implications for cell-based therapies. *Tissue Eng* **2001**, *7*, 211–228, doi:10.1089/107632701300062859.

56. Shaik, S., Hayes, D., Gimble, J., Devireddy, R. Inducing heat shock proteins enhances the stemness of frozen-thawed adipose tissue-derived stem cells. *Stem Cells Dev* **2017**, *26*, 608–616, doi:10.1089/scd.2016.0289.

57. Sarkar, N. Thermal gelation properties of methyl and hydroxypropyl methylcellulose. *J Appl Polym Sci* **1979**, *24*, 1073–1087, https://doi.org/10.1002/app.1979.070240420.

58. Sarkar, N., Walker, L. C. Hydration—dehydration properties of methylcellulose and hydroxypropylmethylcellulose. *Carbohydr Polym* **1995**, *27*, 177–185.

59. Ford, J. L. Thermal analysis of hydroxypropylmethylcellulose and methylcellulose: Powders, gels and matrix tablets. *Int J Pharm* **1999**, *179*, 209–228, doi:10.1016/s0378-5173(98)00339-1.

60. Hirose, M., Kwon, O. H., Yamato, M., Kikuchi, A., et al. Creation of designed shape cell sheets that are noninvasively harvested and moved onto another surface. *Biomacromolecules* **2000**, *1*, 377–381, doi:10.1021/bm0002961.

61. Tate, M. C., Shear, D. A., Hoffman, S. W., Stein, D. G., et al. Biocompatibility of methylcellulose-based constructs designed for intracerebral

gelation following experimental traumatic brain injury. *Biomaterials* **2001**, *22*, 1113–1123, doi:10.1016/s0142-9612(00)00348-3.

62. Li, L., Shan, H., Yue, C., Lam, Y. C., et al. Thermally induced association and dissociation of methylcellulose in aqueous solutions. *Langmuir* **2002**, *18*, doi:10.1021/la020029b.
63. Zheng, P., Li, L., Hu, X., Zhao, X. Sol–gel transition of methylcellulose in phosphate buffer saline solutions. *J Polym Sci B Polym Phys* **2004**, *42*, 1849–1860, https://doi.org/10.1002/polb.20070.
64. Tang, Z., Kikuchi, A., Akiyama, Y., Okano, T. Novel cell sheet carriers using polyion complex gel modified membranes for tissue engineering technology for cell sheet manipulation and transplantation. *React Funct Polym* **2007**, *67*, 1388–1397, https://doi.org/10.1016/j.reactfunctpolym.2007.07.058.
65. Gimble, J. M. Adipose tissue-derived therapeutics. *Expert Opin Biol Ther* **2003**, *3*, 705–713, doi:10.1517/14712598.3.5.705.
66. Halvorsen, Y. D., Franklin, D., Bond, A. L., Hitt, D. C., et al. Extracellular matrix mineralization and osteoblast gene expression by human adipose tissue-derived stromal cells. *Tissue Eng* **2001**, *7*, 729–741, doi:10.1089/107632701753337681.
67. Qureshi, A. T., Monroe, W. T., Dasa, V., Gimble, J. M., et al. miR-148b-nanoparticle conjugates for light mediated osteogenesis of human adipose stromal/stem cells. *Biomaterials* **2013**, *34*, 7799–7810, doi:10.1016/j.biomaterials.2013.07.004.
68. Sen, A., Lea-Currie, Y. R., Sujkowska, D., Franklin, D. M., et al. Adipogenic potential of human adipose derived stromal cells from multiple donors is heterogeneous. *J Cell Biochem* **2001**, *81*, 312–319, doi:10.1002/1097-4644(20010501)81:2<312::aid-jcb1046>3.0.co;2-q.
69. Thirumala, S., Gimble, J. M., Devireddy, R. V. Cryopreservation of stromal vascular fraction of adipose tissue in a serum-free freezing medium. *J Tissue Eng Regen Med* **2010**, *4*, 224–232, doi:10.1002/term.232.
70. Thirumala, S., Gimble, J. M., Devireddy, R. V. Evaluation of methylcellulose and dimethyl sulfoxide as the cryoprotectants in a serum-free freezing media for cryopreservation of adipose-derived adult stem cells. *Stem Cells Dev* **2010**, *19*, 513–522, doi:10.1089/scd.2009.0173.
71. Thirumala, S., Wu, X., Gimble, J. M., Devireddy, R. V. Evaluation of polyvinylpyrrolidone as a cryoprotectant for adipose tissue-derived adult stem cells. *Tissue Eng Part C Methods* **2010**, *16*, 783–792, doi:10.1089/ten.TEC.2009.0552.

72. Wickham, M. Q., Erickson, G. R., Gimble, J. M., Vail, T. P., et al. Multipotent stromal cells derived from the infrapatellar fat pad of the knee. *Clin Orthop Relat Res* **2003**, 196–212, doi:10.1097/01.blo.0000072467.53786.ca.

73. Zuk, P. A., Zhu, M., Ashjian, P., De Ugarte, D. A., et al. Human adipose tissue is a source of multipotent stem cells. *Mol Biol Cell* **2002**, *13*, 4279–4295, doi:10.1091/mbc.e02-02-0105.

74. Mizuno, H., Zuk, P. A., Zhu, M., Lorenz, H. P., et al. Myogenic differentiation by human processed lipoaspirate cells. *Plast Reconstr Surg* **2002**, *109*, 199–209; discussion 210–191, doi:10.1097/00006534-200201000-00030.

75. Awad, H. A., Wickham, M. Q., Leddy, H. A., Gimble, J. M., et al. Chondrogenic differentiation of adipose-derived adult stem cells in agarose, alginate, and gelatin scaffolds. *Biomaterials* **2004**, *25*, 3211–3222, doi:10.1016/j.biomaterials.2003.10.045.

76. Seo, M. J., Suh, S. Y., Bae, Y. C., Jung, J. S. Differentiation of human adipose stromal cells into hepatic lineage in vitro and in vivo. *Biochem Biophys Res Commun* **2005**, *328*, 258–264, doi:10.1016/j.bbrc.2004.12.158.

77. Li, S., Poche, J. N., Liu, Y., Scherr, T., et al. Hybrid synthetic-biological hydrogel system for adipose tissue regeneration. *Macromol Biosci* **2018**, *18*, e1800122, doi:10.1002/mabi.201800122.

78. Pritchard, C. D., O'Shea, T. M., Siegwart, D. J., Calo, E., et al. An injectable thiol-acrylate poly(ethylene glycol) hydrogel for sustained release of methylprednisolone sodium succinate. *Biomaterials* **2011**, *32*, 587–597, doi:10.1016/j.biomaterials.2010.08.106.

79. Jammalamadaka, U., Tappa, K. Recent advances in biomaterials for 3D printing and tissue engineering. *J Funct Biomater* **2018**, *9*, doi:10.3390/jfb9010022.

80. Zaszczyńska, A., Moczulska-Heljak, M., Gradys, A., Sajkiewicz, P. Advances in 3D printing for tissue engineering. *Materials* **2021**, *14*, 3149.

81. Nikolova, M. P., Chavali, M. S. Recent advances in biomaterials for 3D scaffolds: A review. *Bioactive materials* **2019**, *4*, 271–292.

82. K. Naidu Dhanpal, J. A comprehensive review of 3D printing techniques for biomaterial-based scaffold fabrication in bone tissue engineering. *Ann 3D Print Med* **2024**, *13*, 100141, https://doi.org/10.1016/j.stlm.2023.100141.

83. Janarthanan, G., Noh, I. Recent trends in bioinks for 3D printing. *Biomaterials Research* **2018**, *22*, doi:10.1186/s40824-018-0122-1.

84. Xinkun, W., Wenwan, S., Xiaojiang, L., Zhongze, G. Recent advances in 3D-printing-based organ-on-a-chip. *EngMedicine* **2024**, *1*, 100003, https://doi.org/10.1016/j.engmed.2024.100003.

85. Sousa, A. C., Alvites, R., Lopes, B., Sousa, P., et al. Three-dimensional printing/bioprinting and cellular therapies for regenerative medicine: Current advances. *J FunctBiomater* **2025**, *16*, 28.

86. Wu, Z., Liu, R., Shao, N., Zhao, Y. Developing 3D bioprinting for organs-on-chips. *Lab Chip* **2025**.

87. Yang, D.-L., Faraz, F., Wang, J.-X., Radacsi, N. Combination of 3D printing and electrospinning techniques for biofabrication. *Adv Mater Technol* **2022**, *7*, 2101309, https://doi.org/10.1002/admt.202101309.

88. Kilian, D., Ahlfeld, T., Akkineni, A. R., Bernhardt, A., et al. 3D Bioprinting of osteochondral tissue substitutes - in vitro-chondrogenesis in multi-layered mineralized constructs. *Sci Rep* **2020**, *10*, 8277, doi:10.1038/s41598-020-65050-9.

89. Yu, X., Zhang, T., Li, Y. 3D printing and bioprinting nerve conduits for neural tissue engineering. *Polymers (Basel)* **2020**, *12*, doi:10.3390/polym12081637.

90. Harshavardhan, B., Dhakshinamoorthy, S., Swaminathan, S. Embedded 3D bioprinting – An emerging strategy to fabricate biomimetic & large vascularized tissue constructs. *Bioact Mater* **2024**, *32*, 356–384, https://doi.org/10.1016/j.bioactmat.2023.10.012.

91. Davoodi, E., Sarikhani, E., Montazerian, H., Ahadian, S., et al. Extrusion and microfluidic-based bioprinting to fabricate biomimetic tissues and organs. *Adv Mater Technol* **2020**, *5*, doi:10.1002/admt.201901044.

92. Zhenrui, Z., Xianhao, Z., Yongcong, F., Zhuo, X., et al. AI-driven 3D bioprinting for regenerative medicine: From bench to bedside. *Bioact Mater* **2025**, *45*, 201–230, https://doi.org/10.1016/j.bioactmat.2024.11.021.

93. Hilzenrat, G., Gill, E. T., McArthur, S. L. Imaging approaches for monitoring three-dimensional cell and tissue culture systems. *Journal of Biophotonics* **2022**, *15*, e202100380, https://doi.org/10.1002/jbio.202100380.

94. Ding, H., Dole, A., Tourlomousis, F., Chang, R. C. Design of a skin grafting methodology for burn wound using an additive biomanufacturing system guided by hyperspectral imaging. In *Proc ASME 2016 11th Int Manuf Sci Eng Conf*, 2016.

95. Lutzweiler, G., Ndreu Halili, A., Engin Vrana, N. The overview of porous, bioactive scaffolds as instructive biomaterials for tissue regeneration

and their clinical translation. *Pharmaceutics* **2020**, *12*, doi:10.3390/pharmaceutics12070602.

96. Yang, L., Pijuan-Galito, S., Rho, H. S., Vasilevich, A. S., et al. High-throughput methods in the discovery and study of biomaterials and materiobiology. *Chem Rev* **2021**, *121*, 4561–4677, doi:10.1021/acs.chemrev.0c00752.

97. Mehta, N., Shaik, S., Devireddy, R., Gartia, M. R. Single-cell analysis using hyperspectral imaging modalities. *J Biomech Eng* **2018**, *140*, 0208021–02080216, doi:10.1115/1.4038638.

98. Jiangbo, W. Hyperspectral imaging for non-invasive blood oxygen saturation assessment. *Photodiagnosis and Photodynamic Therapy* **2024**, *45*, 104003, https://doi.org/10.1016/j.pdpdt.2024.104003.

99. Mehta, N., Shaik, S., Prasad, A., Chaichi, A., et al. Multimodal label-free monitoring of adipogenic stem cell differentiation using endogenous optical biomarkers. *Adv Funct Mater* **2021**, *31*, doi:10.1002/adfm.202103955.

100. Filippi, M., Buchner, T., Yasa, O., Weirich, S., et al. Microfluidic tissue engineering and bio-actuation. *Adv Mater* **2022**, *34*, 2108427, https://doi.org/10.1002/adma.202108427.

101. Liang, M., Shijie, Y., Xiaodong, X., Sidney, Jing, Z. et al. Application of artificial intelligence in 3D printing physical organ models. *Mater Today Bio* **2023**, *23*, 100792, https://doi.org/10.1016/j.mtbio.2023.100792.

102. Praemer, A., Furner, S., Rice, D. P., American Academy of Orthopaedic, S. *Musculoskeletal conditions in the United States*; American Academy of Orthopaedic Surgeons: Rosemont, IL, 1999.

103. Kruyt, M. C., van Gaalen, S. M., Oner, F. C., Verbout, A. J., et al. Bone tissue engineering and spinal fusion: The potential of hybrid constructs by combining osteoprogenitor cells and scaffolds. *Biomaterials* **2004**, *25*, 1463–1473, doi:10.1016/s0142-9612(03)00490-3.

104. Gupta, A., Kukkar, N., Sharif, K., Main, B. J., et al. Bone graft substitutes for spine fusion: A brief review. *World J Orthop* **2015**, *6*, 449–456, doi:10.5312/wjo.v6.i6.449.

105. Hebert, T. L., Wu, X., Yu, G., Goh, B. C., et al. Culture effects of epidermal growth factor (EGF) and basic fibroblast growth factor (bFGF) on cryopreserved human adipose-derived stromal/stem cell proliferation and adipogenesis. *J Tissue Eng Regen Med* **2009**, *3*, 553–561, doi:10.1002/term.198.

106. James, A. W., Levi, B., Nelson, E. R., Peng, M., et al. Deleterious effects of freezing on osteogenic differentiation of human adipose-derived stromal cells in vitro and in vivo. *Stem Cells Dev* **2011**, *20*, 427–439, doi:10.1089/scd.2010.0082.

107. Johnson, S., Hall, C., Das, S., Devireddy, R. Freezing of solute-laden aqueous solutions: Kinetics of crystallization and heat- and mass-transfer-limited model. *Bioengineering* **2022**, *9*, 540.

108. Costa, M., Cerqueira, M. T., Santos, T. C., Sampaio-Marques, B., et al. Cell sheet engineering using the stromal vascular fraction of adipose tissue as a vascularization strategy. *Acta Biomater* **2017**, *55*, 131–143, doi:10.1016/j.actbio.2017.03.034.

109. Baksh, N., Gallant, N. D., Toomey, R. G. Cell sheet engineering for integrating functional tissue in vivo: Successes and challenges. *MRS Bulletin* **2017**, *42*, 350–355, doi:10.1557/mrs.2017.91.

110. Sudo, Y., Kawai, R., Sakai, H., Kikuchi, R., et al. Star-shaped thermoresponsive polymers with various functional groups for cell sheet engineering. *Langmuir* **2018**, *34*, 653–662, doi:10.1021/acs.langmuir.7b04213.

Chapter 7

Organoids: In Vitro Alternatives to Study In Vivo Events and the Possibility of Using 3D Bioprinting and Hyperspectral Imaging for Better Outcomes

Saumya Rai, Mohankumar B. S., Qudsiya Mohiuddin, Bhawana, Sanya Sharma, and Santosh K. Misra
Department of Biological Sciences & Bioengineering, Indian Institute of Technology Kanpur, Kalyanpur, UP, India
The Mehta Family Centre for Engineering in Medicine, Indian Institute of Technology Kanpur, Kalyanpur, UP, India
skmisra@iitk.ac.in

Understanding various biological in vivo events in a simpler way and pre-screening of interventions have been a major development in modern medical regimens. It has not only ensured more effective therapy in complex diseases but also has helped in cutting down the side effects. But it requires a certain simplification of testing methods for the newly developed therapeutic interventions, which can be performed in laboratory conditions. Among many possible

High-Throughput Imaging Methods: Cells and Tissues
Edited by Manas Ranjan Gartia and Ram V. Devireddy

ISBN 978-981-5352-08-5 (Hardcover), 978-1-003-77417-4 (eBook)
www.jennystanford.com

DOI: 10.1201/9781003774174-7

simplifications, like two-dimensional (2D) cell culture, 3D scaffold-based studies, and ex-vivo evaluations; organoids have provided added benefits to researchers. Organoids are 3D structures resembling a particular organ in their miniature version, generally called mini organs. Though it is a promising system for studying various biological processes, generating and imaging organoids has always been difficult and needs special techniques to perform, like 3D-bioprinting and specialized bioimaging, respectively. In this chapter, different bioprinting techniques and varieties of bioinks used for 3D-bioprinting of organoids are discussed with an extension to understand the possible use of hyperspectral imaging in studying and improving the outcome.

7.1 Introduction

Organoids are cell culture models in which cells self-assemble in three-dimensional (3D) conformations, differentiate, and rearrange in such a way that they mimic the events of organogenesis, mimicking the in vivo condition of the organ. Organoids have gained much attention from researchers as various diseases can be studied in their native conditions. These in vitro constructed cell assemblies can even recapitulate the microenvironment of tumors. But to study these entities, advanced imaging tools would be preferential rather than conventional imaging tools. In the recent past, the establishment of new imaging techniques and applications in the field of bioimaging has been significantly hampered by the limited propagation of light within tissue. As light travels through biological tissue, it experiences several scattering and absorption processes (Patterson et al., 1991). The optical characteristics of biological tissues may vary spatially and show variations in refractive index that cause different extents of scattering. Cells and extracellular proteins in surrounding tissues scatter due to small-scale inhomogeneities and large-scale structural differences (Joel and Tuan, 2003). Organoids are expected to be a heterogeneous cellular mass composed of different cell types and extracellular proteins to enumerate the functional and structural characteristics of a particular organ. In such cases, the imaging and characterization of heterogenous organoid structures are challenging. In such a scenario, hyperspectral (HSpec) imaging (HSI) has been found to have enormous potential for non-intrusive disease diagnosis and surgical guidance, which can be effectively utilized for the characterization of the organoids, as well. Hyperspectral imaging collects fluorescence spectral data for each pixel of an image (Govender et al., 2007), separating each pixel into several wavelength components and linking its chemical composition to its spectral composition. HSI can distinguish between chemical species like retinol and retinoic acid using overlapping spectra, because it provides higher-resolution 3D datasets as compared to multispectral imaging using overlapping spectra (Gao and Smith, 2015) and the phasor method (Browne et al., 2017).

There are various techniques used for the in-situ characterization of organoids, like light microscopy, fluorescence microscopy, confocal microscopy, and electron microscopy such as scanning electron microscopy (SEM) and transmission electron microscopy (TEM). Molecular biology techniques like western blotting, RNA sequencing, and qRT-PCR can be used to characterize the cells from organoids after lysis, to investigate the organ-specific protein and gene expression. Flow cytometry can also be used to study the cell surface-specific markers on the organoids (McKinnon et al., 2018). Immunohistochemistry and immunocytochemistry can be done to determine protein localization and expression of cell surface and intracellular markers using fluorescence and confocal microscopy, but these techniques require several steps of sample preparation, so it is most likely that we can lose specific signals (Hawes et al., 2009). SEM and TEM can be used to study the ultrastructure of organoids, but they are associated with disadvantages, like SEM, which only allows us to study the surface features, while TEM imaging requires the need for preparing a very thin section of the tissue sample. On the other hand, hyperspectral imaging (HSI) is noninvasive, has quick data acquisition ability, and requires minimal sample preparation steps; thus, it gives less variability in results and fewer chances of human error. HSI gives detailed information about biological processes in organoids and helps in understanding the origin of diseases. HSI can be used to distinguish different chemical species in cells, like retinol accumulation in retinal organoids, where it can give information about metabolic changes at different stages of organoid development, and these results can also be validated by immunohistochemistry, qPCR, and single-cell RNA sequencing (sRNA-seq) (Xue et al., 2019). Thus, these features of HSI ensure better bioimaging of organoids to study possible biological processes occurring to reveal the fate of the therapy or strategy under investigation for future in vivo studies and eventual human use.

7.2 Types of Organoids and Optimized Applications

Organoids are the new generation 3D construct models that are close to in vivo organs with most of the biological characteristics retained in vitro. Organoids mimic the characteristics of the organ they represent (Zhao et al., 2021). Different organoids have been developed so far, as depicted in Fig. 7.1. Some of the important organoids, based on the organs they mimic, are discussed in this section.

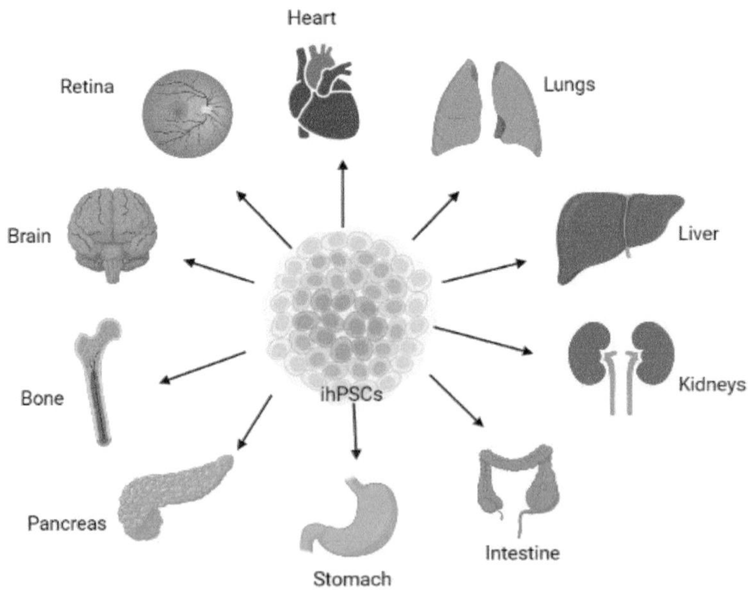

Figure 7.1 Different types of organoid models developed so far from human-induced pluripotent stem cells (hiPSCs). 3D organoids can better recapitulate the structure and functions of the organs in vitro.

7.2.1 Cardiac Organoid

The heart is an important organ in the human body, comprising of different types of cells, like cardiomyocytes, endothelial cells (critical for cardiomyocyte survival and reorganization), vascular smooth muscle cells (responsible for vascular network maintenance), fibroblasts (responsible for constructing the extracellular matrix), and immune cells like lymphoid cells. Cardiomyocytes account for ~25 to 35% of all the cell population in the heart, which provide contraction ability to the heart. All cardiac cells are adjoined by a 3D network of extracellular matrix (ECM) composed of collagen, cell adhesion proteins like laminin, fibrillin, elastin, and other components like cytokines, and other growth hormones (Filippo Buono et al., 2020). The versatility of human cardiac organoids (CO) is demonstrated by their capacity to generate contractile, functional tissue with biological properties like the native functional heart (Zhao et al., 2021). Organoids, also termed scaffold-free designed cardiac tissues, were created using 3D-bioprinting and hiPSCs in a work by Kawai et al. After developing hiPSC-COs and shaping them into spheres, the cells were put onto the needle array and grown to encourage fusion (Kawai et al., 2022).

In another study, cardiac organoids were developed by inducing the differentiation of human pluripotent stem cell (hPSC) aggregates embedded in matrigel via modulating the Wnt pathway (Drakhlis et al., 2021). In different studies, ECM proteins and hiPSCs were used to develop organoids with internal chambers. Briefly, two heart chambers and two flow exits were 3D printed using ECM-based bioinks, containing hiPSCs. The resulting chambers were shown to have strong musculature, pumping function, and electrical connectivity, in addition to demonstrating pressure and volume changes (Kupfer et al., 2020).

Due to the complex and hierarchical nature of the myocardium, engineering it remains challenging. Incorporating blood vessels into the engineered tissue presents yet another challenge. Zhang et al. reported that by 3D bioprinting endothelial cells within a microporous scaffold and then seeding cardiomyocytes, an endothelialized myocardial formed, and the resulting heart displayed synchronous contraction. They used a bioink made from ethane sulfonic acid-dissolved alginate, gelatin methacrylate, and photoinitiator Irgacure. Endothelial cells were bioprinted using composite ink and then implanted into microfibrous hydrogel scaffolds. Endothelial cells were shown to migrate to the microfibers' perimeter, where they formed a confluent layer (Zhang et al., 2016). Bioprinting technology for pliable materials, known as FRESH (freeform reversible embedding of suspended hydrogel) (Colly et al., 2021). Alginate, collagen, and fibrin are all examples of soft hydrogels that can be printed in thin layers with no distorted results (Thomas and Adam, 2016). Soft hydrogels can now be printed with precise spatial control using this method. The bioinks are held in place while they dry after being extruded on a stress support bath (Shiwarski et al., 2021). Lee et al. bioprinted a human heart at the neonatal scale, along with a heart ventricle, a tri-leaflet heart valve, a multi-scale vasculature, and a human vascular network using pH-driven gelation control. An MRI-based computer-aided design (CAD) model was used to construct all of these structures (Lee et al., 2019).

By using cardiomyocytes derived from hiPSCs, Jacqueline et al. printed a contractile heart tube using FRESH 3D bioprinting. The heart tube that forms during the embryonic stages to supply blood and nutrients to the developing embryo served as inspiration for the simplified heart model that was produced using bioprinting. They used acidified type 1 collagen bioink to create the first bioprinted collagen heart tube. Due to the support bath of gelatin's pH-driven neutralization, collagen gelated as a result. The printed collagen structure was released after the supporting bath melted at 37°C, which further caused the collagen to gel. The collagen tube was examined using brightfield microscopy and optical coherence tomography (OCT). The collagen tubes seeded with cells were cultured in PDMS wells. In order to create a cellularized heart tube, collagen gels containing cardiomyocytes and cardiac fibroblasts in a 9:1 ratio were cast around the collagen tube (Bliley et al., 2022). Additionally, full-size models of the human heart have been bioprinted utilizing FRESH technology. Mirdamati et al. bioprinted a full-scale alginate model of the human heart using MRI information from a patient. Since alginate can replicate the elastic modulus of the heart, its mechanical characteristics are similar to those of heart tissue (Mirdamadi et al., 2020).

Cardiac organoids have potential for use in many different medical applications (Fig. 7.2). Researchers have produced developmental cardiac organoids to better comprehend congenital heart disease. If these heart defects appear within the first eight weeks of the embryo's development, then it may be fatal to the foetus. It has been demonstrated that the in vitro-created organoids accurately reproduce different cardiogenic events. Numerous signaling molecules and pathways, such as WNT3A and BMP/WNT signaling, play a significant role in heart development. Organoid models can therefore be used to simulate developmental flaws (Pang et al., 2021). Organoids have also been shown to model genetic cardiomyopathy. In a study by Buono et al., cardiac organoid models for genetic cardiomyopathy were developed from the hiPSCs-derived cardiomyocytes, from cardiomyopathic patients. The authors also developed organoid models from the hiPSCs-derived cardiomyocytes of a healthy individual to compare the healthy organoid model with the cardiomyopathic model (Buono et al., 2020). Other diseases, like myocardial infarction, heart failure, genetic heart disease, tissue fibrosis, and arrhythmia, can also be studied with the help of organoid models (Zhao et al., 2021).

7.2.2 Characterization of Heart Organoids

Human heart organoids have been developed and characterized using techniques like patch-clamp, OCT, confocal microscopy to visualize cells, microelectrode array, etc. (Drakhlis et al., 2021; Lewis-Israeli et al., 2021a, b). Apart from these techniques, calcium-sensitive dye has also been used to conduct measurements of calcium fluorescence intensity and calcium mapping to assess the contractibility of the bioprinted structures. Characterization of heart organoids can be done using different techniques like SEM for cross-sectional imaging, histochemical analysis, immunohistochemistry, and RNA sequencing (Lee et al., 2019).

HSI allows researchers to examine the chemical makeup of heart organoids. In a study by Kato et al., different biomolecules, including proteins, carbohydrates, lipids, and phosphate-containing protein-rich regions, were visualized by hyperspectral imaging using mid-infrared photothermal microscopy (MIP). HSI can be used to visualize the spatial distribution of various chemical components, which can aid in understanding the biological process and the origin of diseases that could be linked to the heart (Kato et al., 2022). Hyperspectral imaging was used in a different study to identify advanced atherosclerotic plaques in aorta tissue and characterize them in vitro. This opens up new opportunities for characterizing atherosclerotic lesions in arteries of heart organoids when the goal is to recreate a disease model and investigate its pathogenesis in vitro (Larsen et al., 2011). Additionally, arteries and veins have been detected and distinguished using HSI, which can be used as an important tool for the characterization of cardiac organoids. Many anatomical structures and physiological traits in cardiac organoids can be studied using hyperspectral imaging (Akbari et al., 2009). Brunner et al. have used hyperspectral imaging to quantify inflammatory cells. They found that myocardium tissues taken from myocarditis patients contained inflammatory cells that were CD3 and CD45 positive. This opens up new avenues for the application of heart organoids, as the detection of inflammatory cells may be a key indicator of the efficacy of any potential therapeutic approach that makes use of heart organoids (Brunner et al., 2022).

7.2.3 Application

Cardiac disease modeling: The utilization of cardiac organoids can be highly advantageous in examining the progression of cardiovascular diseases and elucidating their underlying pathophysiology, hence facilitating a comprehensive recapitulation of the physiological aspects associated with these conditions. Several recent reports have been published that employ cardiac organoids as a means to investigate various illnesses, including myocardial infarction (Richards et al., 2020), ischemic heart disease (Voges et al., 2017), arrhythmia (Shinnawi et al., 2019), and heart failure (Zhao et al., 2021), etc.

Drug prediction: The preclinical stages of drug development sometimes witness a significant proportion of therapeutic candidates failing to progress to clinical trials. This can be attributed to the utilization of 2D cell culture or small animal models for initial testing and target validation. The recent advancements and emerging prospects in the domain of organoid development have facilitated the utilization of patient-specific induced pluripotent stem cells (iPSCs) for drug screening purposes, as well as the differentiation of cells afflicted by various diseases (Xuan et al., 2022). Mills et al. employed mature human cardiac organoids as a means of conducting drug screening in the context of cardiac dysfunction induced by SARS-CoV-2 infection (Mills et al., 2021).

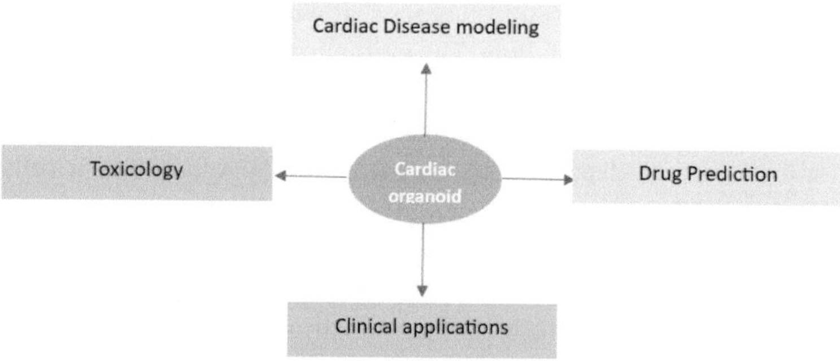

Figure 7.2 Various applications of cardiac organoids. It is used for disease modeling, drug prediction, clinical application, and toxicology.

Toxicology: Currently, the utilization of a cardiac organoid model in conjunction with a preclinical model has shown promise in accurately replicating pathological processes and heart development, as well as identifying potential toxicity and adverse reactions of drugs. Cardiac organoids generated in laboratory settings fail to achieve the desired physiological form and size observed in vivo. Additionally, these organoids do not accurately replicate multifactorial diseases due to the absence of neurological and immunological systems. However, these organoid models present encouraging opportunities for preclinical treatment (Zhao et al., 2021).

Clinical applications: The rapid emergence of iPSCs has significantly expedited advancements in the generation of organoids intended for transplantation into specific organs. The transplantation of organoids that are specific to a particular tissue has been shown to result in enhanced tissue function (Xuan et al., 2022). Varzideh et al. conducted a study wherein they successfully generated cardiac organoids obtained from hiPSCs for transplantation. They developed an approach to promote the maturation of human cardiomyocytes in a laboratory setting. This was achieved by constructing 3D cardiac organoids using human embryonic stem cell (ESC)-derived cardiac progenitor cells, endothelial cells, and mesenchymal stem cells (MSCs) (Varzideh et al., 2019).

7.3 Lung Organoid

There are over 40 different cell types that work together to form the 3D structure of a human lung. The epithelium that lines the airways of the lungs consists primarily of ciliated and secretory cells, such as goblet and club cells. Alveoli are tiny air sacs in the lungs that are lined by two distinct types of cells: alveolar type I and II cells (AEC1 and AEC2, respectively). AEC2 cells secrete surfactants, while AEC1 cells provide surface area for the exchange of gases with the capillaries in the surroundings. Lung tissue also contains immune cells such as eosinophils, dendritic cells, mast cells, T cells, and macrophages (Barkauskas et al., 2017). There are two distinct kinds of lung organoids, known as airway and alveolar organoids. Cystic structures, called airway organoids, are formed after basal cells multiply and develop into club, goblet, and ciliated cells. AEC2 cells give rise to bud-shaped alveolar organoids when they undergo differentiation into AEC1 cells. Certain chemicals and growth factors, including BMP-4, FGF-7/10, and Wnt/catenin, play crucial roles in lung organoid development, which is a miniature version of the full lung (Lee et al., 2021).

Dayoon et al. used inkjet printing to create a 3D alveolar barrier model utilizing four types of lung cell lines: AEC1, AEC2, lung endothelial cells, and lung fibroblasts. They began by printing endothelial cell ink on a porous membrane, then collagen ink, and finally fibroblast-laden collagen ink. After collagen crosslinking, AEC1, and AEC2 were deposited equally over the printed layers. They even evaluated this model for influenza infection and the subsequent antiviral response shown in the alveolar model (Kang et al., 2021). Daisuke and his team used chondrocytes and MSCs taken from rats to make a scaffold-free trachea. They prepared multicellular spheroids from mixed cell suspensions of endothelial cells, chondrocytes, and MSCs. The spheroids were seeded in a needle array using a 3D bioprinter. The artificial trachea formed from the aggregation of spheroids was matured in a bioreactor, and the needles were removed. The tracheas were then tested for mechanical strength, followed by implantation in the rats, and were followed up to 23 days, after which they were sacrificed for histological assessment by immunohistochemistry. Glycosaminoglycan (GAG) content was assessed in the tracheas, as the amount of GAG could relate to the material's mechanical strength (Taniguchi et al., 2018). In another study by Sang-Woo-Bae, respiratory epithelial cells and stem cells made from chondrocytes in bone marrow were used to make an artificial trachea. They printed five layers of microporous PCL layer, an alginate layer with MSC, nonporous PCL layer, alginate layer with epithelial cells, and a microporous PCL layer. The characterization of the bioprinted trachea was done by SEM (Bae et al., 2018).

Lung organoids hold potential in various applications, such as research on pulmonary diseases, viral, and bacterial infections. Recently, lung organoids have been used as a great model to study SARS-CoV2 infection. Besides this, lung organoids are used to study the regeneration of lungs after various injuries and traumas (Lee et al., 2021). By using lung organoids, it was shown that there are many progenitor cells when the alveolar epithelium repairs itself after damage (Kong et al., 2021). IL-17 signaling in respiratory infections has been studied using lung organoids. Studies have shown the important role of IL-17R (interleukin-17 receptor) expression on lung epithelial cells in elucidating the innate immune response in response to defense against fungal pathogens of the pulmonary tissue. For studying IL-17R expression, polarized lung epithelium is required, which is best provided by lung organoids, as they can recapitulate the polarity of tissues. Also, these organoids provide an immune-free microenvironment. Apart from infectious diseases, these organoid models are utilized in studying the role of IL-17F in asthma (Paolicelli et al., 2019).

In one of the other cases, human lung cancer cells (A549), lung adenocarcinoma-associated fibroblasts, and human umbilical vein endothelial cells (HUVECs) were 3D bioprinted together in a laminin and collagen-rich stromal environment to create lung cancer assembloids in a work by Nath et al. First, spheroids were formed using these cells, and then the spheroids were mixed with a custom bioink consisting of collagen and laminin, and 3D constructs were then made using a bioprinter (Nath et al., 2021). Lung organoids are utilized in studying the developmental pattern of the lung during embryonic development, which is important as several diseases have a developmental origin. Such disorders could arise due to various reasons, like exposure to toxins, nutritional deficiency, and mutations. These in vitro lung models have the potential to serve as a testing ground for pharmaceuticals and to evaluate the effectiveness of various medications. Moreover, the efficacy of cell-based anticancer therapies, like CAR-T cell therapy, has been tested on 3D lung tumor models. The toxicological effects of cigarettes and tobacco can be studied on lung organoids. Another application is to study fibrosis, in which there is an accumulation of excessive ECM that leads to scarring of the tissue in the lungs (Sacchi et al., 2020).

7.3.1 Characterization of Lung Organoids

In the study by Nath et al., the cancer cells of the generated cancer spheroids were genetically encoded with GFP, allowing the movement and fusion of cancer spheroids inside the matrix to be observed using fluorescence microscopy. Immunostaining was used to examine the expression of cell surface markers and intracellular markers in 3D-printed lung cancer tissue (Nath et al., 2021).

Histopathological analysis of lung tissues for their characterization is most commonly done with Hematoxylin and Eosin (H&E) staining, but it creates problems while performing automated image analysis of samples. The hyperspectral images of lungs have been stained virtually using adversarial generative networks. Such a kind of virtual staining of hyperspectral images could become a fast and cost-effective way of characterization of lung organoids (Bayramoglu et al., 2017). Another way by which hyperspectral imaging can be applied in lung organoids is by developing spectral signatures of healthy cells and upon disease induction, it can then be used for diagnosis of diseased lung tissue from healthy lung tissue, as presented in a study where they differentiated a healthy lung tissue from a smoker's lung tissue by seeing the spectral signatures of absorbance and emission in a range of wavelengths of light (Lee and Won, 2009). In another study, green fluorescent protein (GFP)-expressing cells could be quantified in highly auto-fluorescent lung tissue even when the intensity of GFP was very low. Hence, HSI is employed for the sensitive detection of fluorescence signals in highly auto-fluorescent cells of organoids as well. Studying the localization of a particular protein in the cells of lung organoids can be done by expressing GFP-tagged proteins and then further detection and characterization by HSI (Leavesley et al., 2012).

7.3.2 Application

Disease modeling: Lung carcinoma is a complex disease that encompasses various molecular and histological subtypes, including non-small cell lung cancer (NSCLC) and small cell lung cancer (SCLC) (Hirsch et al., 2017). Shi et al. devised a culture methodology that enables the generation of NSCLC organoids for varying durations. These durations include shorter timeframes ranging from 1 to 3 months and 1 to 9 passages, as well as longer periods exceeding 3 months and more than 10 passages. The organoids under investigation were subjected to a controlled culture environment, resulting in a notable success rate. Specifically, short-term maintenance of the organoids was achieved with a success rate of 72% (47 out of 65), while long-term maintenance exhibited a success rate of 15% (10 out of 65). The histological characteristics of both the patient and the patient-derived xenograft (PDX) tumor were precisely replicated by the NSCLC organoids, regardless of whether they were generated for short-term or long-term culture periods. The comprehensive examination of genetic material using whole exome and RNA sequencing techniques revealed that NSCLC organoids maintained the inherent genomic diversity observed in the original lung cancer (Shi et al., 2020). Choi et al. successfully generated patient-derived tumor organoids from individuals diagnosed with SCLC. The organoids successfully replicated the distinct structural features of the original tumor tissues. Furthermore, thorough attempts were made to augment the growth of SCLC organoids for an extended period. The long-term viability of SCLC organoids was facilitated by the addition of WNT3A or R-spondin1 to the culture medium. The results of this study underscore the significance of activating the Wnt/β-catenin pathway in facilitating the prolonged expansion of SCLC organoids. In the same study, it was observed that SCLC organoids, after an extended period of expansion, exhibited molecular attributes, genetic profiles, and morphological configurations that were similar to the original tumor tissues (Choi et al., 2021).

Infectious disease modeling: Zhou et al. successfully developed optimized conditions for the cultivation of 2D monolayer cultures of differentiated airway organoids. The primary objective of their study was to examine the dynamics of influenza infection within these organoids. Both the refined 2D differentiated airway organoids and the previously documented 3D model have demonstrated their ability to faithfully replicate the structural and functional characteristics of the human airway epithelium. Furthermore, both models are susceptible to infection by the influenza A virus (IAV). Significantly, the IAV strain, known for its high infectivity, demonstrated vigorous replication, resulting in a significantly higher viral titer in comparison to the less severe strain. The aforementioned observation underscores the considerable promise of utilizing airway organoids as a means to rapidly evaluate the infectivity of newly emerging respiratory viruses within human populations (Zhou et al., 2018). The feeder-free, long-term, 3D culturing technique for cultivating primary human lung tissue-derived alveolar type 2 epithelial cells was introduced by Youk and colleagues. Remarkably, this particular system exhibited a heightened vulnerability to SARS-CoV-2 infection in comparison to previously documented methodologies (Youk et al., 2020). In addition, Tindle et al. introduced a highly advanced adult lung organoid model that encompasses both the proximal airway and distal alveolar cell components. These components are known to have significant involvement in viral infection and the subsequent immune responses of the host (Tindle et al., 2021). In a similar context, the researchers Tiwari et al. effectively produced lung organoids using hiPSCs. These organoids consisted of a combination of alveolar type 1 and type 2 epithelial cells. According to Tiwari et al., it has been observed that these organoids exhibit the presence of crucial host cell receptors, namely angiotensin-converting enzyme 2 and transmembrane serine protease 2. Furthermore, it has been determined that these organoids are susceptible to infection by the SARS-CoV-2 virus. The aforementioned statement highlights the significant value that these entities possess in the realm of investigating infection mechanisms and exploring potential therapeutic interventions for respiratory viruses (Tiwari et al., 2021).

Drug testing: Jabs and colleagues established an automated microscopy-based assessment to distinguish drug-induced cell death from proliferative inhibition. This method has been used to evaluate lung cancer therapeutics (Jabs et al., 2017). In another attempt, Kim et al. created a collection of 80 lung cancer organoids from five subtypes to assess anticancer medicines. An organoid with a BRCA2 mutation was sensitive to olaparib, one with an EGFR mutation was responsive to erlotinib, and one with both EGFR mutation and MET amplification was efficacious with crizotinib. Compared to organoids with the BRCA2 p.W2619C mutation, organoids with the BRCA2 p.M965I mutation had greater olaparib IC50 values due to its non-pathogenicity (Kim et al., 2019). High-throughput screening assessed FDA-approved medicines, such as imatinib, mycophenolic acid, and quinacrine dihydrochloride, for SARS-CoV-2 entrance inhibition (Han et al., 2021). Lieber et al. tested molnupiravir against several SARS-CoV-2 strains using human airway epithelial organoids. This shows how organoid models can evaluate therapies for emerging virus strains (Lieber et al., 2022).

7.4 Brain Organoid

Current studies of the human brain generally rely on postmortem samples and animal models, which include the brains of primates. As stem cell technology has improved, hPSCs have become important tools for modeling human diseases in vitro (Sakaguchi et al., 2015). When the appropriate signaling cues and microenvironment are provided, hPSCs can be differentiated into a variety of cell types in the human body. This property is used to make organoids for the brain. Brain organoids are 3D suspension cultures derived from stem cells that resemble the human brain in cellular makeup and developmental pathways.

In this process, embryoid bodies (Ebs) derived from hPSCs are frequently cultivated using a rotary bioreactor in matrigel to promote tissue proliferation and neuronal differentiation in order to develop brain organoids (Qian et al., 2016). Chemical cues and growth factors are commonly supplied during culture to encourage hPSCs for the formation of specific structures in various brain areas (Qian et al., 2019). Patterning of the neural tube during development establishes the identity of various brain areas by the activity of developmental signaling pathways triggered or suppressed by signaling molecules/

morphogens released in vivo. Organoids of different parts of the brain can be fabricated using a miniaturized multi-well spinning bioreactor and hiPSCs, which resemble the human brain structurally and functionally at the molecular, cellular, and organ system levels (Qian et al., 2018).

Existing organoid techniques are generally constrained by cellular death and inner hypoxia brought on by insufficient diffusion, making them unable to create architecture that resembles the later embryonic phase. The sliced neocortical organoid (SNO) system avoids cell death in long-term cultures by bypassing the diffusion limit (Qian et al., 2020). Even though cerebral organoid methods can create tissues that resemble many interconnected brain regions, the spatial organization of these tissues is highly variable and unexpected. Several groups developed new approaches simultaneously to model interregional interactions, hPSCs were initially differentiated into diverse brain region-specific organoids before being fused in a controlled manner to generate organoids with multiple distinct area identities (Bagley et al., 2017). For instance, it has been demonstrated that united dorsal and ventral forebrain organoids produce an "assembloid" with two different, yet interacting, domains (Birey et al., 2017). Organoid fusions can be employed to provide tests for cellular activities, including neural migration and the creation of new connections between neurons. It is expected that the subsequent extensions of these fusion technologies will aid in the development of more intricate and neuronal connections using non neural organoids or source tissues.

More types of cells, such as oligodendrocytes, have been employed to make organoids with advanced technology. In addition to protecting neuronal axons for impulse transmission and providing nourishment and metabolic support to neurons, oligodendrocytes are crucial for brain growth. Single-cell sequencing studies, on the other hand, show that typical cortical organoids are devoid of oligodendrocyte progenitor cells (Quadrato and Arlotta, 2017). To address these problems, produced organoids were subjected to oligodendrocyte growth factors in cortical spheroids, which resulted in the production of oligodendrocyte progenitors and myelinating oligodendrocytes (Madhavan et al., 2018).

Traditional methods for producing brain organoids typically lack vasculature, which is thought to be deleterious for the organoids.

The lack of circulation limits oxygen transfer and nutrition to the innermost sections of the organoids in conditions of prolonged culture, resulting in cell death in core regions (Lancaster and Knoblich, 2014a; Yin et al., 2016). For vascularization, the cerebral organoids were cultured with vascular cells. The hiPSCs-derived endothelial cells seeded in Matrigel were used to successfully vascularize human brain organoids. The brain organoids developed numerous penetrating capillaries after 34 days of coculture, and these capillaries produced platelet endothelial cell adhesion molecule, an endothelial marker (Pham et al., 2018).

7.4.1 Characterization of Brain Organoids

Cortical or brain organoids have a sophisticated organization that resembles that of a developing brain. The intricacy of brain organoids, on the other hand, presents significant challenges in terms of characterization. On sectioned organoids, immunostaining has been utilized to examine the structural makeup of organoids. Brain organoids include apical progenitors that express radial glial markers (SOX2 and PAX6) in the ventricle (Kadoshima et al., 2013; Qian et al., 2016; Xiang et al., 2017). Single-cell-based genomics methods are another fast-growing avenue for resolving cellular heterogeneity within organoids (Polioudakis et al., 2019). Single-cell RNA sequencing of complex tissues is done to establish cellular individuality (Tang et al., 2009). The electrical attribute of neurons is generally essential for determining the level of maturity and identification of neurons (Tambalo and Lodato, 2020). Nonetheless, the duration it takes to build the neural networks must be considered. Organoids that were four months old had no neural activity, whereas organoids that were eight months old had reflex activity and excitatory synapses (Quadrato et al., 2017). Various experimental methodologies are currently being used to validate the electrical function of brain organoids (Poli et al., 2019). Electrophysiological recordings were made on cells selected randomly from a sliced section of organoids. Patch clamp is also time-consuming and ineffective in determining the interconnectedness of the brain organoid. As a result, brain organoid investigations have employed the microelectrode arrays technique as a functional test (Giandomenico et al., 2019; Trujillo et al., 2019). In addition to conventional methods of characterization, we can employ HSI, which is a noninvasive, label free, and convenient technology that has the potential to allow direct monitoring of morphological, functional, metabolic, and developmental changes in the brain organoid (Rashmiparvathi et al., 2022, Luca et al., 2018). FADH2 and NADH are the most aimed markers of mitochondrial function in the cerebral cortex due to their autofluorescence properties. As a result, their detection can be employed to monitor the brain organoid's metabolic activity. HbO2 and HHb detection can aid in monitoring brain tissue oxygenation. Using this technique will eventually aid in the improvement of culture conditions to recapitulate the in vivo development (Giannoni, Luca, et al., 2018).

7.4.2 Application

Brain evolution: Organoids provide the opportunity to investigate the origins and evolution of the human brain. Organoids produced using iPSCs derived from large apes and organoids made from human cells can be compared to learn more about what makes humans different from nonhuman primates (Fig. 7.3).

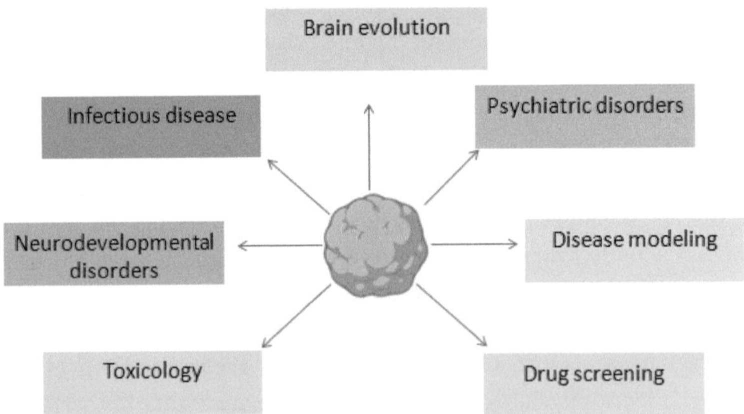

Figure 7.3 Various applications of brain organoids. They have been used for studying neurodevelopmental disorders, toxicology studies, drug screening, disease modeling, psychiatric disorders, and evolution of the brain.

Disease modeling: The potential of cerebral organoids produced from hPSCs, particularly iPSCs derived from patients to simulate neurodevelopmental brain diseases has been intensively investigated (Chen et al., 2018). They have been proven especially good at recreating disease-related phenotypes in circumstances where structural abnormalities appear early in the embryonic stage. Premature differentiation, cell cycle disruption, and decreased proliferation, all of which can be accurately examined in cerebral organoids, are usually linked to defective progenitor cell control in such illnesses. Some of the disease models are summarized in Table 7.1.

Table 7.1 Different disease model developed in brain organoids

Modeled Disease	Type of Brain Organoid	References
Alzheimer's disease (AD)	Dorsal forebrain	Papaspyropoulos et al., 2020
Primary microcephaly	Cerebral organoid	Zhang et al., 2019; Lancaster et al., 2013
Autism spectrum disorder	Dorsal forebrain	Wang, P. et al., 2017
Zika virus infection (ZIKV)	Forebrain	Qian et al., 2017

Drug screening: The capacity to create large quantities of brain organoids with great consistency opens up the potential for chemical validation and testing using brain organoids. Such advancements could be accelerated by the creation of automated high-throughput systems.

Infectious disease modeling (Zika virus infection): The viral infection caused by the Zika virus is the most researched since it coincided with a rise in the number of microcephalic neonates, raising suspicions of a causative link. Cerebral organoids have been frequently employed to imitate this viral infection and examine the cellular mechanisms behind it. Research on neurotropic viruses, such as the Zika virus, has benefited from the development of brain organoids, because they propose a solution to mammalian and 2D cell culture models of the viral infection (Antonucci and Gehrke, 2019).

7.5 Bone Organoid

Bone is a multifunctional tissue that provides mechanical support for mobility, protection of the interior organs, and hematopoiesis, among other things. A variety of cells and an ECM comprise bone, which is mechanically stable and flexible to perform these various functions in the human body (Scheinpflug et al., 2018). Repairing and regrowing bone defects caused by tumor resection, trauma, degeneration, and so on has always been a key concern. Bone has a high capacity for regeneration, and in recent decades, bone tissue engineering has produced a variety of useful materials, including scaffolds and cells seeded in it for bone restoration, which has progressively resolved many clinical complexities. However, the process of metabolism of bone is complex, and the study takes a long time and is tough, limiting the advancement of bone regeneration and restoration research.

7.5.1 Composition of Bone Organoid

Organoids derived from stem cells can be programmed to display the targeted phenotype. Different types of bone-related cells, such as osteoblasts, osteoclasts, and osteocytes, can be used to govern the functioning of bone organoids. In order to study bone diseases, osteoblasts are employed to create a simplified bone model because of their important roles in bone development. Human-periosteum-derived cells (hPDCs), hiPSCs, multipotent MSCs, and ESC-derived osteoblast/osteocyte population are examples of autologous or allogeneic stem cells with tremendous potential for bone construction (Chen et al., 2022).

7.5.2 Characterization of Bone Organoid

Using a lubrication-based model, the fluid shear stress was estimated for three different positions in space along the bottom of the well. Fluorescence microscopy was used to examine cellular morphology on day seven. Second Harmonic Generation pictures were used to visualize deposited collagen filaments (Delaine-Smith et al., 2012). For the woven bone model, the osteocytes were entirely entrenched in their ECM, volumetric imaging using a focused ion beam scanning electron microscope (FIB/SEM) volume imaging with 3D reconstruction. Immunohistochemical examination revealed the non-collagenous proteins deposited extracellularly. The presence of matrix mineralization under biological control was confirmed using a multiscale imaging approach. The spectral signature of growing bone was revealed by Raman micro spectroscopy of the ECM, confirming the mineral's co-localization with collagen (Akiva et al., 2021). Surface characterization is usually done using SEM. HSI is used to visualize the progress of bone mineralization. In newly regenerated bone, the phosphate and mono hydrogen phosphate gradients are clearly defined, with the highest concentration seen in trabecular packets (Timlin et al., 1999). The formation and preservation of collagen using HSI can help in studying the ECM (Akiva et al., 2021). In 2019, Fauch et al. proposed the investigation of 2D distributions of bone features using narrowband autofluorescence, applicable to various bone imaging perspectives, which provided enhanced images for bone analysis, and this could be extrapolated for the characterization of bone organoids by utilizing HSI (Fauch et al., 2019).

7.5.3 Application

Disease modeling: In comparison to classic animal bone disease models, in vitro development of bone organoids has the capability to more accurately imitate the pathological conditions of our body, allowing researchers to better understand the mechanism and diagnosis of disease. One of the most common and widely recognized bone illnesses is osteoporosis. In such conditions, resorption of bone occurs at a faster rate than its replacement, resulting in a loss of bone mineral density. The microenvironment of osteoporosis will be mimicked by creating a bone osteoporotic model in the form of an organoid, which will increase the proportion and activity of osteoclasts (Chen et al., 2022). The bone defect model allows researchers to use a consistent, non-load-bearing orthotopic setup to test bone tissue and biomaterials production processes (Fig. 7.4) (Spicer et al., 2012).

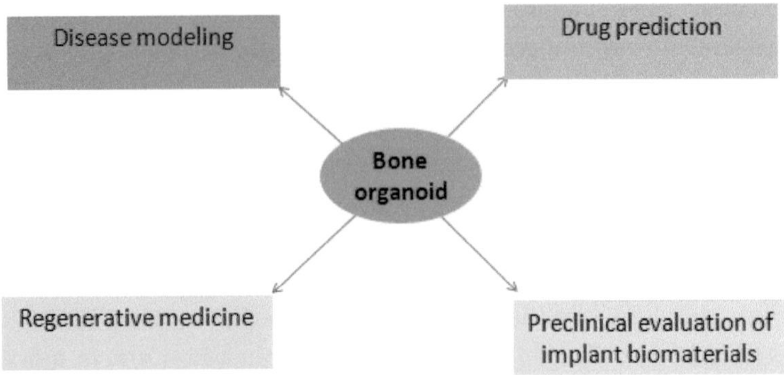

Figure 7.4 Various applications of bone organoids. It is used for disease modeling, drug prediction, and for preclinical evaluation of implant biomaterial.

Regenerative medicine: The primary objective of regenerative medicine is the in vitro generation of replacement tissue for diseased or dysfunctional organs, thereby avoiding lifelong anti-rejection medication and achieving non-immunosuppression, disease-free, and toxicity reduction. Bone organoids can grow with identical genetic tissue for use in autotransplantation, and they can be used as renewable resources in organ replacement systems. For instance, to attain scalable production of callus organoids derived from microspheroids, cells obtained from human periosteum could be employed (Gabriella et al., 2019).

Preclinical examination of implant biomaterials: In order to acquire data, preclinical testing and evaluation of implant biomaterials is now largely done at the cost of a high proportion of animal models. Furthermore, in large-scale therapeutic trials, the information collected by sacrificing these animals is sometimes ambiguous due to species differences.

Drug prediction: Before clinical approval, maximum clinical medications for treatment of bone diseases, namely, growth-promoting drugs, anti-inflammatory therapies, and so on, must go through a lengthy testing process in vitro and in vivo. During phase I trials, these medicines are frequently discontinued. Two key drawbacks are the extended cycle and organ toxicity. The development of a human bone organoid will significantly reduce the amount of time required to detect drugs and provide an accurate method for predicting drug toxicity.

7.6 Liver Organoid

The liver is a major organ in the human body, and it is located in the upper right abdominal cavity. It is one of the important organs for lipid, amino acid, and xenobiotic metabolism. It also aids in the regulation of blood lipid titers and the maintenance of glucose levels. Several proteins that circulate in the blood are produced by the liver, including albumin, which collectively accounts for 55–70% of total protein in plasma. In addition, it serves as a major repository for fat-soluble vitamins, glycogen, and other enzymes. The human liver comprises four major cell types: hepatocytes, kupffer, stellate, and sinusoidal endothelial cells; among which hepatocytes account for around 80% of the total cell population (Jacek et al., 2016). The hepatic lobules (1 mm) are the smallest functional unit that is made up of the hepatocytes, non-hepatocytes, sinusoidal lumen, biliary canaliculi, space of Disse, stellate cells, and kupffer cells. Because of the complex architecture, excellent cell–cell adhesion, and high cell density of the liver, culturing primary hepatocytes along with other cell types of the liver remains very problematic in 2D settings (Yixin Chen et al., 2012). To overcome this problem, 3D hepatocyte spheroid is utilized for drug screening and drug testing (Vorrink et al., 2018). Improved bioprinting technology may be a useful tool for creating 3D liver organoids. 3D bioprinting stands out among the several other strategies for fabricating biomimetic liver tissues. The bioink used to fabricate organoids should mimic the ECM components of the liver. A study by Chang and colleagues used syringe-based, layered bioprinting to create a micro-organ device with HepG2 cells encased in alginate hydrogel. The bioprinted organoid was planned to be used by NASA for assessing drug metabolic investigations in the planetary environment (Chang et al., 2010). In another study by Matsusaki and group, they used HepG2 and HUVECs (human umbilical vein endothelial cells)-laden fibronectin along with gelatin to 3D bioprinted hierarchical liver tissue micro-arrays using inkjet bioprinting technique. The hierarchical sandwich architectures displayed strong cell–cell interaction that was comparable to the native liver structure, and it showcased hepatic functions such as albumin and CYP3A4 secretion (Matsusaki et al., 2013). Ma et al. have shown that the activity of liver organoids was due to the 3D aggregation of hepatocytes with supportive cells. They developed a hydrogel-based triculture hepatic model by 3D bioprinting technology using the hepatic progenitor cells derived from hiPSCs (hiPSC-HPCs), adipose-derived stem cells, and HUVECs in a microscale hexagonal layout. The bioprinted triculture model showed prominently improved liver-specific gene expression, enhanced cytochrome P450 production, enhanced morphological organization, and increased liver-specific metabolic product secretion after 20 days of culture (Ma et al., 2016). Wang et al. developed a 3D structure composed of gelatin hydrogel and hepatocytes utilizing rapid prototyping technology. The technique used pressure-activated syringe to deposit up to 30 layers of the laminated cell-biomaterial mixture. They showed that the laminated hepatocytes were viable and capable to perform biological functions for up to two months, and thus this technology has the potential to develop organoids (Wang et al., 2016). Nguyen et al. have utilized micro-extrusion 3D bioprinting technology to create a distinctive bioprinted human micro liver tissue made of primary human hepatocytes, HUVEC cells, and HSC in the hydrogel to further investigate in vitro drug-induced toxicity. The albumin level, ATP synthesis, cytochrome P450 expression, and drug-induced enzymatic activities were reported in the liver tissue even after 28 days (Nguyen et al., 2016). The liver organoids can be utilized as a model to examine drug-induced toxicity, drug discovery, disease model, and drug metabolism. Skardal et al. printed particular primary liver spheroids using a multifunctional hyaluronic acid (HA) and gelatin-based hydrogel to fabricate in vitro liver constructions with increased cell survival, urea and albumin production (Skardal et al., 2016). Grix and colleagues have used stereolithography to bioprint 3D liver organoids using HepaRG and stellate cells. The bioprinted organoid demonstrated the basic characteristics of the liver (Grix et al., 2018). Song et al. created arrays of liver lobules using a predefined extrusion bioprinting technology. Pre-set cartridge was first fabricated and different materials were injected into the cartridge from the nozzle of the extrusion printer to bioprint a heterogenous, multifunctional liver organoid (Song et al., 2020).

Bouwmeeste et al. used extrusion bioprinting to fabricate hepatic organoids using hepatocyte-like cells and gelatin-based hydrogel (GelMA) as a bioink. Sacrificial hydrogel Pluronic-127 was co-printed with cell-laden GelMA to form porous structures. Printed constructs were made up of 600 to 1,000 μm printed strands with 200 to 400 μm pores. The organoids immediately differentiated as liver tissue utilizing the differentiation medium. This kind of organoids can be produced in vitro, in large numbers, so that complex biologics like clotting factors and biologics can be outsourced (Bouwmeeste et al., 2021).

7.6.1 Characterization of Liver Organoid

The morphology of the organoid was examined 10 days after cultivation using H&E staining. Fluorescence imaging of live and dead cells also confirmed that the number of viable cells remains stable over time when compared to controls. Protein and gene expression profiles were studied as a characterization tool for organoid constructs. Gene expression profiling for some of the liver-specific genes, such as ATP-binding cassette superfamily G member 2 (ABCG2), glucose-6-phosphatase catalytic subunit (G6PC), bile salt export pump (BSEP), and cytochromeP4503A4(CYP3A4), were done for characterization. Cytokeratin 18 (CK18), a cytoskeleton marker, found in the liver can also be checked for characterization by immunofluorescence analysis. Expression of some of the liver specific markers such as HNF4 (Hepatic Nuclear Factor 4-alpha), arginine-succinate synthase (ASS), and albumin are also checked. Glycogen storage is an important function of hepatocytes. Bioprinted organoids demonstrate glycogen buildup when stained with periodic acid-Schiff for glycogen, indicating liver activity (Bouwmeeste et al., 2021). In addition to these characterization techniques, HSI can also be used for the purpose. It can be used to monitor tissue pathology, based on spectral characteristics. HSI has been successfully used to monitor oxygenation and to evaluate the levels of hepatic ischemia (Felli et al., 2021). Therefore, we can use HSI to predict liver viability in the 3D bioprinted liver organoid to assess its functionality. Similarly, we can use HSI in the liver organoid to study different diseases associated with the liver-like fatty liver disease. Okubo et al. used HSI to monitor lipid distribution to diagnose fatty liver severity and liver cancer associated with fatty liver disease by quantifiable mapping of the total lipid content in mouse liver (Okubo et al., 2021). HSI can also be used for real-time proper identification of tissues during intraoperative procedures. HSI shows the highest precision for the detection of the liver and other nearby organs, and thus it is a promising approach to image-guided surgery (Cervantes et al., 2021). Also, we can optimize the HSI software on the 3D bioprinted organoids so that it can help provide us with highly precise information regarding the disease state of their respective organs and can be a useful tool in treating disease conditions.

7.6.2 Application

Drug screening: Liver organoid models accurately display the pathophysiology of human liver diseases; thus, they can be used for developing patient-specific drugs for clinical use. They can be used for drug screening as most of the drugs screened through animal experiments fail in clinical trials on humans. Prior et al. established that primary liver cancer (PLC)-derived organoids can be used as a model for identifying novel targets and can be used for testing drugs (Prior et al., 2019).

Disease modeling: Liver organoids can play an important role in understanding liver metabolic disorders like alcoholic liver disease (ALD) and nonalcoholic fatty liver disease (NAFLD) and help in the development of their therapies. Wang et al. developed a liver organoid model to study ALD and steatohepatitis. They developed the model using ESCs and human fetal liver mesenchymal cells by co-culturing them together. The organoid model successfully mimicked the disease's developmental stages and its pathophysiology (Wang et al., 2019). NAFLD is caused by obesity, a sedentary lifestyle, a high-energy diet, hyperlipidemia, and diabetes. Soret et al. exposed liver organoids with a free fatty acid, and they observed that the hepatocytes showed increased levels of collagen and pro-inflammatory cytokines, along with ballooning and steatosis, thus mimicking nonalcoholic steatohepatitis (NASH) characteristics (Soret et al., 2020). Viral hepatitis caused by hepatitis B virus (HBV) leads to necrosis and inflammation in hepatocytes. The infection of

HBV has different clinical outcomes like either infection cirrhosis or hepatocellular carcinoma in some cases, and the therapy response to the same drug also varies in different patients. The complicated pathophysiology associated with it leads to difficulty in understanding the replication cycle of the HBV virus. The organoid model produced by co-culturing hiPSCs with endothelial and mesenchymal cells generates a better organoid model to study HBV infection (Nie et al., 2018). Some of the genetic diseases associated with liver are Alagille syndrome (ALGS), Citrullinemia type 1 (CTLN1), and mitochondrial DNA deficiency syndrome (MDS), which are also studied in the liver organoid model to better understand the disease condition (Liu et al., 2022).

Regenerative medicine: Liver has a remarkable capability of regeneration; therefore, it is an appropriate candidate for regenerative therapy. Liver transplantation is the most efficient treatment for hepatic failure, however, there is a sparsity of healthy donors for transplantation. Liver organoids can be an alternative source for transplantation. Sampaziotis et al. made biliary organoids from cells of the extrahepatic biliary tree, and the organoid was then transplanted in a mouse model, where it was able to reconstruct the walls of the gall bladder and repair the damaged biliary epithelium (Sampaziotis et al., 2017).

7.7 Kidney Organoid

The kidney is a bean-shaped organ located in the lower abdomen, one on each side in the retroperitoneal space. The kidney's main function is to cleanse the blood plasma by filtering harmful metabolic end products, such as urea, uric acid, and creatinine. It also helps in maintaining the overall body fluid balance, helps in RBC production, keeps the electrolyte balance in the blood, and releases hormones that regulate blood pressure. Nephrons are the kidney's structural and functional unit. There are around 1 to 1.2 million nephrons in each kidney and they are arranged in a consistent pattern (Harth et al., 1983).

Metanephric mesenchyme (MM) cells are successfully used in nephron-related disease modeling and medication screening, and thus make up the majority of kidney organoids. They could be used in regenerative medicine, modeling renal disorders, medication screening, and toxicity testing of chemicals on nephrons. Cultured kidney organoids do not enable the development of a mature renal structure with a complex vascular system. The cultured organoid does not closely mimic the native kidney microenvironment, as different types of cells are present to form a functional kidney (Nishinakamura, 2019). Also, the internal part of kidney organoids becomes necrotic as they mature, and thus, they cannot grow beyond the millimeter level.

Taguchi et al. developed a procedure to harvest MM from mouse ESCs and hiPSCs cultivated into 3D spheres by employing Wnt agonists, retinoic acid, and other agents to stimulate mesoderm growth. It resulted in tubular epithelial and Bowman's capsule cells (Taguchi et al., 2014). Human ESCs were used to produce kidney cells in 3D spheroids by Takasato et al. They created hiPSC-derived kidney organoids with cells of different types like endothelial, stromal, Bowman's capsule, and uterine bud, by first subjecting 2D plane-induced culture and then allowing stem cells to assemble in 3D to form organoids. However, kidney organoids made with these procedures have low reproducibility and a lot of inter-group heterogeneity (Takasato et al., 2016). However, 3D bioprinting technology has made it easier to create functional kidney organoids and use them in clinical settings (Pregosin et al., 2021). In a different study, Singh and colleagues generated an in vitro model using HUVECs and renal PTECs implying, core-shell bioprinting to 3D bioprinted kidney organoids. The core of this structure contains pluronic F127 and the external shell kidney-derived cells containing bioink.

The printed proximal tubule was cultured for 28 days. Albumin resorption, epithelium, and endothelium permeability assay, western blot assay, immunofluorescence, and immunohistochemical staining show that the bioprinted organoid mimics the in vivo microenvironment of renal tubules. The renal tubes were implanted into the renal subcapsular space of immunocompromised mice. The graft was harvested 8 weeks post-operation, and the explant maintained its structure by the expression of functional kidney markers. (Singh et al., 2020). Homan et al. constructed a 3D proximal tubule of the nephron. Firstly, a silicon gasket was printed on a glass slide, which marks the outer border. ECM composed of gelatin-fibrin hydrogel was deposited on the gasket. Fugitive ink was used to print the proximal tubules, which were removed from the final construct to form the tubular structure. Two types of bioinks were used in this process: first, a silicone ink mixed with a catalyst was used to print the chip gasket, and the fugitive ink, composed of pluronic F127 and thrombin, was used for bioprinting the proximal tubule. Gene expression analysis, cytokine analysis, albumin uptake, and drug toxicity testing using cyclosporin A diffusional permeability measurement was done on the bioprinted organoid along with electron microscopy and immunostaining to characterize the bioprinted organoid (Homan et al., 2016). Lin et al. have generated a 3D vascularized proximal tubule model having epithelium and endothelium cells lined in adjacent conduits, which are being embedded in the ECM using the sacrificial bioprinting method. The 3D bioprinted proximal tubule performs efficient renal absorption due to the exchange between tubules and vascular channels. This 3D bioprinted model exhibited excellent solute reabsorption. This model was used to quantify glucose and albumin reabsorption over time. This organoid model can be utilized in vitro to examine kidney function and kidney disease, and for testing drug toxicity. (Lin et al., 2019). In another study by Shelby and the group, they developed a 3D bioprinted proximal tubule model using primary human proximal tubule cells, endothelial cells, and renal fibroblasts suspended in thermo-responsive bioink. The bioprinted organoid showed a microvascular network and expressed renal influx and efflux transporters on the epithelial cells. The bioprinted organoid was used to study nephrotoxicity due to cisplatin, and the proximal tissues showed loss of epithelial tissue viability in a dose-dependent manner (Shelby et al., 2017).

7.7.1 Characterization of Kidney Organoid

The bioprinted proximal tubule could be characterized using light and electron microscopy techniques like confocal microscopy, SEM, and TEM. The proximal tubule endothelial cells (PTECs) assembled to form densely packed columnar renal tubules that could be evidently seen in TEM images of the tubule cross-section (Homan et al., 2016). Single-cell RNA sequencing and gene expression studies could also be done to characterize the kidney organoids. Sucher et al. used HSI for real-time assessment of allo-transplanted human kidney function and for checking donor kidney quality. Images of components of the kidney allograft were analyzed using HSI acquisition software, which was capable of providing information regarding tissue oxygen saturation level, water concentration, and hemoglobin concentration, thus allowing viability assessment of the kidney (Sucher et al., 2022). Similarly, HSI could be used as a characterization technique for real-time assessment of kidney organoid functioning and to develop a closely mimicking organoid model, which can be utilized for studying disease conditions, and for drug testing and drug discovery. Best et al. have shown the use of HSI for renal tissue oxygen level mapping to monitor renal oxygenation during nephrectomy. Visible light of 520–645 nm wavelength, corresponding to the oxyhemoglobin spectrum, was used to illuminate the kidney, and the reflectance images captured were processed digitally to ascertain the oxyhemoglobin percentage of each pixel. The renal oxygenation levels at baseline were compared with postoperative renal oxygenation levels, thus helping in predicting the renal insufficiency of patients undergoing partial nephrectomy (Best et al., 2013). Therefore, these characteristics of HSI can also be used for the assessment of 3D-printed renal organoids and to help maintain the tissue-specific micro-environment of the bioprinted organoids.

7.7.2 Application

Disease modeling: The chronic kidney disease (CKD) is caused due to lifestyle preferences and aging. Around 25% of CKD patients have a hereditary predisposition, and mutations in about 200 genes are associated with CKD. Recently, several researchers have started using kidney organoids to study disease pathophysiology associated with CKD. Jansen et al. studied COVID-19-associated CKD and acute kidney injury (AKI) in kidney organoids and showed that kidney cells are infected by SARS-CoV-2, which leads to the activation of profibrotic signaling pathways, causing fibrosis in the kidney (Jansen et al., 2022). Fabry disease is caused by a mutation in the galactosidase alpha (GLA) gene. It is an X-linked lysosomal disorder that leads to lipid deposition in tissues. Kim et al. modeled galactosidase alpha knockout (GLAKO) organoids to study the disease. They used hiPSCs to derive GLAKO kidney organoids, and the organoids showed less expression of tubular and podocyte markers, apoptosis, increased oxidative stress, and deformation in nephrons (Kim et al., 2021).

Drug and toxicity screening: It is easier to control the experimental conditions in organoid models as compared to animal models, therefore, organoids increase the chances of reproducing the drug screening results. Levitt et al. studied Mucin 1 kidney disease (MKD) in kidney organoids to check the role of BRD4780 (small molecule) in treating the disease. They showed that the response of BRD4780 in kidney organoids was similar to that in the mouse model. It reversed proteinopathy and cleared mucin from the kidney (Levitt et al., 2019). Lawrence et al. derived kidney organoids from HMOX1-mCherry-reporter hiPSCs and studied drug-induced toxicity. They identified HMOX1 as a stress marker that is expressed in kidney organoids in response to oxidative stress. They showed that the organoids responded precisely to toxicants in a dose-dependent way. The intensity of m-Cherry expression determines the stress level in organoids (Lawrence et al., 2022). These types of reporter organoid models are very helpful in screening drugs that cause nephrotoxicity.

Regenerative medicine: Kidney diseases such as polycystic kidney disease (PKD) and acute or chronic renal failure are responsible for the global burden on the healthcare system. The treatment strategy for these problems includes kidney transplantation and dialysis, but, due to the lack of donors and the use of immunosuppression drugs after transplantation raise the challenges in the treatment of these kidney diseases. Regenerative medicine aims to completely establish damaged human tissues or organs and to restore their normal structure and function. Human stem cell-derived kidney organoids, when transplanted into animal models, showed that after a few days host blood vessels were observed in organoids even without adding any external vascular growth-promoting substances. The study indicated that transplantation of kidney organoids into kidney capsules of rats showed remarkable improvement in renal morphogenesis (Bantounas et al., 2018).

7.8 Retinal Organoid

The retina consists of photoreceptor (PR) cells (rods and cones) and a layer of neurons of 0.25 mm thickness present at the posterior part of the eye. It is responsible for receiving and organizing visual cues and sending the signals to the brain for interpretation (Masland, 2012). Rods and cones are light-sensitive cells present in the eye, which convert light into neural signals; rods are responsible for dim light vision, and cones enable color vision. The different types of neuronal cells present in the retina are bipolar, retinal ganglion, horizontal, and amacrine cells. The bipolar cells are responsible for transmitting the signals from the PR to the ganglion cells, while the retinal ganglion cells relay the signals from amacrine and bipolar cells to the brain through the optic nerve, whereas the horizontal cells are responsible for regulating the signals that are emerging from the rods and cones. Moreover, the amacrine cells reach several bipolar cells, regulating the signals reaching the retinal ganglion cells. There is a specialized monolayer of pigment epithelium of cells, known as retinal pigment epithelium (RPE) cells, present under the retina. RPE cells provide physical support, nutrients, and growth factors to the retina, thus providing a sustainable microenvironment for retinal cells (Grossniklaus et al., 2015). Defects or degeneration in the retina can lead to different types of diseases, like glaucoma, retinitis pigmentosa (RP), and age-related macular degeneration (AMD). These retinal diseases can ultimately cause blindness. Emerging 3D bioprinting technology can be utilized to print scaffolds matching the characteristics of the native retina. This approach can help reverse the native retinal condition. In a study by Wang et al., a 3D bioprinted retinal organoid was fabricated to mimic the native retinal environment. They used HA hydrogels, which were chemically altered to attain the compressive modulus matching that of the native retina. The retinal scaffold was bioprinted using HA, and the retinal progenitor cells (RPCs) co-differentiated into PRs. The RPE cells supported the culture system due to their regeneration potential (Wang et al., 2018).

The retina is structurally very complex tissue having several layers with different types of cells. 3D bioprinting can help in providing a precise scaffold, and RPCs can be seeded on them to achieve the spatial arrangement of cells in the retina. Lorber et al.

have investigated a piezoelectric inkjet bioprinter's possibility of bioprinting RGCs and retinal glial cells, and the impact of inkjet bioprinting on cell growth and survival. Their study concluded that RGCs and glial cells can be successfully bioprinted with an inkjet bioprinter, as the viability of these cells was not affected during jet ejection of cells from the printer nozzle and drop formation (Lorber et al., 2013). Thus, RGCs can be bioprinted using an inkjet printer on the desirable scaffold to fabricate retinal organoids.

In another study by Worthington et al., they used two-photon polymerization to fabricate scaffolds with variable pore sizes, hatching types, hatching, and slicing distances. These parameters play a significant role in determining the vertical pore diameter and geometry of the horizontal pore. The scaffolds were seeded with hiPSC-derived RPCs. The cells seeded in the pores with larger diameters revealed long neuronal processes parallel to vertical pores. The study's findings highlight the use of hiPSC-derived RPCs as a strategy for reviving vision in patients suffering from retinal degenerative diseases (Worthington et al., 2017). In the field of tissue engineering of the retina, the development of pluripotent stem cell (PSC)-derived retinal organoids in the culture dish using robotics and 3D bioprinting has provided new insights, addressing the requirement to maintain retinal integrity in order to restore vision (Nair et al., 2021). Cell sources for retinal organoids can be fetal RPCs, hiPSCs, or human ESCs (Singh et al., 2018). The patterning of the early ocular field is regulated by networks of signaling channels and transcription factors that are finely tuned by networks of signaling routes and transcription factor regulatory nodes.

Specification of forebrain neuroectoderm leads to retinogenesis. To form the optic cups, the rostral neuroectoderm undergoes significant morphological changes. To generate retinal progenitor lineage, extrinsic chemical agents are used to regulate particular signaling pathways for in vitro retinal cell differentiation (Arthur et al., 2022). For retinal tissue engineering, both natural biomaterials, like collagen, alginate, gelatin, and laminin, and synthetic biomaterials, like poly (lactic-co-glycolide acid), poly (l-lactic acid), polydimethylsiloxane, etc., can be employed as a scaffold. On the other hand, synthetic scaffolds can be regulated to have better mechanical qualities, and a longer shelf life (Nair et al., 2021). Chichagova et al. have described a protocol that can be used to develop retinal organoids from hiPSCs. These organoids are sensitive to light and contain all of the significant retinal cell types. Single-cell suspensions of hiPSCs were seeded on 96-well plates and formed spheroids, which then transformed into retinal laminae over time. To ensure proper maturation and organization of retinal cell types, retinal organoids were cultured for 22 weeks. The anticipated alterations in retinal organoids were monitored using morphological observations and immunocytochemistry to determine protein localization. The organoids formed a thick layer of neuroepithelium comprising PRs, bipolar, amacrine, horizontal, Muller, and retinal ganglion cells after 22 weeks in culture (Chichagova et al., 2019).

7.8.1 Characterization of Retinal Organoid

Yuntian Xue and colleagues employed hyperspectral imaging to characterize structural and molecular alterations in retinal organoids. Retinol accumulation is a characteristic of PR cells that are functioning. Consequently, Hspec, which is capable of distinguishing between distinct chemical species, was utilized to monitor the alterations in the retinal organoids, which were validated by single-cell RNA sequencing (sRNA-seq), qPCR, and immunohistochemistry (Xue et al., 2019). Browne et al. used HSI to study the structure and function of hiPSC-derived retinal organoids. HSI can provide information regarding the metabolic function and structure of the retinal organoid during different stages of growth. HSpec is used for the characterization of retinal organoids, because it enables monitoring of the metabolic changes due to the differentiation of PRs. It detected free NADH, retinoic acid, and retinol accumulated during PR genesis in the organoid. Thus, HSI facilitates continuous evaluation of the retinal organoid's structure and function (Browne et al., 2017). Harvey and colleagues also used HSI for the detection of retinal disease in patients (Harvey et al., 2002). Similarly, we can use HSI to monitor the degeneration of PRs in the 3D bioprinted retinal organoid.

7.8.2 Application

Drug screening: Retinal organoids are used for drug screening and for developing strategies for the treatment of retinal diseases.

Retinal organoids derived from hiPSCs are a better model for drug screening, as they can be generated in large numbers, and they are more similar to human retinas. Dorgau et al. developed retinal organoids from hPSC-derived cell lines and validated the presence of bipolar and retinal ganglion cells through single-cell RNA sequencing. These organoids were exposed to drugs such as thioridazine, sildenafil, and digoxin, resulting in cell death of PRs, whereas all other cell types of the retina, including muller glia cells, were affected by digoxin and thioridazine. Dendrites-like structure emerged from the neuroepithelium, indicating astrocyte activation. The number of active retinal ganglion cells decreased in numbers due to drug exposure, but the potential of the retinal organoids to respond to light stimuli was conserved. The study indicates that retinal organoids are suitable model for drug toxicity studies (Dorgau et al., 2022).

Inherited retinal disease model: Retinal organoids derived from human PSCs serve as a model for most of the inherited retinal diseases. It helps in understanding the mechanism of these diseases and in developing their related therapies. Retinal organoids derived from a patient's PSC are a powerful model to study Leber's congenital amaurosis (LCA-2), which is a rare inherited retinal disease caused by a deficiency in the RPE65 gene, RP, and some other ocular diseases, such as Stargardt disease, microphthalmia, glaucoma, macular telangiectasia type 2, X-linked juvenile retinoschisis, and retinoblastoma (Zhang et al., 2021).

Retinal cell therapy: Retinal organoid development is now a convenient strategy and a retina-like structure with cell–cell connections is formed between different types of cells. These therapies can be more patient-specific due to the use of hiPSCs from the patients, thus making them more physiologically relevant as compared to animal models. Iraha et al. transplanted retinal organoids into immunodeficient mouse models with end-stage retinal degeneration. The immunostaining studies suggested that the retinal organoid integrates and forms synapses with the host retina. It also displayed its retinal function as confirmed by recording the light response from ganglion cells using a multi-electrode array set-up (Iraha et al., 2018).

Retinal gene therapy: Retinal organoids are used as a reliable model for preclinical gene therapy. The viral mode of gene therapy mediated by AAV has gained a lot of success in the recent past after the augmentation of the RPE65 gene for treating LCA-2. The viral mode of gene therapy involves the infection of the virus in the targeted cells and the efficient expression of the gene upon infection. Retinal organoids can be utilized to study the effectiveness of AAV-mediated infection in the retina. In addition to this, retinal organoids can also be used to optimize the delivery of genes of interest by the non-viral gene delivery method (Juttner et al., 2019).

7.9 Types of Bioink and Its Role in Generating Organoids for Different Applications

The phrase "bioink" was originally used for organ printing in 2003, coupled with the term "biopaper," which was later described as a hydrogel. Initially, bioprinting was referred to as the printing of biopaper (hydrogel) by seeding the cells or spheroids into it and culturing them in a 3D orientation (Groll et al., 2019). By definition, bioink is "a formulation of cells suitable for processing by an automated biofabrication technology that may also contain biologically active components and biomaterials" (Groll et al., 2019). Whitford and Hoying defined bioinks as a fluid deposited by 3D bioprinters containing the matrix or the nutrients needed for cell growth and survival (Whitford and Hoying, 2016). An ideal bioink should consist of polymer hydrogels or other biomaterials, cells, biomolecules, and additives that promote cell growth in the matrix. Bioinks are the raw materials of the bioprinting process, which are loaded with live cells (Hospodiuk et al., 2017). There are many biomaterials that are printable and then subsequently loaded with cells; these biomaterials are collectively termed biomaterial inks (Groll et al., 2019) with properties like biocompatibility, bioactivity, cytocompatibility, etc. (Murphy et al., 2014; Hölzl et al., 2016; Gopinathan et al., 2018). On the basis of the materials used, bioinks are broadly divided into two broad categories: cell-scaffold-free bioinks and cell-scaffold-based bioinks (Kaushik et al., 2016).

7.9.1 Cell-Scaffold-Free Bioinks

Cell-scaffold-free bioinks generate cell mass or aggregates without the addition of any other foreign substance, so they are highly biocompatible, and they secrete the ECM with proper cell–cell interactions. They also facilitate the migration or assembly of cells in a 3D microenvironment that closely mimics the tissue microenvironment (Yut et al., 2016; Alblawi et al., 2020). Using bioinks made of cellular aggregates in the shape of sheets, spheroids, pellets, or tissue strands, scaffold-free bioprinting creates complex biological structures (Ji and Guvendiren, 2017). The major advantage of using cells and bioinks is that it reduces the time for tissue fabrication, probably due to cell adaptation, proliferation, and migration. Cell–cell interaction plays a significant role in the organization of the cells in this method (Ovsianikov et al., 2018). Different types of cell-scaffolds free bioinks are described in Table 7.2.

7.10 Cell-Scaffold-Based Bioink

Bioinks based on cell scaffolds, predominately consist of polymeric and hydrogel materials, and have the tendency to immobilize the live cells for getting 3D-printed. Hydrogels are 3D hydrated structures synthesized from water-soluble synthetic and natural polymers (Lei and Wang, 2016) and contain a highly porous structure that mimics the natural tissue environment. These hydrogels are found to retain a significant amount of water and thus serve as an ideal candidate for application as bioinks in bioprinting (Raza et al., 2018; Bella et al., 2015; Zhou et al., 2019). The hydrogels which are used frequently in tissue engineering applications, generally fall into three major categories based on different types of polymers employed: natural hydrogels, synthetic hydrogels, and a combination of the two for generating the third type (Buwalda et al., 2014). In cell-scaffold-based bioinks, synthesis is performed by blending the high-density cell suspensions in the precursor polymer solution, so that the cells get entrapped within the polymer network, which is either crosslinked before bioprinting or sometimes cross-linked during the process of bioprinting. So, the hydrogels with live cells entrapped in them are called cell-laden hydrogels or cell-laden scaffolds, which can be used for bioprinting. Some of the important categories of cell-scaffold-based bioinks are generated with different individual biomacromolecules or their mixtures, as discussed here.

Table 7.2 Different types of cell-scaffolds-free bioinks

S. No.	Type	Definition	Cell used	References
1.	Tissue spheroids	Densely packed cellular aggregates	MSCs	Lewis et al., 2016
	Magnetic levitation method	External magnetic force is used to trigger the cell spheroid formation	NIH 3T3 mouse fibroblasts and HCC827 non-small-cell lung cancer cells	Türker et al., 2018
	Hanging drop method	Surface tension and gravitational force principles are used for forming cell spheroids	Pancreatic cells	Ware et al., 2016
			Primary buffalo and sheep hepatocytes	Shri et al., 2017
	Hydrogel microarrays	Cells are initially seeded onto the hydrogel microwells which get transformed into cell spheroids	Human adipose-derived stem cell	Tsai et al., 2019
			Human breast adenocarcinoma cells and glioblastoma cells	Seyfoori et al., 2018
	Spinner flask	Culturing of the cells is done in a spinner flask with fixed mechanical spinning, which in turn generates the cell spheroids	Bovine articular chondrocytes (bACs) and rabbit bone marrow-derived mesenchymal stem cells	He et al., 2019
			Umbilical cord tissue-derived mesenchymal stromal cells.	Santos et al., 2015

	Microfluidic systems	A combination of microfluidic and 3D bioprinting cells are directly fabricated in controlled conditions even at complex architectures	Stem cells, carcinoma cells, and fibroblasts cells	Patra et al., 2013
			Primary lung cancer epithelial cells and primary pericytes	Ruppen et al., 2015
2.	Cell pellet techniques	Pelleting down of the cell suspension using centrifugation and the resulting cell pellet is transferred into the suitable mold using 3D bioprinting where cell–cell interaction happens	Bone marrow stem cells and Schwann cells	Owens et al., 2013
3.	Kenzan method	Kenzan arrays of stainless-steel needles are used as support to form cell aggregates	Human-induced PSCs	LaBarge et al., 2019; Gil et al., 2017
			Cardiomyocytes, endothelial and fibroblast cells	Ong et al., 2017
4.	Tissue strand-based bioink	In this method, the highly dense population is injected into an alginate tube, which has predefined mechanical properties and porosity. This alginate tube is digested to get a cell aggregate or tissue strands which are in turn used as bioinks	Human adipose-derived stem cells	Wu et al., 2019
			Human umbilical vein smooth muscle cells	Yu et al., 2014

7.10.1 Collagen-Based Bioink

Collagen is the most prevalent protein in mammals, constituting between 25 and 30% of the mammalian body mass, and is a principal component of the ECM (Rodriguez-Pascual and Slatter, 2016; Ferreira et al., 2012). Collagen is known to be highly biocompatible and biodegradable and induces less immunogenicity in vivo (Parenteau-Bareil et al., 2010; Glowacki et al., 2010). There are around 20 different types of collagens identified so far, but only collagen type I is commonly employed in tissue engineering (Abraham et al., 2008; Ahn et al., 2010; Park et al., 2014). Collagen has been reported to increase cell growth, attachment, and adhesion and qualifies as an excellent scaffold material. Smith et al. experimented with bioprinting of aortic endothelial cells of bovine origin by suspending them in collagen gels (Smith et al., 2004). Cultured primary human dermal fibroblasts and epidermal keratinocytes cells were added to a solution of collagen and used as a bioink (Lee et al., 2013). Though collagen has lots of advantages, it has its limitations due to its slow gelation tendency. It takes 30 mins at 37°C for it to complete crosslinking, which makes its application in 3D bioprinting, challenging (Mori et al., 2013).

7.10.2 Alginate-Based Bioink

Alginate is a polysaccharide polymer derived from brown algae that is composed of 1/4-linked b-D-mannuronic acid (M) and its C-5 epimer a-L-guluronic acid (G) (Pawar et al., 2012). The G units present in the alginate are responsible for the gel-forming ability of the alginate (Lai et al., 2016). Most frequently, alginate is employed as natural hydrogel, which is highly biocompatible, economical, and available for different crosslinking entities (Gudapati et al., 2016). Calcium ions facilitate the cross-linking of alginate in the presence of alginate chains. This convenient and simple method of crosslinking makes it an ideal candidate for bioprinting (Jin and Dijkstra, 2010). In a study by Zhang et al., chondrocytes were mixed with alginate gels and used as a bioink to print hollow vessel-like constructs (Zhang et al., 2013). Christensen et al. used sodium alginate gels to entrap mouse fibroblast cells and print them to get vascular-like constructs using calcium chloride crosslinking (Christensen et al., 2015). The major limitation of alginate bioinks is that they do not have an innate mammalian enzyme to degrade them, and they also have limited cell attachment properties (Nair and Laurencin, 2007).

7.10.3 HA-Based Bioink

Hyaluronic acid is a non-sulfated, linear GAG (Zhang et al., 2009; Yoo et al., 2005). Because of its properties such as biocompatibility, biodegradability, nonadhesiveness, and non-immunogenicity, it is of great use in tissue engineering. It can also form flexible hydrogels (Camci-Unal et al., 2013; Luo et al., 2000). Despite these advantages, HA faces some major limitations such as high molecular weight and the formation of high viscous solution, so it cannot be directly used for 3D bioprinting. Generally, it requires chemical modification and combination with another polymer to improve its physical properties and get used as a bioink (Skardal et al., 2010). Methacrylate-based chemically modified HA hydrogels are known to generate bioinks which could increase the osteogenic properties of the seeded cells (Poldervaart et al., 2017). HA-gelatin based hydrogels were fabricated by incorporating ruthenium-based complexes to facilitate visible light-induced polymerization and to increase human adipose stem cell viability and differentiation (Sakai et al., 2017).

7.10.4 Fibrin-Based Bioink

Blood contains the fibrous protein molecule known as fibrin, which plays a critical role in blood clotting. Fibrinogen is enzymatically cleaved by thrombin to get fibrin, which self assembles to form a straight chain product (Nair and Laurencin, 2007). Due to its high biocompatibility and biodegradable nature, it is used for bioink preparation, but it has poor mechanical properties. In one of the cases, a fibrinogen-coated surface was printed with human microvascular endothelial cells after being mixed with a thrombin solution, which quickly created cell-laden fibrin constructs (Cui and Boland, 2009).

7.10.5 Multicomponent Bioink

Multicomponent bioinks are a combination or blends of two or more individual bioinks. Many of the natural polymeric hydrogels, which are used individually as bioinks, may be biocompatible and biodegradable, but most of them suffer from one or more limitations, such as poor mechanical properties, poor cell attachment, proliferation, and immunogenicity, etc. (Hospodiuk et al., 2017; Fedorovich et al., 2008). In order to improve the mechanical and other biological properties, a single material-based bioink is upgraded to a multicomponent bioinks. Table 7.3 explains different multicomponent bioinks that have been used in bioprinting.

7.11 Types of 3D-Bioprinting

The generation of organoids to the best possible optimization has been achieved by 3D bioprinting technology. It involves integrating living cells with biomaterials in order to generate drop-by-drop or layer-by-layer deposition of cells to achieve the structural properties of multifunctional tissues (Ramadan et al., 2021). 3D bioprinting enables additive manufacturing of living tissues with the help of CAD. This technique enables the controlled fabrication of 3D structures (Ma et al., 2018). 3D Bioprinting allows the use of varying combinations of biomaterials, cell types, as well as supporting growth factors. The bioprinted organoids can be used for regenerative therapy, drug discovery model, and for toxicology studies for newly discovered drugs (Murphy et al., 2014). The general approach of bioprinting is biomimicking, independent self-assembly, and aggregation of small tissue-based methods (Bishop et al., 2017). The biomimetic approach utilizes a single component of tissue, such as the sheer volume and native function of cellular interactions. There are also various other factors, such as types of cells, ligands, and structural components of tissue, and various types of environmental factors, such as pressure, temperature, and electrical components, etc., which affect the outcome.

3D bioprinting is majorly categorized into three steps: pre-processing, processing, and post-processing. The first step in pre-processing is to analyze the anatomical structure of a tissue using imaging methods like CT and MRI, and then store the results in an STL file using CAD. The organ blueprint is obtained from these imaging techniques. Then, a tissue construct is obtained using various bioprinting methods, such as laser-assisted 3D bioprinting, inkjet 3D bioprinting, micro-extrusion 3D bioprinting, or stereolithography-based bioprinting (Cadena et al., 2020). The processing step includes the addition of bioink comprising growth factors and cells. In post-processing, the constructed tissue is obtained with the help of in vivo environmental conditions in bioreactor maturation (Fig. 7.5). Bioprinted tissue can be used in regenerative medicine, such as organ transplantation or tissue repair (Ramadan et al., 2021).

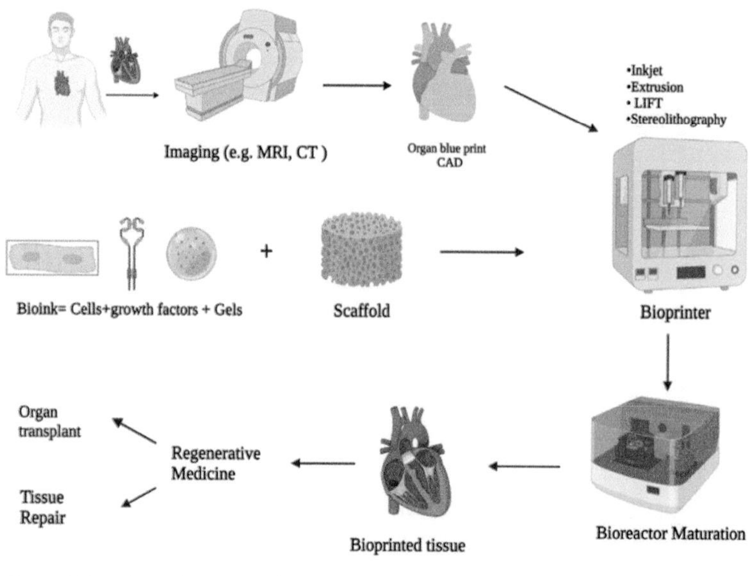

Figure 7.5 Overview of 3D bioprinting of organoids. Imaging of the tissue of interest is done to derive the CAD model. The bioink contains a combination of biomaterials, cells, and growth factors to bioprint the organoids. The scaffold is bioprinted using the CAD model and desired bioink.

Table 7.3 Multicomponent bioinks used in bioprinting

S. No.	Types	Multicomponent bioink	Cells used	Properties	References
1.	Natural-natural polymers	Alginate-gelatin	Primary myoblast	Better cell growth with better mechanical properties	Chung et al., 2013
2.	Natural-natural polymers	Alginate-gelatin	Human mesenchymal cells	Faster crosslinking and long-term crosslinking	Wüst et al., 2014
3.	Natural-synthetic polymers	Alginate-polycaprolactone	Osteoclast and chondrocytes	Improved mechanical properties	Shim et al., 2012
4.	Natural-natural polymers	Alginate-gelatin-agarose	SaOS-2 cells	High cell proliferation and increased cell mineralization	Neufurth et al., 2014
5.	Natural-synthetic polymers with inorganic component	Alginate-polyvinyl alcohol-hydroxyapatite	MC3T3 cells	Optimal rheological properties and increased cell viability	Bendtsen et al., 2017
6.	Natural-natural polymer	Silk fibroin-gelatin	Human turbinate mesenchymal stromal cells	Balanced mechanical properties with optimal degradation rates	Das et al., 2015
7.	Natural-natural polymer	Silk fibroin-gelatin	Chondrocytes and mesenchymal progenitor cells	Increased cell viability and proliferation	Chameettachal et al., 2016

8.	Natural-natural polymer	Agarose-Collagen I	MSCs	Improved mechanical property and cell viability	Duarte Campos et al., 2016
9.	Natural-natural polymers	Agarose-collagen with sodium alginate	Chondrocytes	Enhanced mechanical properties	Yang et al, 2018
10.	Natural-natural polymers	Agarose - alginate-carboxymethyl-chitosan	hiPSCs or human-derived neural cells	Stable 3D constructs	Gu et al., 2016
11.	Natural-natural polymers	Chitosan-gelatin	Fibroblast skin cells	Enhanced 3D shape fidelity and good room-temperature printability	Ng et al., 2016
12.	Natural-natural polymers	Nanocellulose/alginate	Chondrocyctes	Proper curing and increased shape fidelity	Siqueira et al., 2017
13.	Natural-natural polymers	Nanocellulose/HA	Adipocytes	Increased cell viability	Henriksson et al., 2017
14.	Natural-synthetic polymers	Alginate-GelMA	HUVECs	Enhanced mechanical strength	Liu et al., 2017
15.	Natural-synthetic polymers	GelMA-alginate-PEG acrylate	HUVECs and human smooth muscle cells	Enhanced mechanical strength, cell growth, and stability	Pi et al., 2018
16.	Synthetic polymer with inorganic molecules	Poly (ethylene glycol) methacrylate, bioactive glass nanoparticles, and hydroxyapatite	Human MSCs	Higher cell viability and high compressive modulus	Gao et al., 2014
17.	Combination of natural polymers	Decellularized ECM	Human kidney cells	Provide tissue specific microenvironment, so improved cell viability and proliferation	Ali et al., 2019
18.	Combination of natural polymers	Decellularized extra cellular matrix	Cardiac progenitor cells	Improved vascular formation	Jang et al., 2017

7.11.1 3D-Bioprinting Techniques

Bioprinting technology to fabricate different types of tissues for eventual clinical use (Kengla et al., 2015). Some of the major types of 3D-bioprinting methods are: inkjet 3D bioprinting, laser-assisted 3D bioprinting, micro-extrusion 3D bioprinting, and stereolithography-based bioprinting (Fig. 7.6). Each method requires specific characteristics of the bioinks which are to be used for the construction of 3D tissue. Shear rheology and cross-linking are the physiochemical properties of the bioink, which influence the printability, whereas nozzle properties decide the shape and size resolution of printability (Kengla et al., 2015).

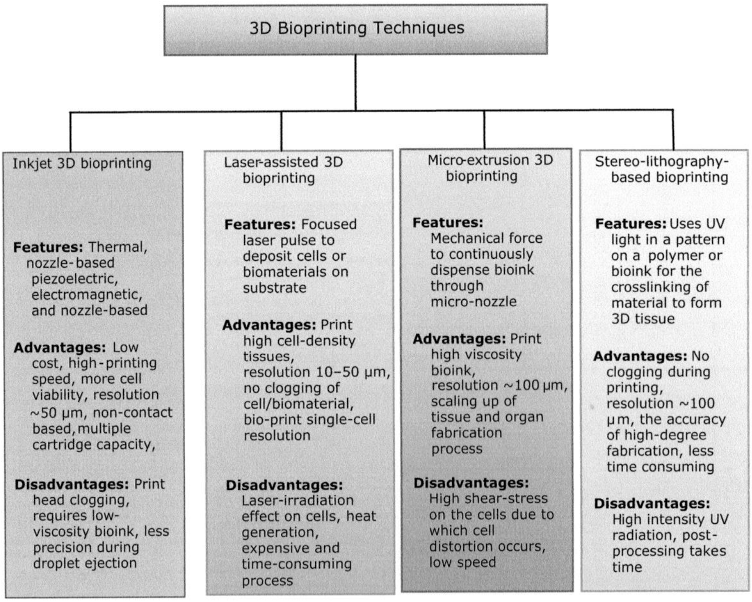

Figure 7.6 Different types of 3D bioprinting technologies used for bioprinting organoids along with their features, advantages, and disadvantages (Rawal et al., 2020; Bishop et al., 2017).

In 3D bioprinting of organoids, cell viability, printing speed, and surface resolution are the most important factors for biological materials. The 3D bioprinting technologies, which are described in Fig. 7.6, have different printing processes, printing speeds, resolution, and post-printing cell viability (Venkataraman et al., 2018).

7.11.2 Inkjet 3D Bioprinting

Inkjet 3D bioprinting is an improvised version of 2D ink-based printing in which ink is replaced by bioink and consist of cells, growth factors, and gels. The ink used in the inkjet 3D bioprinter is called bioink, and it consists of living biological materials, such as various cell types, growth factors, and polymers (natural or synthetic). This method is also known as droplet-based bioprinting, as droplets of bioink contain cells (10,000–30,000), and energy is provided either by the heater or piezoelectric through a non-contact nozzle. The resolution is this process is generally about 50 μm. It is known to have 3D freedom fabrication and multiple cartridge capability. The key features include high speed and low cost. At the end of the process, more than 85% of cells are found to be viable. Murphy et al. used inkjet bioprinting to show in situ regeneration of functional cartilage and skin (Murphy et al., 2014). Inkjet bioprinting has some limitations, including the requirement of low viscosity bioink and the biological sample in liquid form to enable droplet formation (Trivedi et al., 2021), whereas nozzle clogging and cellular distortion are always possibilities in this technique.

7.11.3 Laser-Assisted 3D Bioprinting

Direct laser pulses are used in laser-assisted 3D bioprinting (LAB), which delivers controlled volumes of liquid to a predetermined location through a ribbon containing bioink. The ribbon is supported by a metal layer (Ti or Au), which absorbs and aids in the transfer of energy to the ribbon (Bishop et al., 2017). It is a non-contact and nozzle-free printing process, having a resolution in the range of 10-100 μm. More than 95% of cells are generally found to be viable post-printing process. It has shown successful compatibility with DNA as well as cells. The performance of laser-assisted 3D bioprinting depends on different factors such as substrate wettability, surface tension, thickness, and viscosity of biological materials (Trivedi et al., 2021). The major limitations of this technology are the laser irradiation effect on cells, along with heat generation.

7.11.4 Micro-Extrusion 3D Bioprinting

Micro-extrusion 3D bioprinting involves printers that are robotically controlled and can extrude bioink through the nozzle in a continuous stream by using mechanical force. The resolution of extrusion-based bioprinting is 200–500 μm. It is capable of printing bioink with a high viscosity, complex polymers, cell spheroids, and clay-based substrates (Bishop et al., 2017). It enables the printing of more viscous hydrogels, providing strong mechanical support in the final construction of tissue (Ma et al., 2018). It shows good structural integrity and also has the potential to scale up the generation of tissue. This technique is also called direct ink writing (DIW) and is one of the most commonly used methods. Around 40–80% of cells are found viable post-printing by this method. In this method, culture dish (solid) and materials derived from hydrogel or growth medium (liquid) can be used as the substrate. Shear rheology (viscoelasticity) and shear thinning of hydrogel or polymer play an important role in designing 3D models of organoids by this method (Trivedi et al., 2021). This technique has a major limitation, as it causes high stress on cells, which affects cell viability and proliferation with induction of cell death (Ma et al., 2018).

7.11.5 Stereolithography-Based Bioprinting

UV light is focused in a pattern on the bioink or photopolymerizable polymer in stereolithography-based bioprinting, resulting in the polymer's cross-linking into a solidified 3D tissue construct (Ramadan et al., 2021). It consists of laser-based stereolithography apparatus and projection-based digital light processing (Trivedi et al., 2021). Each layer is polymerized, so multiple cycles take place to form a 3D structure of tissue. At the end of the process, ~85% of cells are viable post-printing. This method is generally associated with a high resolution range of ~200 nm–6 μm. It has a high fabrication accuracy, takes less printing time, as well as fewer chances of clogging during printing. This technology employs a laser to activate the polymerization of photocross-linkable molecules (Carpio et al., 2021). This technology also has limitations, such as the cytotoxic nature of the reassortment of cells in precursor suspension for long-term printing using cross-linking materials (Gudapati et al., 2014).

If cells are exposed to intense UV radiation, the process can cause damage to cells (Ramadan et al., 2021).

7.12 Current Challenge in 3D Bioprinting Technology

3D-bioprinting technology includes several challenges despite significant progress. In tissue modeling, it is necessary to overcome obstacles in the pre-printing, printing, and post-printing phases (Table 7.4) (Almela et al., 2021). The pre-printing phase consists of cell isolation, expansion, and differentiation. It is impossible for a single elemental material to possess all the characteristics required to reconstruct the functions of tissue (Mao et al., 2020). This strategy plays a crucial role in improving fabrication efficiency and reducing fabrication costs, although there is a need for the determination of appropriate sites for implantation, connections between adjacent blood vessels or other channels of the body, and long-term integration and function (Murphy et al., 2019). The fabricated model of a 3D-printed organoid is based on protocols to control the differentiation and maturation along the desired lineage. If there is continuous cell proliferation, the material should be degraded to clear the way for cell growth, and the resulting product should have no or minimal side effects.

7.13 Use of Hyperspectral Imaging in Improving 3D Bioprinted Products

Hyperspectral imaging is the combined usage of standard imaging and spectroscopy and offers spatial and spectral data regarding the specimen. In addition to this, the time-dependent behavior of spectra provides information on the chemical changes occurring within a specimen (Edelman et al., 2012). Hyperspectral imaging contributes to noninvasive diseases and clinical interventions. In recent times, emerging technologies with image analysis methods, hyperspectral cameras, and computational technology advancements have included various applications in the medical field. In such procedures, light interacts in the form of absorption, fluorescence, and scattering, and

Table 7.4 Comparison of different properties of 3D bioprinting technology

Properties	Inkjet-based bioprinting	Laser-assisted 3D bioprinting	**Micro-extrusion 3D bioprinting**	**Stereo-lithography-based bioprinting**	**References**
Cost	Low	High	Medium	Low	Zhang et al., 2021
Printing process	Drop-by-drop	Dot-by-dot	Line-by-line	Layer-by-layer	
Resolution	~ 50 μm	10–50 μm	~100 μm	~100 μm	Rawal et al., 2020
Cell density	Low < 10^6 cells/mL	Medium ~10^8 cells/mL	High Cell spheroids	>10 cells/ml	Ren et al., 2021
Post-printing cell viability	>85%	>95%	40-80%	85%	Zhang et al., 2021
Structural integrity	Low	Low	Low	High	Vijayavenkataraman et al., 2018
Bioink viscosity	3.5–12 mPa/s	1–300 mPa/s	30-6 ×10^7 mPa/s	No limitation of viscosity	Zhang et al., 2021
Printing speed	Fast >10^3 droplets/s	Moderate 200–1600 mm/s	Slow 10–700 mm/s	Fast	Zhang et al., 2021

is captured by an imaging spectrometer (Fei et al., 2020). To achieve optimal results, HSI incorporates pre-processing, feature extraction, selection, classification, and unmixing. In the medical field, pixel- and subpixel-based classification are the most common hyperspectral image classification techniques employed. Hyperspectral imaging requires the pre-processing of data such as the calibration of reflectance and the removal of glare. During the calibration of reflectance, the raw data can be transformed into normalized data. The presence of glare does not help in diagnosis, and the analyzed data contains inaccuracies (Lu et al., 2017). The HSI data includes complex matrices with x-, y-, and z-axes with reflected intensity (Fig. 7.7). Imaging diseased models to determine the physiology and biochemistry generally involves the use of a visible range of 400–700 nm and a NIR range of 700–1000 nm (Wahabzada et al., 2017).

There are several advantages of HSI, such as the quick data acquisition ability, fewer chances of human error, and minimal specimen preparation. Especially, this technique is found to help in the detection, visualization, and identification of 3D printed organoids, noninvasively (Edelman et al., 2012). It also helps in accessing proper tissue oxygenation and blood volume maintenance during surgery and provides real-time data (Rehman et al., 2013). Thus, 3D bioprinting techniques in combination with HSI could be explored to access the real-time localization of active pharmaceutical drugs in 3D bioprinted organoids and follow its after-effects, which could inform the requirement of drug dosage according to the needs of the patient (Vakili et al., 2015).

In addition, hyperspectral data along with spatial dimension algorithms can identify a range of patterns, defects, or disease recognition after the bio-printing process. After 3D bioprinting of organoids, HSI can be extended to visualize the organoids at the cell, tissue, and molecular levels and also to investigate the principal components of organoids. It can create a contrast between the specimen and its background. HSI can also help in monitoring the printed organoids to observe abrupt changes during the printing process.

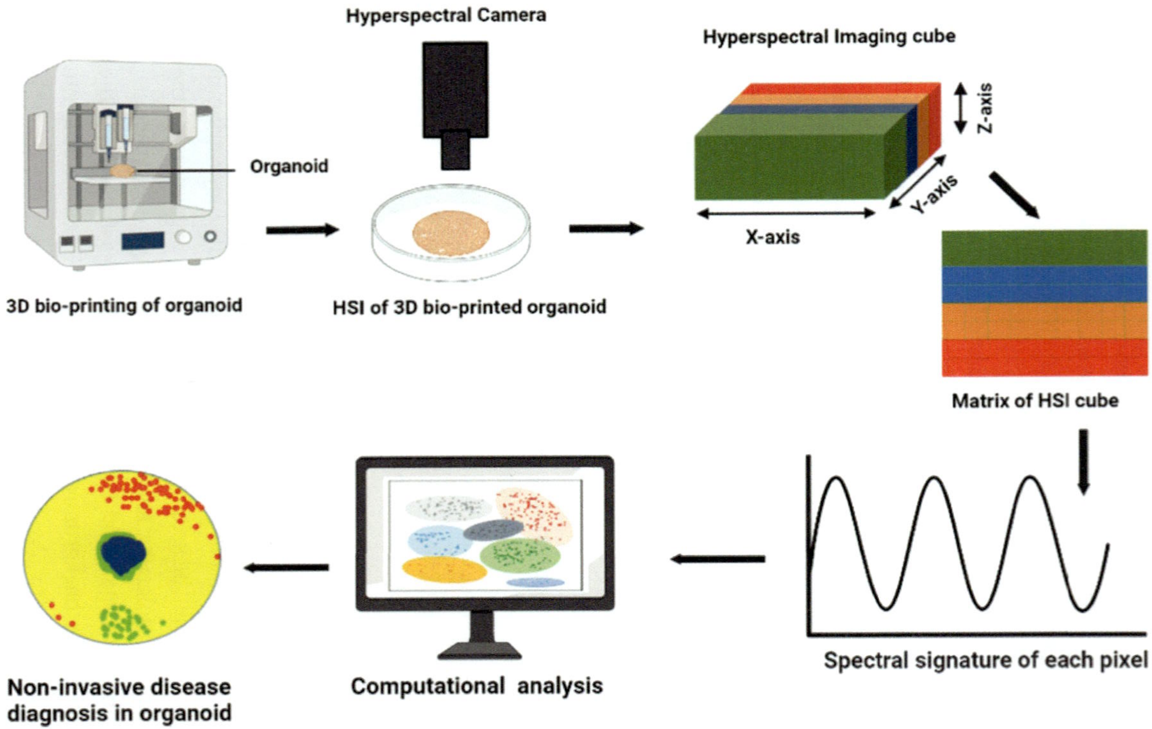

Figure 7.7 Different steps for noninvasive hyperspectral imaging of 3D bioprinted organoids. The hyperspectral camera can be used for mapping different biochemical parameters in the organoid model to generate spectral and spatial information of each pixel to generate a hypercube. Interpretation of the spectral signature of each peak is done through computational analysis.

7.14 Characterization of Organoids

Evaluation of the bioprinted organoid model is necessary to achieve the best outcome. There are several physicochemical and biological techniques employed in previous studies, which can characterize the cells of the bioprinted models either under in situ conditions or after lysing the cells. In situ characterization of cells is done by light microscopy, fluorescence microscopy, electron microscopy, colorimetric methods, tomography, and metabolic flux analysis methods. Techniques that are used to characterize cells after lysing are molecular biology techniques such as western blotting and qRT-PCR to investigate organ-specific gene expression and proteins respectively in the cells of the constructed model. Apart from this, flow cytometry can also be utilized to examine the surface expression of cell-specific markers on organoids (Germain et al., 2022). Some of the important characterization methods are discussed here.

7.15 Scanning Electron Microscopy

The ultrastructure of organoids has been extensively studied using EM techniques such as TEM and SEM (Fei et al., 2022). TEM and SEM analysis of 16 week bioreactor-cultured retinal organoids revealed a layer of PRs composed of inner segments containing mitochondria, connecting cilium (CC), and developing outer segments, as well as an outer limiting membrane (Ovando-Roche et al., 2018). Horsley et al. used SEM to examine distinct multilayered zones of human urothelial organoid formation that had undifferentiated basal cell-like monolayers, disorganized hyperplasia zones, and umbrella-like cells at their surface (Horsley et al., 2018). The scanning electron microscope was employed to assess the liver-derived ECM gels (LEMgel) for essential hydrogel factors such as ultrastructure and pore size (Saheli et al., 2018). The epithelial framework in human airway organoids was visualized in three dimensions using the scanning electron microscope (Sachs et al., 2019) and in kidney tubular epithelial organoids (Schutgens et al., 2019). One of the shortcomings of SEM is that it is only capable of perceiving the surface structure of the sample.

7.16 Transmission Electron Microscopy

TEM excels in its ability to image the intricate details of the structure and stratification of organoids (Fei et al., 2022). Using TEM, Sachs et al. characterized the airway organoids to conform to the polarized, pseudostratified airway epithelial nature containing secretory, multi-ciliated, and basal cells (Sachs et al., 2019). TEM's ultra-fine resolution revealed characteristic structures that optical microscopy could not reveal. Key characteristics of CLN3-JNCL (juvenile neuronal ceroid lipofuscinosis), including an increase in autophagic vacuoles, are reproduced in a human cerebral organoid model (Gomez-Giro et al., 2019). By using TEM, it was possible to see the growth of a ribbon synapse amidst PRs and bipolar cells in the inner plexiform layer, which may indicate that retinal organoids are maturing physiologically (Gonzalez-Cordero et al., 2017; Ovando-Roche et al., 2018). The primary drawback of TEM imaging is the need for extremely thin sections, which are typically between 50 and 100 nm in thickness (Fei et al., 2022).

7.17 Laser Scanning Confocal Microscopy

Laser scanning confocal microscope (LSCM) is the instrument that is most frequently used for imaging organoids (Bremond Martin et al., 2021). This method of characterization creates optical sections of a sample by scanning the sample with a directed laser beam, point-by-point. Pinholes are used to prevent the detection of noise fluorescence (Conchello and Lichtman, 2005). As a result, it is possible to image slices of organoids based on their penetration depth of less than one hundred micrometers and to perform 3D imaging. Human urothelial organoids were characterized by immunolabelling them with different urinary tract-specific antigens and imaging them with high-resolution laser scanning confocal microscopy, which revealed the spatial expression of several markers in particular zones, indicating the differentiation and maturation of the cells in the organoids (Schutgens et al., 2019). Yakoub and Sadek established the presence of three distinct layers of the cerebral organoid, namely the MAP2high/Pax6- layer (preplate/CP), MAP2low/Pax6+ layer (IZ/SVZ), and MAP2-/Pax6+ layer (SVZ/VZ by cryosectioning the organoids followed by immunolabelling and imaging them in confocal microscopy (Yakoub and Sadek, 2018).

7.18 Super-Resolution Fluorescence Microscopy

Abbe's law places a limit on the spatial resolution of standard fluorescence microscopy, with the best resolution being only about the axial and lateral dimensions of 200–300 nm and 500–800 nm, respectively (Abbe, 1873). Super-resolution fluorescence microscopy, on the other hand, can get around the diffraction barrier and achieve nanoscale resolution when imaging living cells (Huang et al., 2009). When this strategy is used for organoid imaging, the spatial resolution increases by an order of magnitude, allowing researchers to observe previously unobserved features of cell structure and even examine single-molecule-level biological processes (Fei et al., 2022).

7.18.1 Stimulated Emission Depletion (STED) Microscopy

LSCM is the foundation of the STED super-resolution imaging technology (Blom and Widengren, 2017). STED microscopy superimposes depletion light on excitation light, causing the two laser sources to overlap and enabling fluorescence depletion, resulting in a significantly smaller effective excitation spot size. STED microscopy has a lateral resolution of approximately 50 nm and an axial resolution of approximately 100 nm. Imaging the intensity distribution of fluorescent samples is STED microscopy's most fundamental application. This method has been utilized to successfully visualize the intensity imaging and quantification of fluorescent nanoparticles and fluorescently labeled organoid samples, as well as the signal intensity of calcium (Sloan et al., 2018) and microglial marker proteins in cerebral organoids (Ormel et al., 2018).

7.18.2 Single-Molecule Localization Microscopy

Stochastic optical reconstruction microscopy and photoactivated localization microscopy (PALM) are the two crucial techniques for single-molecule localization super-resolution imaging (STORM). Specific fluorophores must be used for both of these methods. Using PALM and STORM, the clustering of receptors at the nanoscale (Wnt receptor, LRP6, and EGFR) in colonic organoids is one example of the extracellular components (Michael Salinas et al., 2020).

7.19 Flow Cytometry

Flow cytometry analyzes single cells suspended in buffer solution as they pass through laser. Each particle is analyzed for visible light scatter and fluorescence. Two directions are used to measure visible light scatter: forward scatter (FSC) and 90° side scatter (SSC). The FSC measures cell size, whereas the SSC measures cell complexity or granularity. Fluorescence of the system does not affect light scatter (McKinnon et al., 2018). Organoids from colorectal cancer liver metastases were characterized using flow cytometry, which demonstrated that the organoids displayed widespread occurrence of the epithelial marker EpCAM and adenocarcinoma marker CEA CAM1. The cells in the organoids also expressed stem cell surface markers like CD24 and CD44 (Buzzelli et al., 2018). Flow cytometric analysis was performed on retinal organoids in order to distinguish between developing rods (CD133/CD73 co-labeled cells) and mature PRs. This was done by checking for CD73 and CD133-positive cells in the retinal organoids (Ovando-Roche et al., 2018).

7.20 Biochemical Characterization

7.20.1 Real-Time PCR (RT-PCR)

Real-time polymerase chain reaction (PCR) is frequently the first technique employed to obtain a quick, simple, and quantitative readout of marker genes, indicating cell identity, such as transcription factors and differentiation markers. Quantitative RT-PCR (qRT-PCR) was used to analyze the expression levels of CYP3A4, CYP3A7, TAT, and ALB, which are considered to be the cell maturation markers in liver organoids, and they were found to be upregulated in liver-derived ECM gel organoids (Saheli et al., 2018). Human cerebral organoids were assessed for neuronal maturation markers, such as the glutamatergic receptors (AMPA and NMDA receptors) and transporters (vGluT1 and vGluT2) and the GABAergic receptor (GABBR1, GABA B receptor 1) and transporter (vGAT, vesicular GABA transporter) using RT PCR (Yakoub and Sadek, 2018).

7.20.2 Gene and Transcriptome Profiling

To determine if the organoids are successful, they must contain the desired cell types and closely mimic the functions of the corresponding tissue in vivo (Zhao et al., 2022). Both high-throughput whole-genome transcriptome analysis and low-throughput gene expression validation are typically used to achieve this. It is hypothesized that organoids will exhibit gene expression patterns comparable to the tumor from which they were derived. It has been demonstrated by RNA-seq in different types of tumors, most notably esophageal (Li et al., 2018) and bladder cancer (Yu et al., 2021). Brain organoids in exposure to graphene oxide (50 μg/mL) led to significant upregulation of 80 genes, significantly downregulated 121 genes, and 58,624 genes found to be unaltered analyzed using gene and transcriptome expression profiling (Liu et al., 2022).

7.20.3 Metabolic Activity

Some of the criteria that can be used to assess organoid functionality include accurate response to external stimuli, efficient secretion of cytokines or hormones, and the generation of mature cells (Zhao et al., 2022). The ability of the liver derived ECM gel organoid (LEMgel) to carry out hepatocyte-specific tasks was crucial to evaluate. During the 2nd day of culture, the LEM gel organoid showed a dramatic increase in alpha-1 antitrypsin (AAT) secretions and urea production, using ELISA and colorimetric assays, respectively (Saheli et al., 2018). Alcian blue and periodic acid-Schiff staining revealed a relatively thick mucus layer in functional intestinal organoids, which was the result of mucus secretion from mature goblet cells (Nakamura, 2018; Jung et al., 2018; Son et al., 2020).

7.20.3.1 Characterization of scaffold material

Characterization of cultured organoids can make use of the methods described in previous sections. The following techniques could be used to characterize the bioinks, polymeric hydrogels, etc., that are used to construct the organoids. SEM can be used to analyze the surface architecture, porosity, and extent of crosslinking (Kopeček et al., 2007; MohanKumar et al., 2022). An atomic force microscopy analysis can be performed to measure the surface topography (Pal et al., 2009). The hydrogel's chemical and functional groups are identified using Fourier-transform infrared spectroscopy (FTIR), which is employed to confirm their polymerization and copolymerization (Dhanya et al., 2013). XRD analysis determines crystallinity and confirms the formation of hydrogel cross-linked network gel structures (Ricciardi et al., 2004).

7.21 Challenges in Bioprinting Organoids and Limitations of HSI in Improving the Outcomes

Although research has reached advanced stages, still many challenges are faced in the bioprinting of organoids. The size of organs, vascularization, precise structure and architecture (spatially and temporally), and reproducibility remain major challenges in the bioprinting of organoids (Ren et al., 2021). The generation of some of the organoids has faced challenges because of the cell types and their properties, like a low proliferative rate, as in the case of cardiomyocytes. This eventually creates difficulty in achieving high cell densities in cardiac organoids. Moreover, the handling of mature cells of cardiac muscle remains another challenge (Kupfer et al., 2020). Some major concerns need to be considered for advancements in bioprinting techniques, like the requirement of next-generation bioinks for better protection and support for cells to grow post printing. Technologically advanced printers with better printing approaches can be developed so that cells can be transferred through the nozzle with high precision and thus the survivability of the cells in the organoids can be increased. Effective and efficient crosslinking techniques need to be developed so that the integrity and stability of the bioink are maintained before and after bioprinting. Microfluidic techniques can be integrated along with bioprinting techniques so that the physiological cell culture environment can be maintained for the long term (Sun et al., 2020). Finding a suitable printing material having bioink with better printability, biocompatibility, biodegradability, and desired mechanical properties is another set of challenges for 3D bioprinting technology toward generating better organoids. Another challenge that needs to be considered is the inconsistency of batch-to-batch reproducibility. Human organs are extremely complex structurally and functionally. Hence the search for new generation novel materials is much needed (Mao et al., 2020). In such a process, a real-time imaging technique like HSI could help to a greater extent, but as it can examine the tissues or organoids only close to the surface (3.57 mm at 850 nm and 0.28 mm at 550 nm (Fei et al., 2020)), improvisation would be required.

7.22 Future Prospects

Despite various difficulties, there are some exciting perspectives for this technology of developing organoid bioprinting. As bioprinting processes and biomaterials are fine-tuned on a regular basis to achieve the perfect balance, the 3D printed organoids, as a tool for drug testing or as the therapeutics itself, have immense potential to simulate development of the disease. Further efforts in the future will undoubtedly bring them closer to realizing their full potential. Scaling up the creation of organoids and tissue production is a possibility with 3D bioprinting. More advancement is needed in the hydrogel-based bioinks research for extended success of organoid bioprinting. Furthermore, a new bioink technology with high-resolution capability must be created to increase printability. With the advancement of modern CRISPR/Cas9-like gene editing tools, it is now possible to create mimetic tissue straight from healthy organoids via editing of the gene. With the advancement in organoid bioprinting technology, the generation of individual organoids has also become a possibility that can be further connected through a vascular flow system to mimic the human organ system. This can also be extended to personalized organoids by using specific cell composition and physical and chemical cues. In such improvizations, bioimaging techniques can play a very important role. An assessment of individual/personalized organoids in terms of chemical composition, drug response, and disease cure profile, such bioimaging tools can give inputs for improvizations. HSI is a noncontact imaging modality that uses statistical analysis to process and categorize data. The samples that are specifically connected to biochemical alterations are subsequently given a "molecular fingerprint" and can be used to follow the variations. The setting of a calibrating standard and approving method design of a compact camera system for the characterization of the organoids can surely improvise HSI-mediated outcomes in the 3D-bioprinting based generation of organoids (Pallua et al., 2021).

7.23 Conclusion

3D-bioprinting technology is an emerging and fascinating topic in the discipline of regenerative medicine. It can be used to develop potent fabrication methods to produce micro- and macroscale 3D organoid models for different biomedical applications. These bioprinting methodologies are used to 3D bio-print vascularized tissues and 3D models to imitate skin, kidney, heart, liver, and other tissues. Organoids focus on the rebuilding of biological functions, whereas bioprinting focuses on the reconstruction of tissue architecture. Merging both of these advancing technologies led to the emergence of bioprinted organoids with structure and function comparable to physiology of human organs. Despite the fact that organoid bioprinting is still in its early stages, this technology had showcased remarkable progress in developing bioprinted organoids used for various applications. Bioprinting is a new and intriguing field, and it is used to study experimental biology, pharmacology/toxicology, and tissue engineering. The bioprinted organoids can be used as screening tools, or as a construct for regenerative medicine. Bioprinting techniques have shown significant promise for the bio-fabrication of organoid models, but still, various obstacles are perceived in the field like appropriate biomaterial for bioink formation. The bioprinted organoids have restricted functionality post-printing. To achieve completely functional organoids, more technological and biological advancements are still required. Bioprinted organoids are critical for drug toxicity testing, drug discovery, and disease modeling. Accurate bioprinting is still challenging due to the intricacy of tissues and organs, such as multi-scale and multi-cell bioprinting needs a lot of improvement in the future as tissues or organoids are comprised of different cell types which may vary in size and structure (like smooth muscle cells, fibroblast cells, and endothelial cells. Additional effort is made by including the computational analysis in the bioprinting approach to help in the diagnosis of the diseased and non-diseased area by giving the spectral fingerprint of each pixel. In the field of regenerative medicine, organoids and bioprinting are both fascinating research areas that can be further extended to the field of hyperspectral imaging for disease diagnosis and drug delivery.

References

Abbe, E. (1873). Beiträge zur Theorie des Mikroskops und der mikroskopischen Wahrnehmung. *Archiv für mikroskopische Anatomie* 9(1):413–468.

Abraham, L. C., Zuena, E., Perez-Ramirez, B., and Kaplan, D. L. (2008). Guide to collagen characterization for biomaterial studies, *J. Biomed. Mater. Res. B Appl. Biomater.* 87B(1):264–285.

Adam, E., Mutanga, O., and Rugege, D. (2010). Multispectral and hyperspectral remote sensing for identification and mapping of wetland vegetation: A review, *Wetlands Ecol. Manage.* 18(3):281–296 (2010). 10.1007/s11273-009-9169-z.

Afromowitz, M. A., et al. (1988). Multispectral imaging of burn wounds: A new clinical instrument for evaluating burn depth, *IEEE Trans. Biomed. Eng.* 35(10):842–850. 10.1109/10.7291.

Ahn, S. H., Yoon, H., Kim, G. H., Kim, Y. Y., et al. (2010). Designed three dimensional collagen scaffolds for skin tissue regeneration, *Tissue Engineering Part C Methods* 16(5):813–820.

Akbari, H., Kosugi, Y., Kojima, K., and Tanaka, N. (2009). Blood vessel detection and artery-vein differentiation using hyperspectral imaging, 2009. Annual International Conference of the IEEE Engineering in Medicine and Biology Society, Minneapolis, MN, USA, 2009, pp. 1461–1464, doi: 10.1109/IEMBS.2009.5332920.

Akiva, A., Melke, J., Ansari, S., Liv, N., et al. (2021). An organoid for woven bone. *Adv. Funct. Mater.* 31(17):2010524. https://doi.org/10.1002/adfm.202010524.

Alblawi, A., Ranjani, A. S., Yasmin, H., Gupta, S., et al. (2020). Scaffold free: A developing technique in field of tissue engineering, *Comput. Methods Progr. Biomed.* 185:105148.

Ali, M., Pr, A. K., Yoo, J. J., Zahran, F., et al. (2019). A photo-crosslinkable kidney ECM-derived bioink accelerates renal tissue formation, *Adv. Healthc. Mater.* 8(7):e1800992.

Almela, T., Tayebi, L., and Moharamzadeh, K. (2021). 3D bioprinting for in vitro models of oral cancer: Toward development and validation. *Bioprinting* 22:e00132. doi: 10.1016/j.bprint.2021.e00132. PMID: 34368488; PMCID: PMC8341396.

Amigo, J. and Grassi, S. (2019). Configuration of hyperspectral and multispectral imaging systems. doi:10.1016/B978-0-444-63977-6.00002-X.

Antonucci, J. and Gehrke, L. (2019). Cerebral organoid models for neurotropic viruses. *ACS Infect. Dis.* 5(12):1976–1979. https://doi.org/10.1021/acsinfecdis.9b00339

Arthur, P., Muok, L., Nathani, A., Zeng, E. Z., et al. (2022). Bioengineering human pluripotent stem cell-derived retinal organoids and optic vesicle-containing brain organoids for ocular diseases. *Cells* 11(21):3429. https://doi.org/10.3390/cells11213429.

Bagley, J. A., Reumann, D., Bian, S., Lévi-Strauss, J., et al. (2017). Fused cerebral organoids model interactions between brain regions. *Nat. Methods* 14(7):743–751. https://doi.org/10.1038/nmeth.4304.

Bantounas, I., Ranjzad, P., Tengku, F., Silajdžić, E., et al. (2018). Generation of functioning nephrons by implanting human pluripotent stem cell-derived kidney progenitors. *Stem Cell Rep.* 10(3):766–779.

Barreiro Carpio, M., Dabaghi, M., Ungureanu, J., Kolb, M. R., et al. (2021). 3D bioprinting strategies, challenges, and opportunities to model the lung tissue microenvironment and its function. *Front. Bioeng. Biotechnol.* 24(9):773511. doi: 10.3389/fbioe.2021.773511. PMID: 34900964; PMCID: PMC8653950.

Barrs, R. W., Jia, J., Silver, S. E., Yost, M. (2020). Biomaterials for bioprinting microvasculature. *Chem. Rev.* 120(19):10887–10949. doi: 10.1021/acs.chemrev.0c00027. Epub 2020 Sep 1. PMID: 32867470; PMCID: PMC7810139.

Bayramoglu, N., Kaakinen, M., Eklund, L., and Heikkilä, J. (2017). Towards virtual H&E staining of hyperspectral lung histology images using conditional generative adversarial networks, 2017 IEEE International Conference on Computer Vision Workshops (ICCVW), Venice, Italy, 2017, pp. 64–71, doi: 10.1109/ICCVW.2017.15.

Bendtsen, S. T., Quinnell, S. P., Wei, M. (2017). Development of a novel alginate-polyvinyl alcohol-hydroxyapatite hydrogel for 3D bioprinting bone tissue engineered scaffolds, *J. Biomed. Mater. Res. A* 105(5):1457–1468.

Best, S. L., Thapa, A., Jackson, N., Olweny, E., et al. (2013). Renal oxygenation measurement during partial nephrectomy using hyperspectral imaging may predict acute postoperative renal function. *J. Endourol.* 27(8):1037–1040.

Birey, F., Andersen, J., Makinson, C. D., Islam, S., et al. (2017). Assembly of functionally integrated human forebrain spheroids. *Nature* 545(7652): 54–59. https://doi.org/10.1038/nature22330.

Bishop, E. S., Mostafa, S., Pakvasa, M., Luu, H. H., et al. (2017). 3-D bioprinting technologies in tissue engineering and regenerative medicine: Current and future trends. *Genes Dis.* 4(4):185–195. doi: 10.1016/j.gendis.2017.10.002. Epub 2017 Nov 22. PMID: 29911158; PMCID: PMC6003668.

Blom, H. and Widengren, J. (2017). Stimulated emission depletion microscopy. *Chem. Rev.* 117(11):7377–7427.

Bouwmeester, M. C., Bernal, P. N., Oosterhoff, L. A., van Wolferen, M. E., et al. (2021). Bioprinting of human liver-derived epithelial organoids for toxicity studies. *Macromol Biosci.* 21(12):e2100327. doi: 10.1002/mabi.202100327. Epub 2021 Oct 5. PMID: 34559943.

Brahatheeswaran D., Yasuhiko Y., Toru M., Sakthi Kumar, D. (2011). Polymeric scaffolds in tissue engineering application: A review, *Int. J. Polym. Sci.* vol. 2011, Article ID 290602, 19 pages. https://doi.org/10.1155/2011/290602.

Brémond Martin, C., Simon Chane, C., Clouchoux, C., and Histace, A. (2021). Recent trends and perspectives in cerebral organoids imaging and analysis. *Front. Neurosci.* 15, 629067.

Browne, A. W., Arnesano, C., Harutyunyan, N., Khuu, T., et al. (2017). Structural and functional characterization of human stem-cell-derived retinal organoids by live imaging. *Invest. Ophthalmol. Vis. Sci.* 58(9):3311–3318.

Browne, A. W., Arnesano, C., Harutyunyan, N., Khuu, T., et al. (2017). Structural and functional characterization of human stemcell-derived retinal organoids by live imaging. *Invest. Ophthalmol. Vis. Sci.* 58:3311–3318. doi: 10.1167/iovs.16-20796.

Brunner, A., Schmidt, V. M., Zelger, B., Woess, C., et al. (2022). Visible and near-infrared hyperspectral imaging (HSI) can reliably quantify CD3 and CD45 positive inflammatory cells in myocarditis: Pilot study on formalin-fixed paraffin-embedded specimens from myocardium obtained during autopsy. *Spectrochimica Acta Part A: Molecular and Biomolecular Spectroscopy* 274:121092.

Buwalda, S. J., Boere, K. W. M., Dijkstra, P. J., Feijen, J., et al. (2014). Hydrogels in a historical perspective: from simple networks to smart materials, *J. Control. Release* 190(2014):254–273.

Buzzelli, J. N., Ouaret, D., Brown, G., Allen, P. D., et al. (2018). Colorectal cancer liver metastases organoids retain characteristics of original tumor and acquire chemotherapy resistance. *Stem Cell Res.* 27:109–120.

Cadena, M., Ning, L., King, A., Hwang, B., et al. (2021). 3D bioprinting of neural tissues. *Adv. Healthc. Mater.* 10(15):e2001600. doi: 10.1002/adhm.202001600. Epub 2020 Nov 16. PMID: 33200587; PMCID: PMC8711131.

Camci-Unal, G., Cuttica, D., Annabi, N., Demarchi, D., et al. (2013). Synthesis and characterization of hybrid hyaluronic acid-gelatin hydrogels, *Biomacromolecules* 14(4):1085–1092.

Carlos, K., Anthony, A., Sang, J. L. Chapter 15 - Bioprinting of Organoids, https://doi.org/10.1016/B978-0-12-800972-7.00015-3.

Carrasco, O., et al. (2003). Hyperspectral imaging applied to medical diagnoses and food safety, *Proc. SPIE* 5097:215–221. doi: 10.1117/12.502589.

Cervantes-Sanchez, F., Maktabi, M., Köhler, H., Sucher, R., et al. (2021). Automatic tissue segmentation of hyperspectral images in liver and head neck surgeries using machine learning. *Artif. Intell. Surg.* 1(1): 22–37.

Chameettachal, S., Midha, S., and Ghosh, S. (2016). Regulation of chondrogenesis and hypertrophy in silk fibroin-gelatin-based 3D bioprinted constructs, *ACS Biomater. Sci. Eng.* 2(9):1450–1463.

Chang, R., et al. (2010). *Biofabrication* 2:045004.

Chen, E. P., Toksoy Z., Davis, B. A., and Geibel, J. P. (2021). 3D bioprinting of vascularized tissues for in vitro and in vivo applications. *Front. Bioeng.*

Biotechnol. 13(9):664188. doi: 10.3389/fbioe.2021.664188. PMID: 34055761; PMCID: PMC8158943.

Chen, S., Chen, X., Geng, Z., and Su, J. (2022). The horizon of bone organoid: A perspective on construction and application. *Bioact. Mater.* 18:15–25. https://doi.org/10.1016/j.bioactmat.2022.01.048.

Chen, Y., Wong, P. P., Sjeklocha, L., Steer, C. J., et al. (2012). Mature hepatocytes exhibit unexpected plasticity by direct dedifferentiation into liver progenitor cells in culture. *Hepatology* 55(2):563–574. doi: 10.1002/hep.24712. PMID: 21953633; PMCID: PMC3268884.

Chichagova, V., Dorgau, B., Felemban, M., Georgiou, M., et al. (2019). Differentiation of retinal organoids from human pluripotent stem cells. *Curr. Protoc Stem Cell Biol.* 50(1):e95. https://doi.org/10.1002/cpsc.95.

Choi, S. Y., Cho, Y. H., Kim, D. S., Ji, W., et al. (2021). Establishment and long-term expansion of small cell lung cancer patient-derived tumor organoids. *Int. J. Mol. Sci.* 22(3):1349. doi: 10.3390/ijms22031349.

Christensen K., Xu C., Chai W., Zhang Z., et al. (2015). Freeform inkjet printing of cellular structures with bifurcations. *Biotechnol Bioeng.* 112(5):1047–1055.

Chung, J. H., Naficy, S., Yue, Z., Kapsa, R., Quigley, A., et al. (2013). Bio-ink properties and printability for extrusion printing living cells, *Biomater. Sci.* 1(7):763–773.

Cintron Pregosin, N., Bronstein, R., Mallipattu, S. K. (2021). Recent Advances in Kidney Bioengineering. *Front. Pediatr.* 9.

Conchello, J. A. and Lichtman, J. W. (2005). Fluorescence microscopy. *Nat. Methods* 2(12):910–919.

Cui X. and Boland T. (2009). Human microvasculature fabrication using thermal inkjet printing technology. *Biomaterials* 30(31):6221–6227.

Das, S., Pati, F., Choi, Y.-J., Rijal, G., et al. (2015). Bioprintable, cell-laden silk fibroin–gelatin hydrogel supporting multilineage differentiation of stem cells for fabrication of three-dimensional tissue constructs, *Acta Biomater.* 11:233–246.

Delaine-Smith, R. M., MacNeil, S., and Reilly, G. C. (2012). Matrix production and collagen structure are enhanced in two types of osteogenic progenitor cells by a simple fluid shear stress stimulus. *Eur. Cells Mater.* 24:162–174. https://doi.org/10.22203/ecm.v024a12.

Dhanya, S., Bahadur, D., Kundu, G. C., and Srivastava, R. (2013). Maleic acid incorporated poly-(N-isopropylacrylamide) polymer nanogels for dual-responsive delivery of doxorubicin hydrochloride. *Eur. Polym. J.* 49(1):22–32.

Di Bella, C., Fosang, A., Donati, D. M., Wallace, G. G., et al. (2015). 3D bioprinting of cartilage for orthopedic surgeons: reading between the lines, *Front. Surg.* 2:39.

Dorgau, B., Georgiou, M., Chaudhary, A., Moya-Molina, M., et al. (2022). Human retinal organoids provide a suitable tool for toxicological investigations: A comprehensive validation using drugs and compounds affecting the retina. *Stem Cells Transl. Med.* 11(2):159–177.

Duarte Campos, D. F., Blaeser, A., Buellesbach, K., Sen, K. S., et al. (2016). Bioprinting organotypic hydrogels with improved mesenchymal stem cell remodeling and mineralization properties for bone tissue engineering, *Adv. Healthc. Mater.* 5(11):1336–1345.

Dvela-Levitt, M., Kost-Alimova, M., Emani, M., Kohnert, E., et al. (2019). Small molecule targets TMED9 and promotes lysosomal degradation to reverse proteinopathy. *Cell* 178(3):521–535.

Edelman, G. J., Gaston, E., van Leeuwen, T. G., Cullen, P. J., et al. (2012). Hyperspectral imaging for non-contact analysis of forensic traces, *Forensic Sci. Int.* 223(1–3):28–39. ISSN 0379-0738, https://doi.org/10.1016/j.forsciint.2012.09.012.

Fatehullah, A., Tan, S. H., and Barker, N. (2016). Organoids as an in vitro model of human development and disease. *Nat. Cell Biol.* 18:246–254. http://dx.doi.org/10.1038/ncb3312

Fatimi, A., Okoro, O. V., Podstawczyk, D., Siminska-Stanny, J., et al. (2022). Natural hydrogel-based bio-inks for 3D bioprinting in tissue engineering: A review. *Gels* 14;8(3):179. doi: 10.3390/gels8030179. PMID: 35323292; PMCID: PMC8948717.

Fauch, L., Palander, A., Dekker, H., Schulten, E. A., et al. (2019). Narrowband-autofluorescence imaging for bone analysis. *Biomed. Opt. Express* 10(5):2367–2382. https://doi.org/10.1364/BOE.10.002367.

Fedorovich, N.E., De Wijn, J. R., Verbout, A. J., Alblas, J., et al. (2008). Three dimensional fiber deposition of cell-laden, viable, patterned constructs for bone tissue printing, *Tissue Eng.* 14(1):127–133.

Fei, K., Zhang, J., Yuan, J., and Xiao, P. (2022). Present application and perspectives of organoid imaging technology. *Bioengineering* 9(3):121.

Felli, E., Al-Taher, M., Collins, T., Nkusi, R., et al. (2021). Automatic liver viability scoring with deep learning and hyperspectral imaging. *Diagnostics* 11(9):1527.

Feng, Y. Z. and Sun, D. W. (2012). Application of hyperspectral imaging in food safety inspection and control: a review, *Crit. Rev. Food Sci. Nutr.* 52(11):1039–1058. doi: 10.1080/10408398.2011.651542.

Ferreira A. M., Gentile P., Chiono V., Ciardelli G. (2012). Collagen for bone tissue regeneration. *Acta Biomater.* 8(9):3191–3200.

Fischer, C. and Kakoulli, I. (2006). Multispectral and hyperspectral imaging technologies in conservation: Current research and potential applications, *Stud. Conserv.* 7:3–16.

Gao, G., Schilling, A. F., Yonezawa, T., Wang, J., et al. (2014). Bioactive nanoparticles stimulate bone tissue formation in bioprinted three-dimensional scaffold and human mesenchymal stem cells, *Biotechnol. J.* 9(10):1304–1311.

Gao, L. and Smith, R. T. (2015). Optical hyperspectral imaging in microscopy and spectroscopy – a review of data acquisition. *J. Biophotonics* 8:441–456. doi: 10.1002/jbio.201400051.

Germain, N., Dhayer, M., Dekiouk, S., Marchetti, P. (2022). Current advances in 3D bioprinting for cancer modeling and personalized medicine. *Int. J. Mol. Sci.* 23:3432. https://doi.org/10.3390/ijms23073432.

Giandomenico, S. L., Mierau, S. B., Gibbons, G. M., Wenger, L., et al. (2019). Cerebral organoids at the air-liquid interface generate diverse nerve tracts with functional output. *Nat. Neurosci.* 22(4):669–679. https://doi.org/10.1038/s41593-019-0350-2.

Giannoni, L., Lange, F., and Tachtsidis, I. (2018). Hyperspectral imaging solutions for brain tissue metabolic and hemodynamic monitoring: Past, current and future developments. *J. Opt.* 20(4):044009. https://doi.org/10.1088/2040-8986/aab3a6.

Glowacki, J. and Mizuno, S. J. B. (2010). Collagen scaffolds for tissue engineering, 89(5):338–344.

Gomez-Giro, G., Arias-Fuenzalida, J., Jarazo, J., Zeuschner, D., et al. (2019). Synapse alterations precede neuronal damage and storage pathology in a human cerebral organoid model of CLN3-juvenile neuronal ceroid lipofuscinosis. *Acta Neuropathol. Commun.* 7(1):1–19.

Gonzalez-Cordero, A., Kruczek, K., Naeem, A., Fernando, M., et al. (2017). Recapitulation of human retinal development from human pluripotent stem cells generates transplantable populations of cone photoreceptors. *Stem Cell Rep.* 9(3):820–837.

Gopinathan, J. and Noh, I. (2018). Recent trends in bioinks for 3D printing. *Biomater. Res.* 22, 11. https://doi.org/10.1186/s40824-018-0122-1.

Govender, M., Chetty, K., and Bulcock, H. (2007). A review of hyperspectral remote sensing and its application in vegetation and water resource studies, *Water* SA 33(2):145–151.

Gowen, A. A., et al. (2007). Hyperspectral imaging—an emerging process analytical tool for food quality and safety control, *Trends Food Sci. Technol.* 18(12):590–598. 10.1016/j.tifs.2007.06.001.

Grix, T., Ruppelt, A., Thomas, A., Amler, A. K., et al. (2018). Bioprinting perfusion-enabled liver equivalents for advanced organ-on-a-chip applications. *Genes* (Basel). 22;9(4):176. doi: 10.3390/genes9040176. PMID: 29565814; PMCID: PMC5924518.

Groll, J., Burdick, J., Cho, D., Derby, B., et al. (2018).A definition of bioinks and their distinction from biomaterial inks, *Biofabrication* 11(1):013001.

Grossniklaus, H. E., Geisert, E. E., and Nickerson, J. M. (2015). Introduction to the retina. Progress in molecular biology and translational science, 134:383–396.

Gu, Q., Tomaskovic-Crook, E., Lozano, R., Chen, Y., et al. (2016). Functional 3D neural mini-tissues from printed gel-based bioink and human neural stem cells. *Adv Healthc Mater.* 5(12):1429–1438.

Gu, Z., Fu, J., Lin, H., and He, Y. (2020). Development of 3D bioprinting: From printing methods to biomedical applications. *Asian J. Pharm Sci.* 15(5):529–557. doi:10.1016/j.ajps.2019.11.003

Gudapati, H., Dey, M., and Ozbolat, I. (2016). A comprehensive review on droplet-based bioprinting: past, present and future, *Biomaterials* 102:20–42.

Gudapati, H., Dey, M., and Ozbolat, I. (2016). A comprehensive review on droplet-based bioprinting: Past, present and future. *Biomaterials* 102:20–42. doi: 10.1016/j.biomaterials.2016.06.012. Epub 2016 Jun 7. PMID: 27318933.

Hagenbuchner, J., Nothdurfter, D., and Ausserlechner, M. J. (2021). 3D bioprinting: novel approaches for engineering complex human tissue equivalents and drug testing. *Essays Biochem.* 10;65(3):417–427. doi: 10.1042/EBC20200153. PMID: 34328185; PMCID: PMC8365325.

Han, Y., Duan, X., Yang, L., Nilsson-Payant, B. E., et al. (2021). Identification of SARS-CoV-2 inhibitors using lung and colonic organoids. *Nature* 589(7841):270–275. doi: 10.1038/s41586-020-2901-9.

Harth, O. (1983). The function of the kidneys. In: Schmidt, R.F., Thews, G. (eds.) *Human Physiology*. Springer, Berlin, Heidelberg. https://doi.org/10.1007/978-3-642-96714-6_27.

Harvey, A. R., Lawlor, J., McNaught, A. I., Williams, J. W., et al. (2002, November). Hyperspectral imaging for the detection of retinal diseases. In *Imaging Spectrometry* VIII (Vol. 4816, pp. 325–335). SPIE.

Hawes, D., SHI, S. R., Dabbs, D. J., Taylor, C. R., et al. (2009). Immunohistochemistry. *Mod. Surg. Pathol.* 48.

He, H., He, Q., Xu, F., Zhou, Y., et al. (2019). Dynamic formation of cellular aggregates of chondrocytes and mesenchymal stem cells in spinner flask, *Cell Prolif* 52:1–10.

Henriksson, I., Gatenholm, P., and H€agg, D. (2017). Increased lipid accumulation and adipogenic gene expression of adipocytes in 3D bioprinted nanocellulose scaffolds, *Biofabrication* 9(1):015022.

Hildebrandt, C., Büth, H., Thielecke, H. (2011). A scaffold-free in vitro model for osteogenesis of human mesenchymal stem cells. *Tissue Cell.* 43(2):91–100. doi: 10.1016/j.tice.2010.12.004. Epub 2011 Feb 16. PMID: 21329953.

Hirsch, F. R., Scagliotti, G. V., Mulshine, J. L., Kwon, R., et al. (2017). Lung cancer: Current therapies and new targeted treatments. *Lancet* 389(10066):299–311. doi: 10.1016/s0140-6736(16)30958-8.

Hölzl, K., Lin, S., Tytgat, L., Van Vlierberghe, S., et al. (2016). Bioink properties before, during, and after 3D bioprinting. *Biofabrication* 8(3):032002. doi: 10.1088/1758-5090/8/3/032002.

Homan, K., Kolesky, D., Skylar-Scott, M. et al. (2016). Bioprinting of 3D convoluted renal proximal tubules on perfusable chips. *Sci. Rep.* 6:34845 (2016). https://doi.org/10.1038/srep34845.

Hospodiuk, M., Dey, M., Sosnoski, D., and Ozbolat, I. T. (2017). The bioink: A comprehensive review on bioprintable materials, *Biotechnol. Adv.* 35(2):217–239.

Huang, B., Bates, M., and Zhuang, X. (2009). Super-resolution fluorescence microscopy. *Annu. Rev. Biochem.* 78:993–1016.

Iraha, S., Tu, H. Y., Yamasaki, S., Kagawa, T., et al. (2018). Establishment of immunodeficient retinal degeneration model mice and functional maturation of human ESC-derived retinal sheets after transplantation. *Stem Cell Rep.* 10(3):1059–1074.

Jabs, J., Zickgraf, F. M., Park, J., Wagner, S., et al. (2017). Screening drug effects in patient-derived cancer cells links organoid responses to genome alterations. *Mol. Syst. Biol.* 13(11):955. doi: 10.15252/msb.20177697.

Jang, J., Park, H.-J., Kim, S.-W., Kim, H., et al. (2017). 3D printed complex tissue construct using stem cell-laden decellularized extracellular matrix bioinks for cardiac repair, *Biomaterials* 112:264–274.

Jansen, J., Reimer, K. C., Nagai, J. S., Varghese, F. S., et al. (2022). SARS-CoV-2 infects the human kidney and drives fibrosis in kidney organoids. *Cell Stem Cell* 29(2):217–231.

Ji, S., and Guvendiren, M. (2017). Recent advances in bioink design for 3D bioprinting of tissues and organs, Front. *Bioeng. Biotechnol.* 5:1–8.

Ji, S., Guvendiren, M. (2021). Complex 3D bioprinting methods. *APL Bioeng.* 11;5(1):011508. doi: 10.1063/5.0034901. PMID: 33728391; PMCID: PMC7954578.

Jin R. and Dijkstra P. J. (2010). Hydrogels for tissue engineering applications. In: Ottenbrite R. M., Park K., Okano T., editors. Biomedical applications of hydrogels handbook. New York: Springer; p. 203–225.

Joel, M. and Tuan, V.-D. (2003). Optical properties of tissue, Chapter 2 in *Biomedical Photonics Handbook*, pp. 1–76, CRC Press, Boca Raton, Florida.

Jung, K. B., Lee, H., Son, Y. S., Lee, M. O., et al. (2018). Interleukin-2 induces the in vitro maturation of human pluripotent stem cell-derived intestinal organoids. *Nat. Commun.* 9(1):3039.

Jüttner, J., Szabo, A., Gross-Scherf, B., Morikawa, R. K., et al. (2019). Targeting neuronal and glial cell types with synthetic promoter AAVs in mice, non-human primates and humans. *Nat. Neurosci.* 22(8):1345–1356.

Kadoshima, T., Sakaguchi, H., Nakano, T., Soen, M., et al. (2013). Self-organization of axial polarity, inside-out layer pattern, and species-specific progenitor dynamics in human ES cell-derived neocortex. *Proc Natl Acad Sci USA* 110(50): 20284–20289. https://doi.org/10.1073/pnas.1315710110.

Kang, D., Hong, G., An, S., et al. (2020). Bioprinting of multiscaled hepatic lobules within a highly vascularized construct. *Small* 16:1905505. https://doi.org/10.1002/smll.201905505.

Kato, R., Yano, T. A., Minamikawa, T., and Tanaka, T. (2022). High-sensitivity hyperspectral vibrational imaging of heart tissues by mid-infrared photothermal microscopy. *Anal. Sci.* 38(12):1497–1503.

Kaushik, S. N., Kim, B., Walma, A., Choi, S. C., et al. (2016). Biomimetic microenvironments for regenerative endodontics. *Biomater Res.* 20:14.

Keshara, R., Kim, Y. H., and Grapin-Botton, A. (2022). Organoid imaging: Seeing development and function. *Annu. Rev. Cell Dev. Biol.* 38:447–466. https://doi.org/10.1146/annurev-cellbio-120320-035146.

Kim, J. W., Kim, H. W., Nam, S. A., Lee, J. Y., et al. (2021). Human kidney organoids reveal the role of glutathione in Fabry disease. *Exp. Mol. Med.* 53(10):1580–1591.

Kim, M., Mun, H., Sung, C. O., Cho, E. J., et al. (2019). Patient-derived lung cancer organoids as in vitro cancer models for therapeutic screening. *Nat. Commun.* 10(1):3991. doi: 10.1038/s41467-019-11867-6.

King Shelby M., Higgins J. William, Nino Celina R., Smith Timothy R., et al. (2017). 3D proximal tubule tissues recapitulate key aspects of renal physiology to enable nephrotoxicity testing. *Front. Physiol.* 8, https://www.frontiersin.org/article/10.3389/fphys.2017.00123.

Kopeček, J. (2007). Hydrogel biomaterials: A smart future? *Biomaterials,* 28(34):5185–5192.

Kupfer, M. E., Lin, W.-H., Ravikumar, V., Qiu, K., et al. (2020). In situ expansion, differentiation, and electromechanical coupling of human cardiac muscle in a 3D bioprinted, chambered organoid. *Circ. Res.* 127:207–224. https://doi.org/10.1161/CIRCRESAHA.119.316155.

Kuula, J., et al. (2012). Using VIS/NIR and IR spectral cameras for detecting and separating crime scene details, *Proc. SPIE* 8359:83590P. 10.1117/12.918555.

LaBarge, W., Morales, A., Pretorius, D., Kahn-Krell, A. M., et al. (2019). Scaffold-free bioprinter utilizing layer-by-layer printing of cellular spheroids, *Micromachines* 10: 1–13.

Lai, K. K., Renneberg, R., and Mak, W. C. (2016). High efficiency single-step biomaterial-based microparticle fabrication via template-directed supramolecular coordination chemistry. *Green Chem.* 18:1715–1723

Lancaster, M. A. and Knoblich, J. A. (2014). Generation of cerebral organoids from human pluripotent stem cells. *Nat Protoc.* 9(10):2329–2340. https://doi.org/10.1038/nprot.2014.158.

Lancaster, M. A. and Knoblich, J. A. (2014). Organogenesis in a dish: Modeling development and disease using organoid technologies. *Science.* 345:1247125.http://dx.doi.org/10.1126/science.1247125

Lancaster, M. A., Renner, M., Martin, C. A., Wenzel, D., et al. (2013). Cerebral organoids model human brain development and microcephaly. *Nature* 501(7467):373–379. https://doi.org/10.1038/nature12517.

Larsen, E. L., Randeberg, L. L., Olstad, E., Haugen, O. A., et al. (2011). Hyperspectral imaging of atherosclerotic plaques in vitro. *J. Biomed. Opt.* 16(2):026011–026011.

Laschke, M. W., Menger, M. D. (2017). Life is 3D: Boosting spheroid function for tissue engineering. *Trends Biotechnol.* 35(2):133–144. doi: 10.1016/j.tibtech.2016.08.004. Epub 2016 Sep 12. PMID: 27634310.

Lawrence, M. L., Elhendawi, M., Morlock, M., Liu, W., et al. (2022). Human iPSC-derived renal organoids engineered to report oxidative stress can predict drug-induced toxicity. *IScience* 25(3).

Leavesley, S. J., Annamdevula, N., Boni, J., Stocker, S., et al. (2012). Hyperspectral imaging microscopy for identification and quantitative analysis of fluorescently-labeled cells in highly autofluorescent tissue. *J. Biophotonics* 5(1):67–84.

Lee, A. R. H. A., Hudson, A. R., Shiwarski, D. J., Tashman, J. W., et al. (2019). 3D bioprinting of collagen to rebuild components of the human heart. *Science* 365(6452):482–487.

Lee, D., Kim, Y., & Chung, C. (2021). Scientific validation and clinical application of lung cancer organoids. Cells 10(11): 3012.

Lee V., Singh G., Trasatti J. P., Bjornsson C., et al. (2013). Design and fabrication of human skin by three-dimensional bioprinting. *Tissue Eng Part C Methods* 20:473–484.

Lei, M. and Wang, X. (2016). Biodegradable polymers and stem cells for bioprinting, *Molecules* 21(5):539.

Lewis, E. E. L., Wheadon, H., Lewis, N., Yang, J., et al. (2016). A quiescent, regeneration-responsive tissue engineered mesenchymal stem cell bone marrow niche model via magnetic levitation, *ACS Nano* 10:8346–8354.

Li, X., Francies, H. E., Secrier, M., Perner, J., et al. (2018). Organoid cultures recapitulate esophageal adenocarcinoma heterogeneity providing a model for clonality studies and precision therapeutics. *Nat. Commun.* 9(1):2983.

Liang, H. (2012). Advances in multispectral and hyperspectral imaging for archaeology and art conservation, *Appl. Phys.* A 106(2):309–323. 10.1007/s00339-011-6689-1.

Liao, J., Han, R., Wu, Y., and Qian, Z. (2021). Review of a new bone tumor therapy strategy based on bifunctional biomaterials. *Bone Res.* 9(1):18. https://doi.org/10.1038/s41413-021-00139-z.

Lin, N. Y. C., Homan, K. A., Robinson, S. S., Kolesky, D. B., et al. (2019). Renal reabsorption in 3D vascularized proximal tubule models. *Proc. Natl. Acad. Sci. U.S.A.* 116:5399–5404. https://doi.org/10.1073/pnas.1815208116.

Liu, Q., Zeng, A., Liu, Z., Wu, C., et al. (2022). Liver organoids: From fabrication to application in liver diseases. *Front. Physiol.* 13:956244.

Liu, W., Heinrich, M. A., Zhou, Y., Akpek, A., et al. (2017). Extrusion bioprinting of shear-thinning gelatin methacryloyl bioinks, *Adv. Healthc. Mater.* 6(12):1601451.

Liu, X., Yang, C., Chen, P., Zhang, L., et al. (2022). The uses of transcriptomics and lipidomics indicated that direct contact with graphene oxide altered lipid homeostasis through ER stress in 3D human brain organoids. *Sci. Total Environ.* 849:157815.

Lorber, B., Hsiao, W. K., Hutchings, I. M., and Martin, K. R. (2013). Adult rat retinal ganglion cells and glia can be printed by piezoelectric inkjet printing. *Biofabrication* 6(1):015001.

Lu, G. and Fei, B. (2014). Medical hyperspectral imaging: A review. *J Biomed Opt.* 19(1):10901. doi: 10.1117/1.JBO.19.1.010901. PMID: 24441941; PMCID: PMC3895860.

Lu, G., Little, J. V., Wang, X., Zhang, H., et al. (2017). Detection of head and neck cancer in surgical specimens using quantitative hyperspectral imaging. *Clin. Cancer Res.* 15;23(18):5426–5436. doi: 10.1158/1078-0432.CCR-17-0906. Epub 2017 Jun 13. PMID: 28611203; PMCID: PMC5649622.

Luo, Y., Kirker, K. R., and Prestwich, G. D. (2000). Cross-linked hyaluronic acid hydrogel films: new biomaterials for drug delivery, *J. Control. Release* 69(1):169–184.

Ma, X., Liu, J., Zhu, W., et al. (2018). 3D bioprinting of functional tissue models for personalized drug screening and in vitro disease modeling. *Adv. Drug Deliv. Rev.* 132:235–251. doi: 10.1016/j.addr.2018.06.011.

Ma, X., Qu, X., Zhu, W., Li, Y. S., et al. (2016). Deterministically patterned biomimetic human iPSC-derived hepatic model via rapid 3D bioprinting. *Proc Natl Acad Sci U S A.* 23;113(8):2206–2211. doi: 10.1073/pnas.1524510113. Epub 2016 Feb 8. PMID: 26858399; PMCID: PMC4776497.

Madhavan, M., Nevin, Z. S., Shick, H. E., Garrison, E., et al. (2018). Induction of myelinating oligodendrocytes in human cortical spheroids. *Nat. Methods* 15(9):700–706. https://doi.org/10.1038/s41592-018-0081-4.

Mao, D., Li, Q., Bai, N., Dong, H., et al. (2018). Porous stable poly(lactic acid)/ethyl cellulose/hydroxyapatite composite scaffolds prepared by a combined method for bone regeneration. *Carbohydr Polym.* 180:104–111. doi: 10.1016/j.carbpol.2017.10.031. Epub 2017 Oct 7. PMID: 29103485.

Mao, H., Yang, L., Zhu, H., Wu, L., et al. (2020). Recent advances and challenges in materials for 3D bioprinting. *Prog. Nat. Sci. Mater. Int., SI: Biomaterials* 30:618–634. https://doi.org/10.1016/j.pnsc.2020.09.015.

Masland, R. H. (2012). The neuronal organization of the retina. *Neuron* 76(2):266–280.

Matsusaki, M., Sakaue, K., Kadowaki, K., and Akashi, M. (2013). Three-dimensional human tissue chips fabricated by rapid and automatic inkjet cell printing. *Adv. Healthc. Mater.* 2:534–539. https://doi.org/10.1002/adhm.201200299.

McKinnon, K. M. (2018). Flow cytometry: An overview. *Curr. Protoc. Immunol.* 120:5.1.1–5.1.11.

Mills, R. J., Humphrey, S. J., Fortuna, P. R. J., Lor, M., et al. (2021). BET inhibition blocks inflammation-induced cardiac dysfunction and SARS-CoV-2 infection. *Cell* 184(8):2167–2182. doi:10. 1016/j.cell.2021.03.026.

MohanKumar, B. S., Priyanka, G., Rajalakshmi, S., Sankar, R., et al. (2022). Hydrogels: Potential aid in tissue engineering—a review. *Polym. Bull.* 79(9):7009–7039.

Mori H., Shimizu K., and Hara M. (2013). Dynamic viscoelastic properties of collagen gels with high mechanical strength. *Mater Sci Eng C.* 33(6):3230–3236.

Murphy, S. V. and Atala, A. (2014). 3D bioprinting of tissues and organs, *Nat. Biotechnol.* 32(8):773–785.

Murphy, S. V., De Coppi, P., Atala, A. (2020). Opportunities and challenges of translational 3D bioprinting. *Nat. Biomed. Eng.* 4(4):370–380. doi: 10.1038/s41551-019-0471-7. Epub 2019 Nov 6. PMID: 31695178.

Nair L. S. and Laurencin C. T. (2007). Biodegradable polymers as biomaterials. *Prog. Polym. Sci.* 32:762–798.

Nair, D. S. R., Seiler, M. J., Patel, K. H., Thomas, V., et al. (2021). Tissue engineering strategies for retina regeneration. *Appl. Sci.* (Basel, Switzerland), 11(5):2154. https://doi.org/10.3390/app11052154.

Nakamura, T. (2019). Recent progress in organoid culture to model intestinal epithelial barrier functions. *Int. Immunol.* 31(1):13–21.

Neufurth M., Wang X., Schröder H. C., Feng Q., et al. (2014). Engineering a morphogenetically active hydrogel for bioprinting of bioartificial tissue derived from human osteoblast-like SaOS-2 cells. *Biomaterials* 35:8810–8819. doi: 10.1016/j.biomaterials.2014.07.002.

Ng, W. L., Yeong, W. Y., and Naing, M. W. (2016). Polyelectrolyte gelatin-chitosan hydrogel optimized for 3D bioprinting in skin tissue engineering, *Int. J. Bioprint.* 2(1):53–62.

Nguyen, D. G., Funk, J., Robbins, J. B., Crogan-Grundy, C., et al. (2016). Bioprinted 3D primary liver tissues allow assessment of organ-level response to clinical drug induced toxicity in vitro. *PLoS ONE* 11(7):e0158674. https://doi.org/10.1371/journal.pone.0158674.

Nie, Y. Z., Zheng, Y. W., Miyakawa, K., Murata, S., et al. (2018). Recapitulation of hepatitis B virus–host interactions in liver organoids from human induced pluripotent stem cells. *EBioMedicine* 35:114–123.

Nilsson Hall, G., Mendes, L. F., Gklava, C., Geris, L., et al. (2019). Developmentally engineered callus organoid bioassemblies exhibit predictive in vivo long bone healing. *Adv. Sci.* (Weinheim, Baden-Wurttemberg, Germany), 7(2):1902295. https://doi.org/10.1002/advs.201902295.

Nishinakamura, R. (2019). Human kidney organoids: Progress and remaining challenges. *Nat. Rev. Nephrol.* 15:613–624. https://doi.org/10.1038/s41581-019-0176-x.

Okubo, K., Kitagawa, Y., Hosokawa, N., Umezawa, M., et al. (2021). Visualization of quantitative lipid distribution in mouse liver through near-infrared hyperspectral imaging. *Biomed. Opt. Exp.* 12(2): 823–835.

Ong, C. S., Fukunishi, T., Nashed, A., Blazeski, A., et al. (2017). Creation of cardiac tissue exhibiting mechanical integration of spheroids using 3D bioprinting, *J. Vis. Exp.* 2017:2–6.

Ormel, P. R., Vieira de Sá, R., Van Bodegraven, E. J., Karst, H., et al. (2018). Microglia innately develop within cerebral organoids. *Nat. Commun.* 9(1):4167.

Ovando-Roche, P., West, E. L., Branch, M. J., Sampson, R. D., et al. (2018). Use of bioreactors for culturing human retinal organoids improves photoreceptor yields. *Stem Cell Res. Ther.* 9:1–14.

Ovsianikov, A., Khademhosseini, A., Mironov, V. (2018). The synergy of scaffold-based and scaffold-free tissue engineering strategies, *Trends Biotechnol.* 36:348–57.

Owens, C. M., Marga, F., Forgacs, G., Heesch, C. M. (2013). Biofabrication and testing of a fully cellular nerve graft. *Biofabrication* 5:45007.

Pal, K., Banthia, A. K., and Majumdar, D. K. (2009). Polymeric hydrogels: Characterization and biomedical applications. *Designed Monomers Polym.* 12(3):197–220.

Pallua, J. D., Brunner, A., Zelger, B., Huck, C. W., et al. (2021). New perspectives of hyperspectral imaging for clinical research. *NIR News*, 32(3-4):5-13.

Papaspyropoulos, A., Tsolaki, M., Foroglou, N., and Pantazaki, A. A. (2020). Modeling and targeting Alzheimer's disease with organoids. *Front. Pharmacol.* 11:396. https://doi.org/10.3389/fphar.2020.00396.

Parenteau-Bareil, R., Gauvin, R., and Berthod, F. (2010). Collagen-based biomaterials for tissue engineering applications, *Materials* 3(3):1863–1887.

Park, J. Y., Choi, J.-C., Shim, J.-H., Lee, J.-S., et al. (2014). A comparative study on collagen type I and hyaluronic acid dependent cell behavior for osteochondral tissue bioprinting, *Biofabrication* 6(3).

Patra, B., Chen, Y. H., Peng, C. C., Lin, S. C., et al. (2013). A microfluidic device for uniform-sized cell spheroids formation, culture, harvesting and flow cytometry analysis, *Biomicrofluidics* 7.

Patterson, M., Wilson, B., and Wyman, D. (1991). The propagation of optical radiation in tissue I. Models of radiation transport and their application, *Lasers Med. Sci.* 6(2):155–168. 10.1007/BF02032543.

Pawar, S. N. and Edgar, K. J. (2012). Alginate derivatization: a review of chemistry, properties and applications, *Biomaterials* 33(11):3279–3305.

Pham, M. T., Pollock, K. M., Rose, M. D., Cary, W. A., et al. (2018). Generation of human vascularized brain organoids. *Neuroreport* 29(7):588–593. https://doi.org/10.1097/WNR.0000000000001014.

Pi, Q., Maharjan, S., Yan, X., Liu, X., et al. (2018). Digitally tunable microfluidic bioprinting of multilayered cannular tissues, *Adv. Mater.* 30(43):1706913.

Poldervaart M. T., Goversen B., de Ruijter M., Abbadessa A., et al. (2017). 3D bioprinting of methacrylated hyaluronic acid (MeHA) hydrogel with intrinsic osteogenicity. *PLoS One* 12(6):e0177628. https://doi.org/10.1371/journal.pone.0177628.

Poli, D., Magliaro, C., and Ahluwalia, A. (2019). Experimental and computational methods for the study of cerebral organoids: A review. *Front. Neurosci.* 13:162. https://doi.org/10.3389/fnins.2019.00162.

Polioudakis, D., de la Torre-Ubieta, L., Langerman, J., Elkins, A. G., et al. (2019). A single-cell transcriptomic atlas of human neocortical development during mid-gestation. *Neuron* 103(5):785–801. https://doi.org/10.1016/j.neuron.2019.06.011.

Prior, N., Inacio, P., and Huch, M. (2019). Liver organoids: from basic research to therapeutic applications. *Gut* 68(12):2228–2237.

Qian, X., Jacob, F., Song, M. M., Nguyen, H. N., et al. (2018). Generation of human brain region-specific organoids using a miniaturized spinning bioreactor. *Nat. Protoc.* 13(3):565–580. https://doi.org/10.1038/nprot.2017.152.

Qian, X., Nguyen, H. N., Song, M. M., Hadiono, C., et al. (2016). Brain-region-specific organoids using mini-bioreactors for modeling ZIKV Exposure. *Cell* 165(5):1238–1254. https://doi.org/10.1016/j.cell.2016.04.032.

Qian, X., Song, H., and Ming, G. L. (2019). Brain organoids: Advances, applications and challenges. Development (Cambridge, England), 146(8), dev166074. https://doi.org/10.1242/dev.166074.

Qian, X., Su, Y., Adam, C. D., Deutschmann, A. U., et al. (2020). Sliced human cortical organoids for modeling distinct cortical layer formation. *Cell Stem Cell* 26(5):766–781. https://doi.org/10.1016/j.stem.2020.02.002.

Quadrato, G. and Arlotta, P. (2017). Present and future of modeling human brain development in 3D organoids. *Curr Opin Cell Biol.* 49:47–52. https://doi.org/10.1016/j.ceb.2017.11.010.

Ramadan, Q. and Zourob, M. (2021). 3D Bioprinting at the Frontier of Regenerative Medicine, Pharmaceutical, and Food Industries. *Front. Med. Technol.* 28;2:607648. doi: 10.3389/fmedt.2020.607648.

Ramadan, Q. and Zourob, M. (2021). 3D bioprinting at the frontier of regenerative medicine, pharmaceutical, and food industries. *Front. Med. Technol.* 28;2:607648. doi: 10.3389/fmedt.2020.607648. PMID: 35047890; PMCID: PMC8757855.

Rawal, P., Tripathi, D. M., Ramakrishna, S., and Kaur, S. (2021). Prospects for 3D bioprinting of organoids. *Bio-Des. Manuf.* 4(3):627–640. doi: 10.1007/s42242-020-00124-1.

Raza, F., et al. (2018). A review on recent advances in stabilizing peptides/proteins upon fabrication in hydrogels from biodegradable polymers, *Pharmaceutics* 10(1):16.

Rehman, A. U. and Qureshi, S. A. (2021). A review of the medical hyperspectral imaging systems and unmixing algorithms' in biological tissues, *Photodiagn. Photodyn. Ther.* 33:102165, ISSN 1572-1000, https://doi.org/10.1016/j.pdpdt.2020.102165.

Ren, Y., Yang, X., Ma, Z., Sun, X., et al. (2021). Developments and opportunities for 3D bioprinted organoids. *Int. J. Bioprinting* 7. https://doi.org/10.18063/ijb.v7i3.364.

Ren, Y., Yang, X., Ma, Z., Sun, X., et al. (2021). Developments and opportunities for 3D bioprinted organoids. *Int J Bioprint.* 28;7(3):364. doi: 10.18063/ijb.v7i3.364. PMID: 34286150; PMCID: PMC8287496.

Ricciardi, R., Auriemma, F., De Rosa, C., and Lauprêtre, F. (2004). X-ray diffraction analysis of poly (vinyl alcohol) hydrogels, obtained by freezing and thawing techniques. *Macromolecules* 37(5):1921–1927.

Richards, D. J., Li, Y., Kerr, C. M., Yao, J., et al. (2020). Human cardiac organoids for the modelling of myocardial infarction and drug cardiotoxicity. *Nat. Biomed. Eng.* 4(4):446–462. https://doi.org/10.1038/s41551-020-0539-4.

Rodriguez-Pascual F. and Slatter D. A. (2016). Collagen cross-linking: Insights on the evolution of metazoan extracellular matrix. *Sci Rep.* 6.

Ruppen, J., Wildhaber, F. D., Strub, C., Hall, S. R. R., et al. (2015). Towards personalized medicine: chemosensitivity assays of patient lung cancer cell spheroids in a perfused microfluidic platform, *Lab Chip* 15:3076–3085.

Sachs, N., Papaspyropoulos, A., Zomer-van Ommen, D. D., Heo, I., et al. (2019). Long-term expanding human airway organoids for disease modeling. *EMBO J.* 38(4):e100300.

Saheli, M., Sepantafar, M., Pournasr, B., Farzaneh, Z., et al. (2018). Three-dimensional liver-derived extracellular matrix hydrogel promotes liver organoids function. *J. Cell Biochem.* 119(6):4320–4333.

Sakaguchi, H., Kadoshima, T., Soen, M., Narii, N., et al. (2015). Generation of functional hippocampal neurons from self-organizing human embryonic stem cell-derived dorsomedial telencephalic tissue. *Nat. Commun.* 6:8896. https://doi.org/10.1038/ncomms9896.

Sakai S., Ohi H., Hotta T., Kamei H., et al. (2017). Differentiation potential of human adipose stem cells bioprinted with hyaluronic acid/gelatin-based bioink through microextrusion and visible light-initiated crosslinking. *Biopolymers.*

Salinas, M., Fuentes, N., Wang, X., Fan, Y. Y., et al. (2020). High-fat diet-induced obesity modulates colonic Lgr5+ stem cell homeostasis by dysregulating plasma membrane organization. *Curr. Dev. Nutr.* 4(Supplement_2):1684–1684.

Sampaziotis, F., Justin, A. W., Tysoe, O. C., et al. (2017). Reconstruction of the mouse extrahepatic biliary tree using primary human extrahepatic cholangiocyte organoids. *Nat. Med.* 23:954–963.

Santos, J. M., Cam~oes, S. P., Filipe, E., Cipriano, M., et al. (2015). Three-dimensional spheroid cell culture of umbilical cord tissue-derived mesenchymal stromal cells leads to enhanced paracrine induction of wound healing, *Stem Cell Res. Ther.* 6:1–19.

Scheinpflug, J., Pfeiffenberger, M., Damerau, A., Schwarz, F., et al. (2018). Journey into bone models: A review. *Genes* 9(5):247. https://doi.org/10.3390/genes9050247

Schutgens, F., Rookmaaker, M. B., Margaritis, T., Rios, A., et al. (2019). Tubuloids derived from human adult kidney and urine for personalized disease modeling. *Nat. Biotechnol.* 37(3):303–313.

Seyfoori, A., Samiei, E., Jalili, N., Godau, B., et al. (2018). Self-filling microwell arrays (SFMAs) for tumor spheroid formation, *Lab Chip* 18:3516–3528.

Shi, R., Radulovich, N., Ng, C., Liu, N., et al. (2020). Organoid cultures as preclinical models of non-small cell lung cancer. *Clin. Cancer Res.* 26(5):1162–1174. doi:10.1158/1078-0432.ccr-19-1376.

Shim J.-H., Lee J.-S., Kim J. Y., and Cho D.-W. (2012). Bioprinting of a mechanically enhanced three-dimensional dual cell-laden construct for osteochondral tissue engineering using a multi-head tissue/organ building system. *J. Micromech. Microeng.* 22 doi: 10.1088/0960-1317/22/8/085014.

Shinnawi, R., Shaheen, N., Huber, I., Shiti, A., et al. (2019). Modeling reentry in the short QT syndrome with human-induced pluripotent stem cell-derived cardiac cell sheets. *J. Am. Coll. Cardiol.* 73(18):2310–2324. https://doi.org/10.1016/j.jacc.2019.02.055.

Shri, M., Agrawal, H., Rani, P., Singh, D., et al. (2017). Hanging drop, a best three dimensional (3D) culture method for primary buffalo and sheep hepatocytes, *Sci. Rep.* 7:1–13.

Singh, N. K., Han, W., Nam, S. A., Kim, J. W., et al. (2020). Three-dimensional cell-printing of advanced renal tubular tissue analogue. *Biomaterials* 232:119734. https://doi.org/10.1016/j.biomaterials.2019.119734.

Singh, R., Cuzzani, O., Binette, F., Sternberg, H., et al. (2018). Pluripotent stem cells for retinal tissue engineering: Current status and future prospects. *Stem Cell Rev. Rep.* 14(4):463–483. https://doi.org/10.1007/s12015-018-9802-4.

Siqueira, G., Kokkinis, D., Libanori, R., Hausmann, M. K., et al. (2017). Cellulose nanocrystal inks for 3D printing of textured cellular architectures, *Adv. Funct. Mater.* 27(12):1604619.

Skardal A., Zhang J., and Prestwich G. D. (2010). Bioprinting vessel-like constructs using hyaluronan hydrogels crosslinked with tetrahedral polyethylene glycol tetracrylates. *Biomaterials* 31:6173–6181.

Skardal, A., Devarasetty, M., Kang, H. W., Seol, Y. J., et al. (2016). Bioprinting cellularized constructs using a tissue-specific hydrogel bioink. *J Vis. Exp.* 21(110):e53606. doi: 10.3791/53606. PMID: 27166839; PMCID: PMC4941985.

Sloan, S. A., Andersen, J., Paşca, A. M., Birey, F., et al. (2018). Generation and assembly of human brain region–specific three-dimensional cultures. *Nat. Protoc.* 13(9):2062–2085.

Smith, C. M., Stone, A. L., Parkhill, R. L., Stewart, R. L., et al. (2004). Three-dimensional bioassembly tool for generating viable tissue-engineered constructs. *Tissue Eng* 10:1566–1576.

Son, Y. S., Ki, S. J., Thanavel, R., Kim, J. J., et al. (2020). Maturation of human intestinal organoids in vitro facilitates colonization by commensal lactobacilli by reinforcing the mucus layer. *FASEB J.* 34(8):9899–9910.

Soret, P. A., Magusto, J., Housset, C., and Gautheron, J. (2020). In vitro and in vivo models of non-alcoholic fatty liver disease: A critical appraisal. *J. Clin. Med.* 10(1):36.

Spicer, P. P., Kretlow, J. D., Young, S., Jansen, J. A., et al. (2012). Evaluation of bone regeneration using the rat critical size calvarial defect. *Nat. Protoc.* 7(10):1918–1929. https://doi.org/10.1038/nprot.2012.113.

Sucher, R., Wagner, T., Köhler, H., Sucher, E., et al. (2022). Hyperspectral imaging (HSI) of human kidney allografts. *Ann. Surg.* 276(1):e48–e55.

Sun, W., Starly, B., Daly, A. C., Burdick, J. A., et al. (2020). The bioprinting roadmap. *Biofabrication* 12:022002. https://doi.org/10.1088/1758-5090/ab5158.

Taguchi, A., Kaku, Y., Ohmori, T., et al. (2014). Redefining the in vivo origin of metanephric nephron progenitors enables generation of complex kidney structures from pluripotent stem cells. *Cell Stem Cell* 14:53–67. https://doi.org/10.1016/j.stem.2013.11.010.

Takasato, M., Er, P., Chiu, H. et al. (2016). Erratum: Kidney organoids from human iPS cells contain multiple lineages and model human nephrogenesis. *Nature* 536:238. https://doi.org/10.1038/nature17982.

Tambalo, M. and Lodato, S. (2020). Brain organoids: Human 3D models to investigate neuronal circuits assembly, function and dysfunction. *Brain Res.* 1746:147028. https://doi.org/10.1016/j.brainres.2020.147028.

Tang, F., Barbacioru, C., Wang, Y., Nordman, E., et al. (2009). mRNA-Seq whole-transcriptome analysis of a single cell. *Nat Methods* 6(5):377–382. https://doi.org/10.1038/nmeth.1315.

Timlin, J. A., Carden, A., Morris, M. D., Bonadio, J. F., et al. (1999). Spatial distribution of phosphate species in mature and newly generated mammalian bone by hyperspectral Raman imaging. *J. Biomed. Opt.* 4(1):28–34. https://doi.org/10.1117/1.429918

Tindle, C., Fuller, M., Fonseca, A., Taheri, S., et al. (2021). Adult stem cell-derived complete lung organoid models emulate lung disease in COVID-19. *eLife* 10. doi:10.7554/eLife.66417.

Tiwari, S. K., Wang, S., Smith, D., Carlin, A. F., et al. (2021). Revealing tissue-specific SARS-CoV-2 infection and host responses using human stem cell-derived lung and cerebral organoids. *Stem Cell. Rep.* 16(3):437–445. doi:10.1016/j.stemcr.2021.02.005.

Trivedi, P., Liu, R., Bi, H., Xu, C., et al. (2021). 3D modeling of epithelial tumors-The synergy between materials engineering, 3D bioprinting, high-content imaging, and nanotechnology. *Int. J. Mol. Sci.* 9;22(12):6225. doi: 10.3390/ijms22126225. PMID: 34207601; PMCID: PMC8230141.

Trujillo, C. A., Gao, R., Negraes, P. D., Gu, J., et al. (2019). Complex oscillatory waves emerging from cortical organoids model early human brain network development. *Cell Stem Cell* 25(4):558–569.e7. https://doi.org/10.1016/j.stem.2019.08.002.

Tsai, C. C., Hong, Y. J., Lee, R. J., Cheng, N. C., et al. (2019). Enhancement of human adipose derived stem cell spheroid differentiation in an: In situ enzyme-crosslinked gelatin hydrogel, *J. Mater. Chem. B* 7:1064–1075.

Türker, E., Demirçak, N., and Arslan-Yildiz, A. (2018). Scaffold-free three-dimensional cell culturing using magnetic levitation, *Biomater. Sci.* 6:1745–1753.

Vakili, H., Kolakovic, R., Genina, N., Marmion, M., et al. (2015). Hyperspectral imaging in quality control of inkjet printed personalized dosage forms. *Int. J. Pharm.* 10;483(1–2):244–249. doi: 10.1016/j.ijpharm.2014.12.034. Epub 2014 Dec 16. PMID: 25527212.

Varzideh, F., Pahlavan, S., Ansari, H., Halvaei, M., et al. (2019). Human cardiomyocytes undergo enhanced maturation in embryonic stem cell-derived organoid transplants. *Biomaterials* 192:537–550. https://doi.org/10.1016/j.biomaterials.2018.11.033.

Vijayavenkataraman, S., Yan, W. C., Lu, W. F., Wang, C. H., et al. (2018). 3D bioprinting of tissues and organs for regenerative medicine. *Adv Drug*

Deliv Rev. 132:296–332. doi: 10.1016/j.addr.2018.07.004. Epub 2018 Jul 7. PMID: 29990578.

Voges, H. K., Mills, R. J., Elliott, D. A., Parton, R. G., et al. (2017). Development of a human cardiac organoid injury model reveals innate regenerative potential. *Development* 144 (6):1118–1127. doi:10.1242/ dev.143966.

Vorrink, S. U., Zhou, Y., Ingelman-Sundberg, M., Lauschke, V. M. (2018). Prediction of drug-induced hepatotoxicity using long-term stable primary hepatic 3D spheroid cultures in chemically defined conditions. *Toxicol Sci*. 163(2):655–665. doi:10.1093/toxsci/kfy058

Wahabzada, M., Manuela, B., Milad, K., Matheus, K., et al. (2017). Monitoring wound healing in a 3D wound model by hyperspectral imaging and efficient clustering. *PLoS One* 12. e0186425. 10.1371/journal.pone.0186425.

Wang, P., Li, X., Zhu, W., Zhong, Z., et al. (2018). 3D bioprinting of hydrogels for retina cell culturing. *Bioprinting* 12:e00029.

Wang, P., Mokhtari, R., Pedrosa, E., Kirschenbaum, M., et al. (2017). CRISPR/Cas9-mediated heterozygous knockout of the autism gene CHD8 and characterization of its transcriptional networks in cerebral organoids derived from iPS cells. *Mol. Autism* 8:11. https://doi.org/10.1186/s13229-017-0124-1.

Wang, S., Wang, X., Tan, Z., Su, Y., et al. (2019). Human ESC-derived expandable hepatic organoids enable therapeutic liver repopulation and pathophysiological modeling of alcoholic liver injury. *Cell Res*. 29(12):1009–1026.

Wang, X., Yan, Y., Pan, Y., Xiong, Z., et al. (2006). Generation of three-dimensional hepatocyte/gelatin structures with rapid prototyping system. *Tissue Eng*. 12(1):83–90. doi: 10.1089/ten.2006.12.83. PMID: 16499445.

Ware, M. J., Colbert, K., Keshishian, V., Ho, J., et al. (2016). Generation of homogenous three-dimensional pancreatic cancer cell spheroids using an improved hanging drop technique, *Tissue Eng. C Methods* 22:312–321.

Wechsler, M. E., Rao, V. V., Borelli, A. N., and Anseth, K. S. (2021). Engineering the MSC secretome: A hydrogel focused approach. *Adv. Healthc. Mater.* 10(7):e2001948. https://doi.org/10.1002/adhm.202001948.

Whitford, W. G., and Hoying, J. B. (2016). A bioink by any other name: Terms, concepts, and constructions related to 3D bioprinting, *Future Sci.* 2(3):FSO133.

Wiśniewski, J. R., Vildhede, A., Norén, A., and Artursson, P. (2016). In-depth quantitative analysis and comparison of the human hepatocyte and hepatoma cell line HepG2 proteomes, *J. Proteomics* 136.

Wolfe W. L. (1997). *Introduction to Imaging Spectrometers*, SPIE Press, Bellingham, Washington.

Worthington, K. S., Wiley, L. A., Kaalberg, E. E., Collins, M. M., et al. (2017). Two-photon polymerization for production of human iPSC-derived retinal cell grafts. *Acta Biomater.* 55:385–395.

Wu, Y., Hospodiuk, M., Peng, W., Gudapati, H., et al. (2019). Porous tissue strands: avascular building blocks for scalable tissue fabrication, *Biofabrication* 11.

Wüst S., Godla M. E., Müller R., and Hofmann S. (2014). Tunable hydrogel composite with two-step processing in combination with innovative hardware upgrade for cell-based three-dimensional bioprinting. *Acta Biomater.* 10:630–640. doi: 10.1016/j.actbio.2013.10.016

Xiang, Y., Tanaka, Y., Patterson, B., Kang, Y. J., et al. (2017). Fusion of regionally specified hiPSC-derived organoids models human brain development and interneuron migration. *Cell Stem Cell* 21(3):383–398.e7. https://doi.org/10.1016/j.stem.2017.07.007.

Xuan, W., Tipparaju, S. M., and Ashraf, M. (2022). Transformational applications of human cardiac organoids in cardiovascular diseases. *Front. Cell Dev. Biol.* 10:936084. https://doi.org/10.3389/fcell.2022.936084.

Xue, Y., Browne, A. W., Tang, W. C., Delgado, J., et al. (2019). Retinal organoids long-term functional characterization using two-photon fluorescence lifetime and hyperspectral microscopy. *Front. Cell. Neurosci.* 15:796903. https://doi.org/10.3389/fncel.2021.796903.

Yakoub, A. M. Sadek, M. (2018). Development and characterization of human cerebral organoids: An optimized protocol. *Cell Transplant.* 27(3):393–406.

Yang, X., Lu, Z., Wu, H., Li, W., et al. (2018). Collagen-alginate as bioink for three-dimensional (3D) cell printing based cartilage tissue engineering. *Mater. Sci. Eng. C.* https://doi.org/10.1016/j.msec.2017.09.002.

Yin, X., Mead, B. E., Safaee, H., Langer, R., et al. (2016). Engineering stem cell organoids. *Cell Stem Cell* 18(1): 25–38. https://doi.org/10.1016/j.stem.2015.12.005.

Yoo H. S., Lee E. A., Yoon J. J., and Park T. G. (2005). Hyaluronic acid modified biodegradable scaffolds for cartilage tissue engineering. *Biomaterials* 26(14):1925–1933.

Youk, J., Kim, T., Evans, K. V., Jeong, Y. I., et al. (2020). Three-dimensional human alveolar stem cell culture models reveal infection response to SARS-CoV-2. *Cell Stem Cell* 27(6):905–919. doi:10.1016/j.stem.2020.10.004.

Yu, L., Li, Z., Mei, H., Li, W., et al. (2021). Patient-derived organoids of bladder cancer recapitulate antigen expression profiles and serve as a personal evaluation model for CAR-T cells in vitro. *Clin. Transl. Immunol.* 10(2):e1248.

Yu, Y. and Ozbolat, I. T. (2014). Tissue strands as "bioink" for scale-up organ printing, in: 2014 36th Annu. Int. Conf. IEEE Eng. *Med. Biol. Soc.* EMBC 2014:1428–1431.

Yut, Y., Moncal, K. K., Li, J., Peng, W., et al. (2016). Three dimensional bioprinting using self-Assembling scalable scaffold-free "tissue strands" as a new bioink, *Sci. Rep.* 6:1–11.

Zhang, J., Ooi, J., Utami, K. H., et al. (2019). Expanded huntingtin CAG repeats disrupt the balance between neural progenitor expansion and differentiation in human cerebral organoids, 2019bioRxiv. https://doi.org/10.1101/850586.

Zhang, J., Wehrle, E., Rubert, M., Müller, R. (2021). 3D bioprinting of human tissues: Biofabrication, bioinks, and bioreactors. *Int. J. Mol. Sci.* 22:3971. https://doi.org/ 10.3390/ijms22083971.

Zhang, L., Hu, J., and Athanasiou, K.A. (2009). The role of tissue engineering in articular cartilage repair and regeneration. *Crit. Rev. Biomed. Eng.* 37:1–57.

Zhang, X., Wang, W., and Jin, Z. B. (2021). Retinal organoids as models for development and diseases. *Cell Regen.* 10:1–10.

Zhang Y., Yu Y., Chen H., and Ozbolat I. T. (2013). Characterization of printable cellular micro-fluidic channels for tissue engineering. *Biofabrication* 5(2):025004.

Zhao, Z., Chen, X., Dowbaj, A. M., Sljukic, A., et al. (2022). Organoids. *Nat. Rev. Methods Primers* 2(1):94.

Zhou, D., Chen, J., Liu, B., Zhang, X., et al. (2019). Bioinks for jet-based bioprinting. *Bioprinting* e00060. doi:10.1016/j.bprint.2019.e00060.

Zhou, J., Li, C., Sachs, N., Chiu, M. C., et al. (2018). Differentiated human airway organoids to assess infectivity of emerging influenza virus. *Proc. Natl. Acad. Sci. U. S. A.* 115(26):6822–6827. doi:10.1073/pnas.1806308115.

Zhu, X., Pei, L., Zhu, R., Jiao, Y., et al. (2018). Preparation and characterization of Sn/La co-doped TiO_2 nanomaterials and their phase transformation and photocatalytic activity. *Sci. Rep* 8(1):12387.

Chapter 8

Integration of Electronics with Biological Systems

Anish Maharjan, Julio Zuazola, and Robert Herbert
Department of Mechanical Engineering, Louisiana State University, Baton Rouge, LA 70803, USA
rherbert@lsu.edu

Soft electronics have emerged as a promising solution for integration with biological systems. With the use of stretchable and flexible designs and materials, electronics can be integrated onto skin, tissues, and tissue models for unobtrusive monitoring. Wearable devices are one prominent application of soft electronics to enable human health monitoring without interfering with daily activities. Within tissue models, soft electronics may be embedded to offer monitoring or stimulation without the use of bulky, rigid devices that would disturb sensitive tissues. In this chapter, the integration of soft electronics with biological systems is summarized. Key requirements for wearable electronics are briefly discussed. Wearable electronics are reviewed with a focus on electrochemical sensors and electrophysiological monitoring. Additionally, the

High-Throughput Imaging Methods: Cells and Tissues
Edited by Manas Ranjan Gartia and Ram V. Devireddy

ISBN 978-981-5352-08-5 (Hardcover), 978-1-003-77417-4 (eBook)
www.jennystanford.com

DOI: 10.1201/9781003774174-8

rapidly rising field of biohybrid electronics, where tissue and electronics are integrated for stimulation or monitoring of tissues, is summarized. Current limitations and future directions of soft electronics for integration with biological systems are highlighted.

8.1 Introduction

The development of soft electronics has enabled the intimate integration of electronics with biological systems [1–5]. Stretchable and flexible materials enable devices to conform to the tissue underneath while maintaining electronic functionality (Fig. 8.1A) [6]. While conventional devices rely on rigid materials that were obtrusive to biological materials, soft electronics enable unobtrusive designs to imperceptibly integrate with tissues and offer monitoring or stimulation. Soft electronics, built from stretchable thin film patterns or intrinsically soft materials, deform with soft biological materials to offer a reliable interface for monitoring and stimulation. Distinct from conventional electronics, soft electronics avoid irritation of tissues and seamlessly integrate onto or into biological materials.

A prominent use case of soft electronics is as wearable devices for human health monitoring (Fig. 8.1B) [7–11]. Soft electronics can be conformally integrated onto the skin and allow for monitoring without affecting daily activities. By laminating sensors on the skin, a wealth of information can be detected and analyzed. Electrochemical sensors, such as those measuring proteins, electrolytes, glucose, uric acid, and more, can be used to detect human health conditions through sweat, saliva, and wounds [12, 13]. While traditional measurement methods offer single data points collected with bulky systems, wearable sensors can allow for painless and continuous detection of key signals. Alongside electrochemical sensors, electrophysiological signals can be detected with wearable electrodes. Signals can include electrocardiography (ECG), electromyography (EMG), electrooculography (EOG), and electroencephalography (EEG) to detect heart, muscle, eye, and brain conditions, respectively [14]. Conventional recording systems are bulky and obtrusively attach to the skin, but soft electronic systems offer daily and continuous monitoring throughout daily activities. Soft electronics have also enabled other use cases, including integration within tissue and disease models (Fig. 8.1C) [15]. By matching the mechanics of tissues, electronics have been embedded into tissue models to offer advanced means of stimulation and sensing.

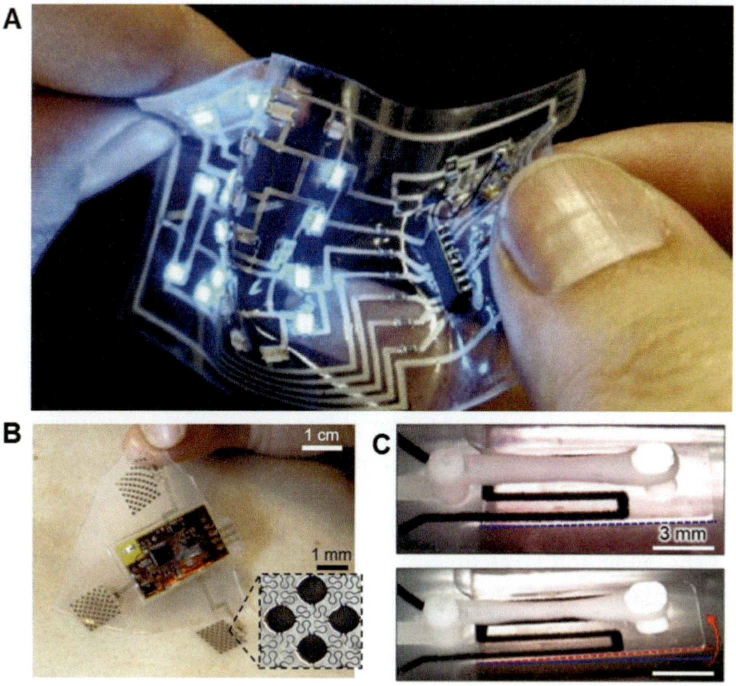

Figure 8.1 Soft electronics. (A) Stretchable circuit [6] (© Springer Nature, 2021). (B) Soft, wearable device for monitoring ECG [7] (© Wiley, 2019). (C) Soft sensor embedded in tissue model for monitoring [15] (© Wiley, 2023).

Collectively, this chapter summarizes the integration of soft electronics with biological systems, including the use cases of wearable devices and biohybrid electronics. First, general requirements for wearable electronics to conformally and reliably interface with the skin are introduced. Next, application areas of wearable electronics for human health monitoring are discussed. Wearable electrochemical sensors are reviewed for the monitoring of sweat, saliva, and wounds. Additionally, the use of wearable devices to measure electrophysiological signals is reviewed. Next, the emerging area of biohybrid electronics, where biomaterials and electronic materials are jointly patterned, is highlighted. Lastly, current limitations and future directions in the integration of electronics with biological systems are summarized.

8.2 Wearable Electronics

8.2.1 Integration with Skin

To integrate electronics with tissues, it is necessary for electronics to mechanically match, adhere to, and be compatible with placement on tissue [16]. Specifically for wearable electronics, electronic devices must be designed to comfortably attach to the skin, which necessitates stretching and deforming with the skin. Secondly, to stay attached to the skin, it is also critical for devices to adhere to the skin. In addition, long-term wearing, such as throughout a day, demands wearable devices to be breathable and avoid irritation to the skin. Overall, there are several key requirements for wearable electronics to successfully integrate with the skin.

A primary requirement for wearable electronics is to match the mechanical properties of the skin and deform during movements of the body and stretching of the skin. This is critical to maintain contact with the skin in both resting and moving conditions. The elastic modulus of the skin ranges from tens to hundreds of kPa, with a stretchability up to 15%. Figure 8.2A shows the range of moduli for tissues along with comparable materials used to build electronics [16]. Conventional materials that have been used for electronics, including metals and silicon, demonstrate moduli orders of magnitude higher than all tissues, including bone. A variety of classes of elastomers have been developed and recently applied toward electronic systems to fit in this design gap, including silicones. These elastomeric materials more closely match many biological soft tissues, including skin. There have been significant strides in the study of hydrogels and hydrogel-based electronic materials, which offer a means to create softer electronics and better integrate with the softest tissues. The use of materials with low moduli, or high stretchability and flexibility, to create wearable electronics is critical to achieve close contact with the skin. As shown in Fig. 8.2B, a conventional wearable electrode loses contact with the skin during normal movements [17]. However, an electrode based on soft materials and patterns maintains contact by deforming with the skin.

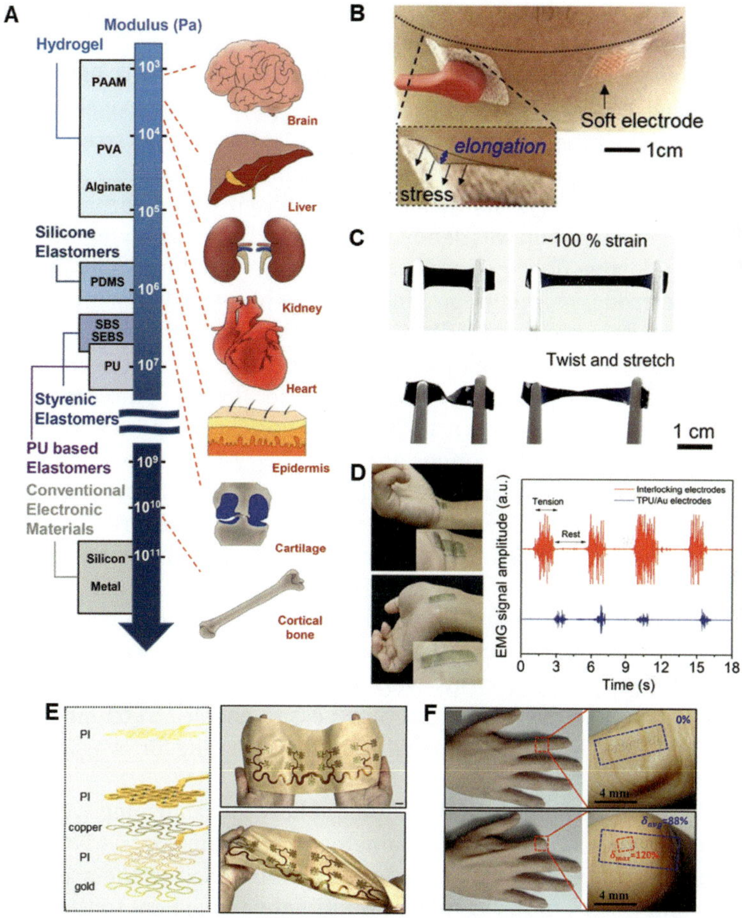

Figure 8.2 Mechanical properties of wearable electronics. (A) Comparison of moduli of tissues and materials for electronics [16] (© Wiley, 2021). (B) On-skin comparison of electrodes fabricated from conventional materials and soft materials [17] (© Springer Nature, 2017). (C) Image of hydrogel electrode stretching and twisting [19] (© Wiley, 2022). (D) Hydrogel electrode worn on skin to collect EMG signals [20] (© Wiley, 2020). (E) Micropatterned polymer and metal for a soft electrode [21] (© Wiley, 2023). (F) Wearable electrode fabricated from a narrow, thin silicon serpentine pattern [22] (© Wiley, 2024).

From a broad perspective, there is a plethora of methods to achieve stretchable and flexible electronics [18]. One option is to use electronic materials that are intrinsically soft, which includes elastomers and hydrogels. With this method, micro- and nano-fillers may be added to a soft matrix to offer electronic functionality while still maintaining stretchability and flexibility. An example of this is the double-network polymer hydrogel consisting of poly(3,4-ethylenedioxythiophene):poly(styrenesulfonate) (PEDOT:PSS) and polyvinyl alcohol (PVA) shown in Fig. 8.2C [19]. The hydrogel, which was demonstrated as an electrode for EMG monitoring, is stretchable up to 150% while demonstrating a conductivity of 10 S cm^{-1}. Figure 8.2D displays another electrode based on a hydrogel [20]. Here, a gold nanofilm was deposited onto electrospun polymer fibers. The conductive fiber network was then embedded into a hydrogel. The electrode, as pictured, conforms to the wrinkles of the skin, owing to its exceptional adhesion and stretchability, and captures EMG signals. An alternative method is to pattern conventional electronic materials into stretchable and flexible patterns. The electrode structure shown in Fig. 8.2E consists of patterned thin films of polyimide (PI), copper, and gold [21]. While copper and gold are not flexible, patterning into microns-thick films allows for high flexibility due to the low bending stiffness. To achieve stretchability, a serpentine or mesh-like pattern is used. This micropatterned electrode displayed less than a 0.11% change in resistance despite stretching to 15%. The use of stretchable patterns and thin films can also yield nearly imperceptible wearable electronics, as displayed in Fig. 8.2F [22]. Silicon strips were patterned into serpentine strips with a 10 μm width and 30 μm thickness to enable stretching up to 100%. The strips showed high durability by lasting for over 50,000 cycles.

While stretchability and flexibility allow for deforming with the skin, adhesion is also a key parameter to maintain contact with the skin. Conformal contact is a term frequently used in the scope of wearable electronics and indicates intimate attachment to the skin. As shown in Fig. 8.3A, a thinner material, which simultaneously increases flexibility, better conforms to the wrinkles of the skin [23]. Whereas gaps between the skin and a wearable device exist for thicker devices, conformal contact is achieved by creating a thinner device with a lower bending stiffness. Conformal contact is also dependent on a material's adhesion, which can be proven analytically and experimentally. Figure 8.3B illustrates conformal and nonconformal contact conditions for an electrode [7]. With a constant thickness, a more adhesive elastomeric layer enables conformal contact, as displayed by an analytical relationship in Fig. 8.3C distinguishing contact conditions. Low modulus and higher adhesion enable conformal contact. The importance of conformal contact from a signal collection perspective is highlighted in Figs. 8.3D and 8.3E. With conformal contact, the impedance between an electrode and the skin drops by over a factor of 3. Additionally, conformal contact and strong adhesion enable more consistent signal collection with a higher signal-to-noise ratio (SNR) during exercise.

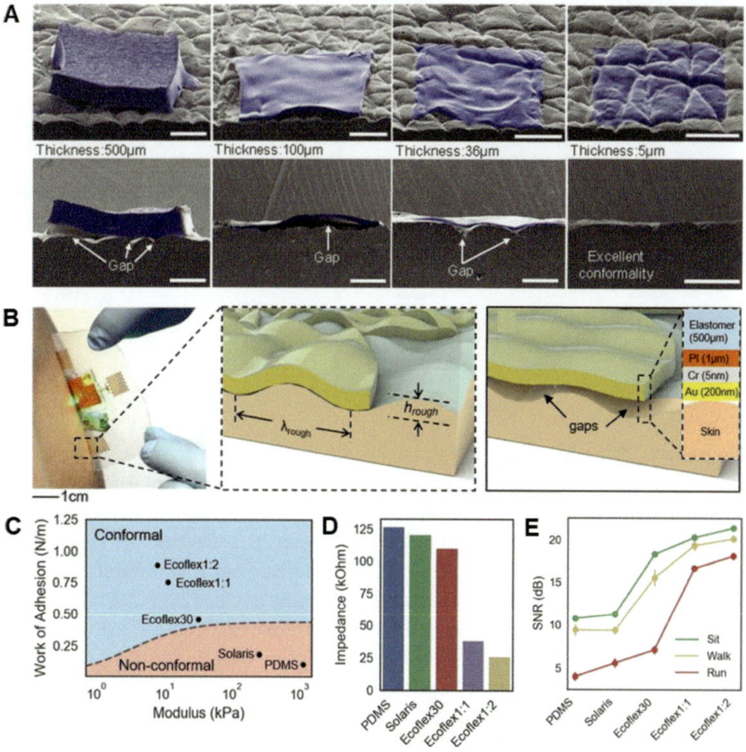

Figure 8.3 Adhesion and conformal contact requirements of soft electronics. (A) Images showing the improvement in conformal contact as the thickness of the substrate decreases [23] (© Wiley, 2013). (B) Image of a wearable ECG device with illustrations demonstrating conformal and nonconformal contact conditions [7] (© Wiley, 2019). (C) Theoretical estimate of required material properties to achieve conformal contact [7] (© Wiley, 2019). (D) Change in impedance correlates strongly with the predicted conformality from the theoretical prediction in (C) [7] (© Wiley, 2019). (E) The corresponding SNR values compares well with the change in impedance and conformal contact [7] (© Wiley, 2019).

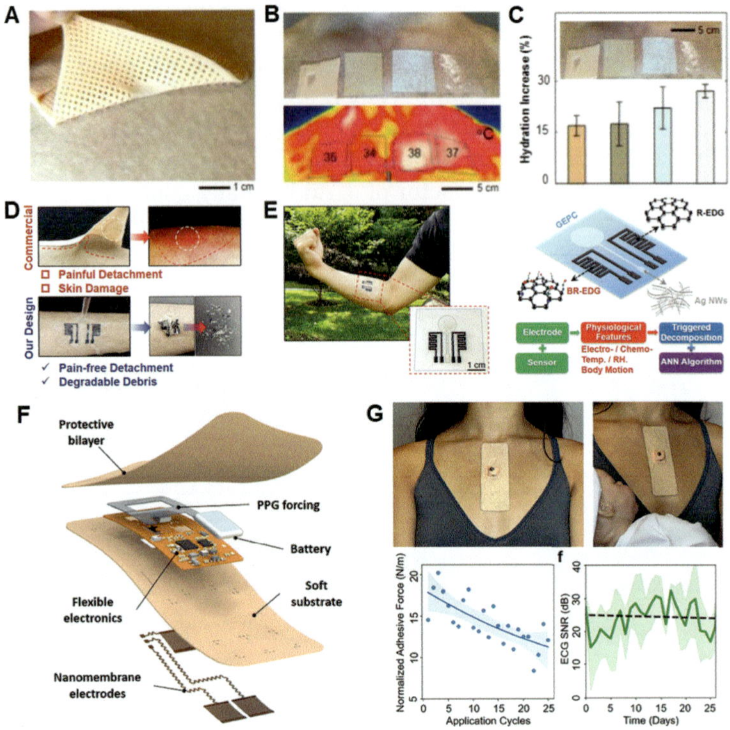

Figure 8.4 Breathability and reusability for wearable devices. (A) A wearable substrate with micropatterned pores to offer higher breathability [24] (© American Chemical Society, 2022). (B-C) The patterned substrate offered (B) better temperature control and (C) decreased collection of sweat [24] (© American Chemical Society, 2022). (D) Comparison of conventional adhesives for skin to a water-degradable wearable electronic device [25] (© Wiley, 2019). (E) Photos and illustration of degradable wearable electronics [25] (© Wiley, 2019). (F) A wearable ECG device for postpartum monitoring [27] (© Wiley, 2024). (G) The device enables daily monitoring for over 25 days, indicating high reusability [27] (© Wiley, 2024).

Along with maintaining contact with the skin, other requirements are considered when creating wearable electronics. Some applications, including daylong use, require the device to be permeable and breathable. Breathable substrates and device designs improve user comfort and decrease irritation of the skin. As shown in Fig. 8.4A, a substrate was created with perforations to enhance breathability [24]. Figures 8.4B and 8.4C demonstrate that the improved breathability allows the skin to maintain a more normal temperature and prevent the accumulation of sweat during exercise. Alongside breathability, detachment from the skin and reusability are increasingly the focus when developing new wearable electronics. While strong adhesion is desired, reversible modes of attachment may be required depending on adhesion strength. As shown in Fig. 8.4D, some commercially available skin-based devices offer strong adhesion but become painful to remove. To avoid difficult removal, one wearable device was created to degrade under running water and comfortably dissolve away from the skin [25]. The electrode shown in Fig. 8.4D is created from a protein-based substrate and a silver nanowire structure that decomposes in water. The device thus enables sensing from the skin before a triggered decomposition with running water (Fig. 8.4E). Another solution for reversible adhesion involves solidified oils within the substrate that are melted at an elevated temperature and reduce adhesion to the skin by more than 75% [26]. Heating is accomplished with an embedded resistive heater to control the release of oil and offer a switchable adhesive. The reusability of wearable and soft electronics is also becoming a focus point to move the field toward widespread usage. A flexible circuit and electrode device, shown in Fig. 8.4F, was successfully used over a month-long, at-home study to measure the ECG of a postpartum patient [27]. The device maintained sufficient adhesion to capture ECG with a high SNR throughout the month (Fig. 8.4G).

8.2.2 Wearable Electrochemical Sensors

The need for in situ examination of bodily fluids, including tears, sweat, and interstitial fluid (ISF), is growing, accelerating the development of wearable biosensors [28–30]. These bodily fluids constitute an abundant source of biomarkers that are critical for the diagnosis of several metabolic disorders, including diabetes, Parkinson's disease, and gout. These biomarkers include ions (Na^+, K^+, Ca^{2+}, and Cl^-); small molecules (glucose, lactate, uric acid, peptides, and ammonia); and proteins [28, 31–34]. The processes include target analyte recognition, transduction, and signal output from electrochemical sensing [35]. Analytes in biofluids are frequently identified using catalysts and enzymes on electrodes.

The detected analyte is transformed into a measurable proportional electrical signal by the electrode or transducer [33]. A detectable electrical signal, which may include changes in resistance, current, or potential, is produced by the sensor's electrochemical reaction with the target analyte [29].

8.2.2.1 Sensor parameters

There are a number of critical performance characteristics that affect the efficacy of electrochemical biosensors for monitoring bodily fluids continuously. The amount of time an electrochemical biosensor may operate efficiently before experiencing degraded performance is referred to as its lifetime. Enzyme stability, electrode fouling, and material deterioration are important factors influencing longevity. Extended lifespans are essential for continuous monitoring applications, particularly for wearable technology, to minimize the need for frequent replacements. Stable enzymes, anti-fouling coatings, and sensor design optimization are a few methods to increase lifespan [36].

A sensor's response time, which typically falls between seconds and minutes, indicates how quickly it can detect changes in analyte concentration and generate a measurable signal. Some parameters that influence response time include the analyte diffusion rate, electrode surface area, reaction kinetics, and sampling procedure. Ensuring real-time tracking of dynamic biomarkers requires rapid reaction times. In critical applications, such as continuous glucose monitoring, it is imperative to enable rapid measurements [34].

For electrochemical biosensors, reusability is an essential feature, especially for daily or continuous monitoring. Reusability is problematic for many electrochemical biosensors because of electrode fouling and biomolecule instability [37]. Reversible immobilization techniques for controlled attachment and release, anti-fouling coatings applied to the surface, and regeneration are some strategies to increase the reusability of electrochemical biosensors. Studies have also investigated methods based on nanomaterials, including graphene or carbon nanotubes, to increase stability and reusability [30, 37].

There is also significant interest, particularly within medical diagnostics, to simultaneously detect several analytes [29, 38]. Several design strategies are used in multi-sensor electrochemical biosensors, such as arrays of selective electrodes, multi-enzyme systems, and aptamer-based platforms for multiple analyte detection. These systems decrease the need for several sensors and provide a more complete health picture. However, creating efficient multi-sensor systems faces obstacles due to possible cross-reactivity between sensors and increased design complexity [30].

8.2.2.2 Sweat sensing

Analytes in readily accessible biofluids can be detected noninvasively with wearable electrochemical sensors, opening a window to the body's dynamic biomolecular state. Though there are a number of biofluids, wearable sensing is not compatible with many. For medical analyte monitoring, blood is considered the gold standard [32, 36, 39]. However, patients who find blood sampling difficult, such as neonates, the elderly, and patients with hemophobia, face significant challenges with repetitive and invasive drawing of blood. Analyte concentrations in tissue fluids and blood are extremely comparable [36]. However, the collection of samples requires the use of a small needle or hypodermic excitation current, both of which have the potential to irritate the dermal tissue [40]. Although urine is frequently utilized as a clinical medical sample, it is incompatible with continuous monitoring [41]. Certain salts, enzymes, proteins, and lipids found in tears can provide information on ocular disorders and diseases [42]. The current techniques for gathering tear samples, however, have the potential to irritate the eyes and induce reflex tears, which could influence comfort and deteriorate sensor signals [34]. Saliva is a rich source of indicators that properly reflect human status, such as hormones, enzymes, antibodies, and antimicrobial agents [43]. However, saliva monitoring is not without its difficulties, because the mouth can harbor a variety of contaminants, including food particles, which can affect data.

Among the many biofluids, sweat provides many benefits for wearable sensing. Sweat performs a range of physiological functions, including immunological defense, moisture retention, electrolyte and pH balance, and thermoregulation [44–47]. It has an abundance of compounds that communicate physiological information and molecular states within the body [41, 48–52]. Also, because sweat glands are located all over the body and can be collected in a noninvasive way, it is perfect for continuous monitoring. Soft, wearable sensors can continually collect and monitor a variety of sweat biomarkers [12]. Sweat contains a wide variety of biomarkers, including electrolytes, metabolites (viz., glucose, alcohol, and lactate); trace elements (viz., iron, copper, and zinc); small molecules (viz., cortisol, urea, and tyrosine); neuropeptides; and cytokines [12, 41, 50]. Sweat mostly consists of water molecules, which constitute approximately 99% of its total composition [53]. It has a pH of 6.3, which is mildly acidic [47]. Measurements of pH can reveal changes in the concentration of different electrolytes in perspiration, which can provide insight into metabolic activity and disease [50]. Wearable sweat sensors have applications across disease detection for medical patients and fitness tracking for athletes [39].

Two methods can be used to collect sweat samples: passive and active. In the passive technique, individuals exercise to produce sweat, such as by jogging or cycling. The active approach involves electrically stimulating a localized region of the skin. Iontophoresis is a popular technique for actively inducing perspiration and enables the collection of sweat while the subject is at rest. By providing a voltage between the iontophoretic electrodes, as illustrated in Fig. 8.5A, a current is created beneath the skin's surface, enabling the delivery of the agonistsuch as molecules of pilocarpine—to the sweat gland at the anode and inducing sweating [39]. Chloride, ethanol, and glucose levels have been monitored using this technique [54–59]. The conventional method of iontophoresis tends to corrode electrodes and irritate the skin [39]. To solve this issue, a periodically applied current source is used to prevent overheating and damage to the skin [54].

To collect and store sweat for on-skin analysis, microfluidic channels are commonly used in wearable devices. Sweat navigates through microfluidic channels and, occasionally, into reservoirs for storage and analysis. To maintain sweat delivery in microfluidic devices, a driving force is necessary. Sweat delivery, when not pressured by additional sweat flowing into the channel, can be triggered within wearable devices via capillary force, osmotic pressure, and evaporation pumps [36]. Figure 8.5B shows a microfluidic channel where sweat enters for analysis and then is removed by capillary force [60]. There are also several methodologies to control the flow and direction of sweat within more complex microfluidic systems. In Fig. 8.5C, the flow and pressure of sweat can be adjusted by incorporating valve structures [61]. To aid in the collection of sweat, a hydrogel may be positioned at the microfluidic inlet [62]. An evaporation-driven micropump can also efficiently create a continuous flow of sweat by creating micropores at the microfluidic exit [86]. While high-throughput production of microfluidic sweat sensors is a current challenge, one fabrication method used a laser-engraved microfluidic channel. It enabled wireless monitoring of sweat-derived tyrosine and uric acid [63].

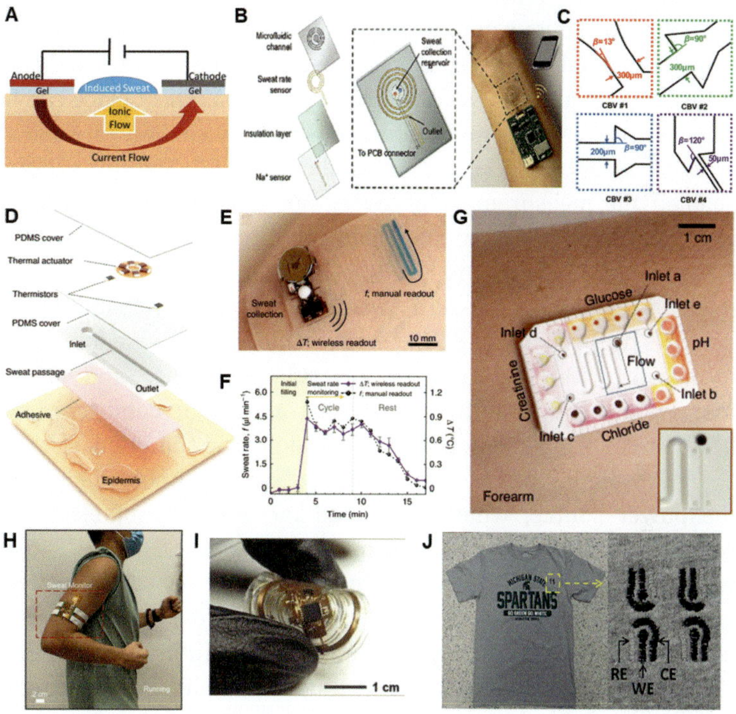

Figure 8.5 Wearable sweat monitors. (A) Schematic of iontophoresis to stimulate sweating [39] (© Elsevier, 2019). (B) Image of microfluidic channels to collect sweat and record signals with a circuit [60] (© American Chemical Society, 2018). (C) Design of capillary bursting valves to control flow of sweat on wearable devices [61] (© American Chemical Society, 2019). (D) Illustration of layers to form a soft electronic system to measure sweat flow rate [64] (© Springer Nature, 2021). (E) Image of on-skin device to wirelessly record sweat rates [64] (© Springer Nature, 2021). (F) Collected sweat rate data recording during exercise [64] (© Springer Nature, 2021). (G) Multimodal monitoring of sweat with a wireless and colorimetric device [64] (© Springer Nature, 2021). (H) Wearable armband for multimodal sweat sensing [65] (© Wiley, 2023). (I) NFC-enabled, battery-free device for sweat sensing [73] (© Wiley, 2018). (J) Electrochemical sensor electrodes woven into textiles [74] (© Royal Society of Chemistry, 2001).

In addition to monitoring chemicals and ions within sweat, wearable devices have been created to monitor sweat rate. A thermal actuator and thermistors were combined in a miniaturized, flexible package to monitor sweat rate (Fig. 8.5D) [64]. Sweat flows down a flexible sweat passage, and the resulting heat transfer can be used to analyze the flow rate in the range of 0 to 5 $\mu L \cdot min^{-1}$. This methodology was combined with a wireless readout scheme to enable continuous monitoring of sweat flow rate (Figs. 8.5E and 8.5F). A manual measurement was used to validate the wirelessly recorded data and showed a strong correlation. This device can be similarly packaged for multimodal sensing of sweat flow rate and an array of sweat biomarkers, as imaged in Fig. 8.5G. The soft package enabled the measurement of sweat rate, glucose, pH, creatinine, and chloride with both wireless and colorimetric readout methods. Colorimetric reagents are used to provide a colorful and visual readout method for on-skin sweat sensing. Another multimodal device is pictured in Fig. 8.5H, where six biosensors were integrated to measure $NH4^+$, Na^+, glucose, pH, skin impedance, and surface temperature [65]. Multimodal sensing allows for comprehensive health monitoring and detection of physiological conditions.

A key limitation of many current soft, wearable devices is the power source, which is needed for a variety of electronic components to process or transmit signals [32, 39, 41, 48]. Some wearable systems integrate energy-harvesting devices to achieve self-sustainable operation, to meet the power demand of wearable electronic devices, and avoid frequent interruptions in charging [52]. One smartwatch-based device used flexible zinc-manganese batteries and flexible photovoltaic cells for energy collection to recharge [66]. Sweat can be used by biofuel cells (BFCs) to convert it into energy to power implants [67]. As a demonstration, a stretchable, textile-based BFC was created to extract energy from sweat. Long-term, reliable power output, and rechargeability were achieved by utilizing a sweat-based BFC to extract energy from lactic acid and store it in an integrated supercapacitor [68]. Perspiration-powered and motion-powered devices have also been used to power wearable devices and detect sweat [69–72]. In addition to energy-harvesting methodologies, wireless powering is of high interest. Near-field communication (NFC) systems have been widely used to create battery-free wearable electronics. Figure 8.5I shows an

NFC device to monitor sweat loss and electrolyte composition. With NFC, miniaturized electronics with long-term signal viability can be realized [73].

Wearable devices must continuously and reliably monitor physiological data without disrupting daily activities. Therefore, materials must comfortably conform to the skin. Figure 8.5J shows one example of enhancing comfort, where electrodes for an electrochemical sensor were embroidered into a shirt [74]. A conductive thread of polyester and carbon ink was created for integration into the textile. Methods to unobtrusively embed electronics into wearable items will continue to be a focus in order to create comfortable and commercially feasible wearable sensors.

8.2.2.3 Wound sensing

Infections, especially in chronic wounds, can severely hinder recovery, and current clinical diagnostic methods have limitations. Thus, new, rapid, and simple techniques are needed [75]. This need has led to the development of wearable sensors integrated into wound dressings to detect biomarkers of infection and wound health, as well as to treat and heal the wound. The most widely utilized biomarkers are uric acid and pH, but lactic acid, oxygenation, inflammatory mediators, microorganisms, and their metabolites have also been used [75–79]. Wearable devices have primarily been created to measure uric acid and pH.

In the topic of wound sensing, a miniaturized, wireless electronic patch was created (Fig. 8.6A) [80]. To interface with the wound, the device used gentle hydrogel and PEDOT:PSS electrodes. Beyond enabling wound temperature and impedance sensing, the device also allowed for electrical stimulation of the wound to enable quicker healing, closure, and better vascularization, and recovery of normal skin properties (Fig. 8.6B). A closed-loop system was applied to offer controlled stimulation based on sensing signals, paving the way towards autonomous wound repair. The device was proven to heal wounds 25% more rapidly than a conventional bandage (Fig. 8.6C). Reusability of electronic components from wound bandages is of interest, as sterile wound dressings are required. With this in mind, a device was created with a detachable and reusable electronic subassembly that offered wireless monitoring (Fig. 8.6D) [81]. This reusable device was connected to a disposable set of bandages with inkjet-printed electrodes that interfaced with the wound. Designs offering reuse of electronics are key to the commercial and long-term viability of wearable electronics.

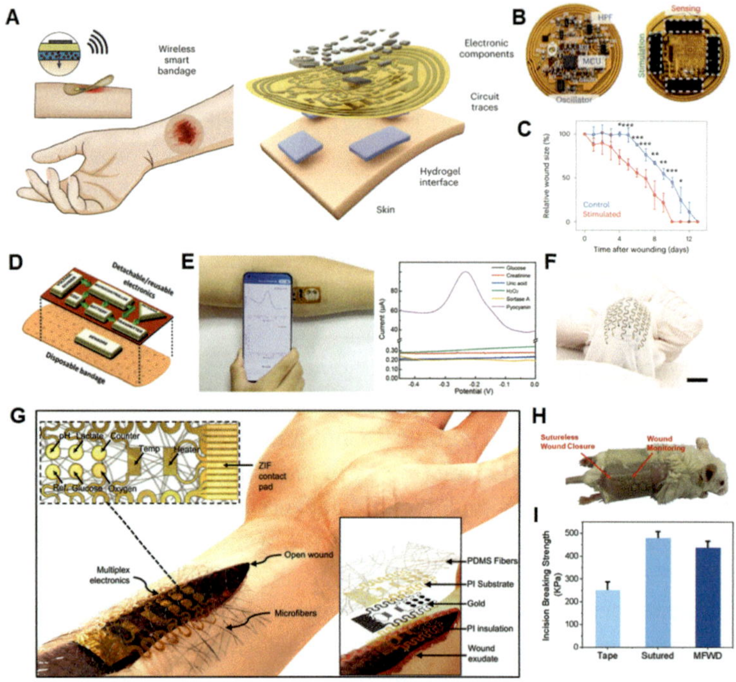

Figure 8.6 Sensing of skin wounds. (A) Illustration of an electronic bandage for wound monitoring [80] (© Springer Nature, 2022). (B) Images of miniaturized electronics for closed-loop sensing and stimulation of skin wounds [80] (© Springer Nature, 2022). (C) Data showing improvement in healing rate due to wearable electronics [80] (© Springer Nature, 2022). (D) Illustration of an electronic device packaging scheme using disposable, printed electrodes and reusable electronic interface [81] (© Springer Nature, 2016). (E) Image of wirelessly recording data from an adhesive, electronic bandage [82] (© Elsevier, 2023). (F) Highly strtetchable pH sensor for wound monitoring [85] (© American Chemical Society, 2017). (G) Schematic of an adhesive-free electronic device for wound monitoring [86] (© American Chemical Society, 2022). (H) A sutureless form of wound closure using soft electronics applied to a rat model [87] (© Wiley, 2021). (I) Data indicating strength of skin is similar when closed with sutures or the soft electronic device [87] (© Wiley, 2021).

The most common cause of hospital-acquired infections, especially in patients with burns or chronic wounds, is the Gram-negative bacterium *P. aeruginosa*, which secretes pyocyanin [82]. To detect this substance, one study developed an electrochemical sensor and integrated it within a smart bandage (Fig. 8.6E) [82]. The pyocyanin sensor showed high selectivity, as observed in Fig. 8.6E. As pyocyanin concentrations increased from 1 μM to 100 μM, peak currents linearly increased. Additionally, an electrode treated with MXenes displayed greater pyocyanin redox peaks than those of the bare electrode. Higher levels of uric acid are also present in chronic wounds, where it plays a role in wound progression. To utilize uric acid as a biomarker for wound infections, a smart bandage was developed with a biosensor integrated into a wound dressing [83]. pH is a second key metric for tracking chronic wounds, as a shift from acidic to alkaline typically indicates bacterial infection [84]. Monitoring the pH of the wound is essential for determining its condition, and early detection of infections or non-healing wounds is helpful for timing the application of medication. A highly stretchable pH sensor, based on a polyaniline conductive layer, for wound monitoring is imaged in Fig. 8.6F [85].

For wound monitoring, the substrate and adhesive covering are of major interest as they affect comfort and infection. To enhance comfort and breathability, an adhesive-free device was created using a permeable, elastomeric fiber matrix to allow for integration with the skin (Fig. 8.6G) [86]. The elastomeric fiber sheet offered high permeability to allow for cellular epithelization while also allowing for sensors to be interfaced with the wound for the sensing of various biomarkers. Another study investigated the use of soft materials with embedded sensors for the sutureless closure of wounds (Fig. 8.6H) [87]. This device used a self-healing and tough elastomer with antibacterial properties to improve wound healing and allow for contraction of the wound. Skin strength was measured after healing, and the soft material device enabled similar strength to that of suture-based closure. Moreover, the device offered wound temperature, pH, and glucose monitoring. Thus, soft materials offer both advanced means to effectively heal wounds while simultaneously enabling sensing.

8.2.2.4 Saliva monitoring

Saliva is an accessible biofluid that contains a variety of components, including protein biomarkers, metabolites, and electrolytes [88, 89].

Salivary analysis gives real-time information on dynamic changes within the body [90]. A variety of electrochemical approaches, including voltammetric, amperometric, potentiometric, and impedance spectroscopy methods, have been investigated for the identification of specific analytes found in saliva [90]. Thus far, mouthguards, retainers, and pacifiers have been developed with sensing capabilities.

A Bluetooth-enabled mouthguard was developed to enable monitoring of salivary uric acid levels (Fig. 8.7A) [91]. To fabricate the electrodes, a silver ink was screen-printed onto polyethylene terephthalate (PET) to interface with saliva from the mouthguard platform. Data were wirelessly tracked from a smartphone [23]. The salivary uric acid levels of hyperuricemia patients were monitored, and a distinction was observed over a multi-hour study (Fig. 8.7B). A mouthguard device was also fabricated for athletic applications as part of a combined saliva sensor and wearable sensor scheme to monitor physiological strain and dehydration. The mouthguard, imaged in Fig. 8.7C, consisted of a flexible circuit and rechargeable battery [92]. The device measures saliva osmolality as a biomarker for dehydration (Fig. 8.7D).

As an alternative to mouthguards, an ultralight, stretchable electronic system was designed for wireless monitoring of salt intake, featuring integrated circuits and serpentine interconnects embedded in a 200-μm-thick, oral elastomer chamber [93]. This device was packaged onto a smooth, microporous dental retainer with a microstructured ion-selective sodium sensor with wireless Bluetooth electronics. The device enabled real-time sodium intake tracking while consuming vegetable juice and other foods. Soft sodium sensors have also been developed, including one based on carbon black and soft elastomer composites [94].

Saliva monitoring is also highly beneficial for infants, particularly for monitoring sodium and potassium levels in newborns in intensive care units. To monitor these ions, a sensor-integrated pacifier was created (Fig. 8.7E) to replace the need to draw blood for tests [95]. Microfluidic channels were embedded into the pacifier to collect saliva and measure electrolytes. The device was evaluated with in vivo studies and successfully demonstrated the monitoring of salivary sodium levels (Fig. 8.7F).

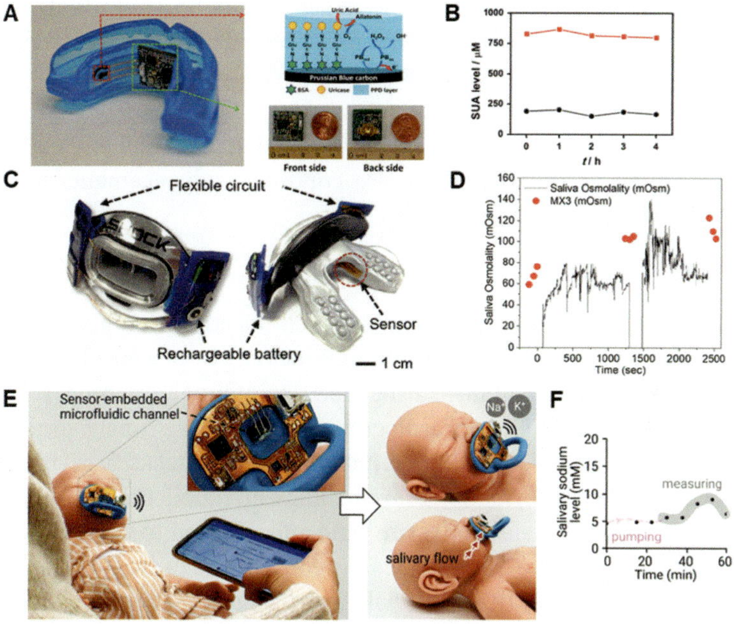

Figure 8.7 Saliva sensing. (A) Mouthguard-based device for sweat sensing of salivary uric acid [91] (© Elsevier, 2015). (B) Data comparison of normal with a hyperuricemia patient [91] (© Elsevier, 2015). (C) A flexible electronics system embedded into a mouthguard with a rechargeable battery [92] (© Wiley, 2024). (D) Wirelessly recorded signals of saliva osmolality changes during exercise [92] (© Wiley, 2024). (E) Microfluidics and electronics integrated into a pacifier for monitoring of newborns [95] (© Elsevier, 2022). (F) Recorded changes in sodium levels from a baby with electronic pacifier [95] (© Elsevier, 2022).

In addition, saliva sensing shows promise for disease detection. One system detected the SARS-CoV-2 nucleocapsid protein with low limit of detection in artificial saliva. The portable system featured Bluetooth communication, allowing it to work with digital health apps and deliver findings on-site in 30 minutes [88].

8.2.3 Biopotential Monitoring

Due to the electrochemical nature of our nerve impulses, there are several biological activities that can be monitored by measuring electric potentials. With the use of surface electrodes on the skin, biopotential sensing is noninvasive and does not require imaging methods. Thus, electrophysiological signals can be comfortably and continuously monitored with the application of wearable electronics. However, there are ongoing efforts to collect more reliable signals and to continue improving the comfort and form factor of devices. From making devices more flexible to miniaturizing them and embedding in clothing accessories, there are continued advancements being studied for biopotential sensing.

8.2.3.1 ECG

Since the heart creates and demands cyclic electrical impulses for its normal functioning, one can monitor its activity by placing electrodes on the chest area and measuring electric potential differences. The technique, called electrocardiography (ECG) is routinely used to diagnose a variety of conditions, including heart attacks, arrhythmia, abnormal heart formations, or insufficient blood flow. Although conventional ECG electrodes are rigid and adhered to the skin with adhesive gels, there have been continuous efforts to create soft, thin sensors. With soft, thin materials and designs, electrodes achieve better electric contact with the skin and minimize signal losses due to gaps in contact [96].

One solution to enhance user comfort with wearable electrode patches is to improve the substrate or adhesive material holding them to the skin. Thinner, more permeable patches improve breathability and decrease contact area, which makes them less irritating to the skin and obtrusive to the user. The material and pattern must endure large strains without rupture to comply with the elasticity of the skin underneath. One strategy is to use origami or kirigami patterns of otherwise rigid materials to make them flexible and stretchable [97]. In this work, a three-layered structure was created in a Kagome pattern (Fig. 8.8A) that unfolds under strain without tearing. The wearable device was used for five days without irritation, and the SNR stayed above 30 dB with discernible QRS complexes (Fig. 8.8B). Notably, there was also a reduction in the self-reported discomfort scale when the wearable device used a kirigami pattern (Fig. 8.8C).

In addition to using stretchable patterns and miniaturization of electronics, there are efforts to use intrinsically soft materials, such as composite elastomers, as electrodes. One study reported an electrode based on an elastomer of styrene–ethylene–butylene–styrene (SEBS) as a flexible scaffold with poly(3,4-ethylenedioxythiophene) and carbon black for conductivity [98]. The goal is to replace the traditionally used Ag/AgCl electrodes that require humid conditions for their gel. The composite electrode resists abrasion and records high-quality ECG signals. In a similar manner, silver nanoparticles were used as an additive to create conductive inks made with SEBS (Fig. 8.8D) [99]. Although its electrical properties are 500× better compared to a composite with carbon black and PEDOT:PSS, it loses electric contact at higher strains. Thus, interconnections with silver nanoparticles were printed to connect to carbon black and PEDOT:PSS electrodes that could better interface with the skin. The materials were patterned onto textile substrates to offer ECG monitoring (Fig. 8.8E). Another option for a soft, conductive material is to use liquid metals, such as eutectic gallium indium (EGaIn). By enclosing liquid metals in an elastomer, a self-healing conduit is formed that is more resistant and elastic than solid electrodes. A wearable device is pictured in Fig. 8.8F [100]. EGaIn and silver microflakes may be mixed to increase conductivity and enable printing of the liquid metal composite. Printed on a thermoplastic polyurethane (TPU) substrate (Fig. 8.8G), the liquid metal matrix maintains electrical connection with electronic components even under larger strains (1000%). The wearable device enables high-resolution monitoring of ECG profiles, as shown in Fig. 8.8H.

While some devices aim to alleviate the effects of strain with highly stretchable electronics, other methods include strain isolation techniques. The strains created when the patient moves can cause not only damage to electronics but also degrade the electrode-to-skin contact. To compensate, a strain isolation layer can be integrated into a wearable patch [101]. The strain-isolated soft bioelectronic (SIS) sensor (Fig. 8.8I) isolates the main detection electrode from the effects of strain, minimizing external interferences to the signal. The device also simultaneously collects acceleration data while recording ECG. While conventional sensors use irritating adhesives, ultrathin devices containing no adhesives are subjected to potential changes due to sudden movement and localized strain effects. Thus, to address this issue, the team also developed a machine learning solution to process data and retrieve the ECG signal despite noise. The patch was compared to a commercially available DEVICE (MAC-ECG), and it had similar performances when the patient is idle, but the proposed SIS performs better during exercise conditions (Fig. 8.8J).

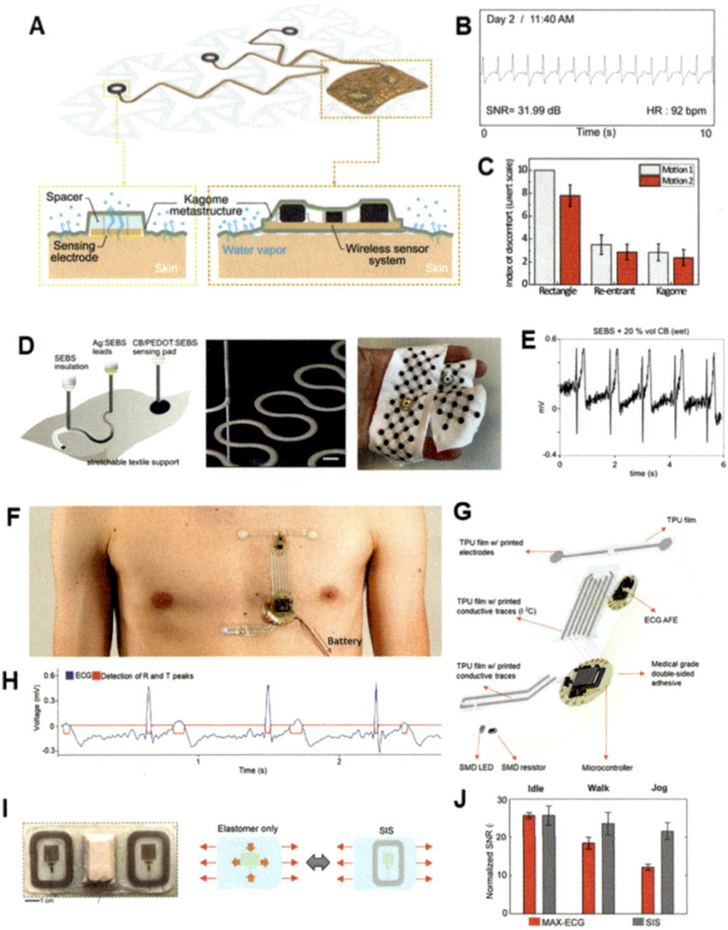

Figure 8.8 Wearable ECG monitors. (A) Kagome pattern with sensor added [97] (© Wiley, 2022). (B) ECG signal with the sensor [97] (© Wiley, 2022). (C) Comparison of the discomfort scale between different cutting patterns on the sensor with two movements [97] (© Wiley, 2022). (D) Schematics of the sensor layers, the shape of the printed sensors, and the resulting band [99] (© Wiley, 2023). (E) Resulting ECG signal [99] (© Wiley, 2023). (F) Sensor added to a patient with indicated battery [100] (© Wiley, 2022). (G) Different components of the TPU-based sensor [100] (© Wiley, 2022). (H) Resulting signal from wearable device [100] (© Wiley, 2022). (I) Complex sensor with strain compensator [101] (© Wiley, 2021). (J) Comparison between complex sensor with the strain isolator and a commercial MAX-ECG [101] (© Wiley, 2021).

8.2.3.2 EMG

EMG is the analysis of the electrical signals either on the studied muscle or between its structure and the corresponding motor nervous system. Most of the aforementioned advancements in ECG can also be applied to EMG to some degree. With the use of a composite consisting of a PEDOT and poly(styrene sulfonate) matrix doped with β-cyclodextrin and citric acid, a highly stretchability patch was made, reaching 700% strain without losing conductivity [102]. The mixture of β-cyclodextrin and citric acid behaves as an ionic liquid contained in the polymeric matrix, allowing for excellent adhesion (Fig. 8.9A) and better compliance with the underlying skin. Another blend made of PEDOT:PSS was doped with polyrotaxanes [103]. This motif allows for more flexibility, considering that cracks started to show at 50% strain rather than 5%.

Hydrogels can also be used as water-based patches with high flexibility and comfort. One study suggests a eutectogel based on polyurethane, tannic acid, and eutectic solvents that is self-adhesive, as imaged in Fig. 8.9B [104]. The electrodes sustained deformation up to 178% and were compared to traditional Ag/AgCl electrodes, showing less noise and better impedance matching (Fig. 8.9C). Other options for skin-integrated electrodes include PI substrates with silicon microneedle electrodes, as shown in Fig. 8.9D [105]. The structure is bioinspired by the thorns of the plant Radix Zanthoxylum. The microneedle electrodes showed a higher amplitude signal (Fig. 8.9E) when compared to conventional wet electrodes.

Additive manufacturing is becoming a common manufacturing technique for prototyping due to the ease with which one can customize shapes and test different materials. In one study, a number of 3D printed electrodes were reviewed and compared [106]. The first mentioned technique is fusion deposition modeling (FDM), with most of the work containing a printed plastic matrix with an embedded conductive wire or strip [107, 108]. Although conductive filaments do exist, they tend to have lower conductivity when compared to metals, but can be used where high deformations are more of a concern. Another method for 3D printing, called direct ink writing (DIW), uses ink droplets for material deposition instead of melting a filament. In one paper, the sensing pads for EMG were piezoelectric and were made using a multi-nozzle system [109].

Another printed electrode used TPU/carbon nanotube composites that are conductive and flexible [110]. Some other project reported a printed array of electrodes with 32 channels (Fig. 8.9F) [111]. The structure was 3D printed with silver channels and used silicone for fixation. The device achieved an SNR nearly 4× that of conventional electrodes.

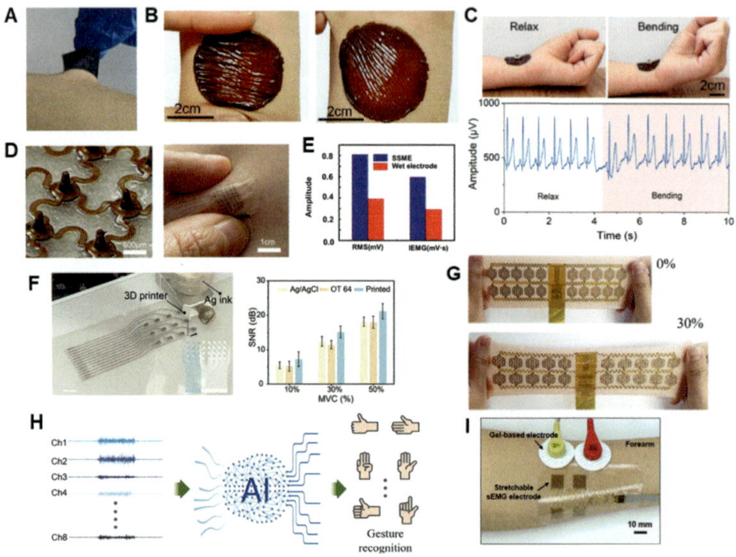

Figure 8.9 Wearable EMG devices. (A) Sensor patch being removed from the skin to demonstrate adherence without irritation [102] (© Springer Nature, 2022). (B) Skin patched in compressed and relaxed positions conforming to the skin [104] (© American Chemical Society, 2021). (C) The same patch added to a moving arm and the resulting EMG pattern, showing higher peaks when in movement [104] (© American Chemical Society, 2021). (D) Microneedle pattern zoomed in and resulting electrode [105] (© Springer Nature, 2023). (E) Comparison of the composite electrode and commercial wet electrode patch's signal intensity [105] (© Springer Nature, 2023). (F) Printed electrode with silver nanoparticle ink and resulting SNR values [111] (© Wiley, 2023). (G) Sensor array in its relaxed state and under strain [112] (© Springer Nature, 2023). (H) Schematics on the data processing from collection to treatment with a convolutional neural network to recognize specific gestures [112] (© Springer Nature, 2023). (I) Installed sensor array and gel-based commercial electrodes for comparison [113] (© Wiley, 2021).

In addition to collecting muscle activity, EMG may be used for pose identification. Due to the noninvasive nature of biopotential sensing, the goal is to incorporate motion sensors in clothes that can automatically determine poses for a variety of applications. A group of researchers used EMG with an array of electrodes for a gesture recognition system (Fig. 8.9G) [112]. Using a self-attention graph neural network (Fig. 8.9H) to interpret the data from the electrodes, 18 different hand positions, four of which were dynamic, were differentiated. The patches were composed of PEDOT:PSS with a Ti/Cu electrode due to commercial availability and high conductivity. After a series of data treatments, the confusion matrix showed over 90% accuracy. In another work, the sensors were tested for sign language recognition and were integrated with strain sensors that compensate for deformations in the underlying skin [113]. The reported device has a stable signal up to 30% strain and, when compared to gel electrodes (Fig. 8.9I), it showed a much stronger signal with a less bulky package.

8.2.3.3 EOG

EOG, another biopotential, may be measured by placing electrodes around the eyes to measure the differences in potential arising from eye movements in the vertical and horizontal directions [114]. Among a variety of applications, it can be used to diagnose Best disease, a genetic condition that creates yolk-like deformities in the eye's macula, or to detect the rapid-eye movement (REM) sleep phase in the polysomnography. There is a slightly different version called electroretinography (ERG) whose procedure is similar except that electrodes are placed on or behind the surface of the eye.

One group of researchers developed a headband with dry electrodes that cause minimal skin irritation than conventional wet, gel electrodes [115]. Made with TPU, the headband (Fig. 8.10A) contains flexible electrodes integrated with a flexible circuit for wireless transmission of EOG data. Using the wearable device, the collected data was processed with a convolutional neural network to perceive eye movement. With an accuracy of 98.3%, the EOG signals were used to control a two-wheeled remote-control car.

With a goal of creating a low-cost electrode, an electrode was fabricated with corn residue [116]. The patch was also made with polylactic acid (PLA) and PEDOT:PSS. To increase stretchability, a Y-shaped kirigami pattern was laser-cut into the material. The resulting patch showed an ~85% area coverage with high

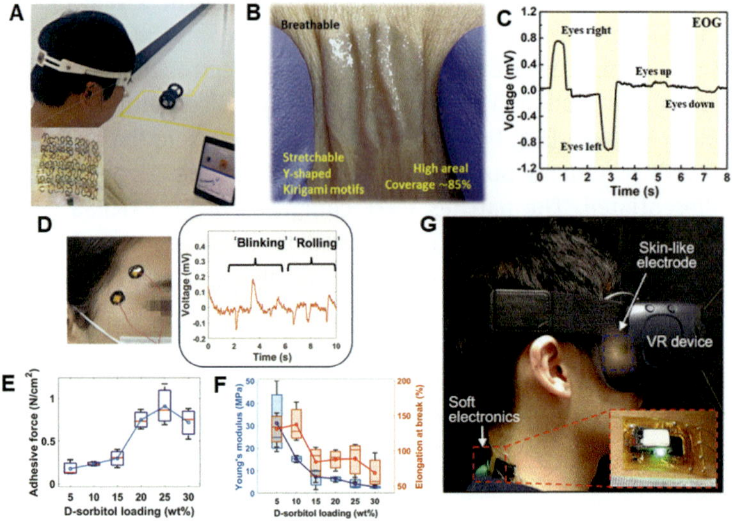

Figure 8.10 On-skin EOG monitoring systems. (A) Remote-control car controlled by EOG signals, with the patch electrode in the detail [115] (© American Chemical Society, 2023). (B) Electrode patch conforming to the skin underneath when pinched [116] (© American Chemical Society, 2021). (C) EOG result showing the electrical impulses for every movement [116] (© American Chemical Society, 2021). (D) Electrode position and the collected EOG showing the pulses for eye movement [118] (© American Chemical Society, 2024). (E) Comparison between D-sorbitol and adhesive force or (F) Young's modulus [118] (© American Chemical Society, 2024). (G) Electrode placement within the VR headset with the zoomed-in sensor in detail [120] (© The American Association for the Advancement of Science, 2020).

transparency and breathability (Fig. 8.10B). The electrodes successfully recorded EOG movements, as shown in Fig. 8.10C. Following the use of vegetable-based materials, the team made a starch-based patch for EOG by making a salt-doped conductive hydrogel [117]. The resulting material has a low Young's modulus (4.4 kPa) and endures strains up to 800% without losing conductivity. The most significant difference is that starch is not only a low-cost material, but it is also biodegradable, so the material can be readily disposed of. A water-soluble electrode was also reported with conductive patches of PEDOT:PSS [118]. Installed around the eyes of the patient (Fig. 8.10D), the main polymer constituting the patch was a mixture of PLA and D-sorbitol, both water-soluble components. As a dry electrode, it presents high adherence to the skin and conforms well due to the humidity absorbed from the air that dampens PLA but does not affect its conductivity. The research includes the influence of the concentration of D-sorbitol on the adhesive force, peaking at 25% wt. (Fig. 8.10E), and on Young's modulus (Fig. 8.10F), peaking at 10% wt. One group developed EOG electrodes for sight tracking [119]. Electrodes were made of multiple layers of silver nanowires and dielectric intermediate layers. Due to the alignment of the nanowires, the material is anisotropic, which makes signals in different directions noticeable. Moreover, it is self-healable and has better performance in transmitting signals due to the dielectric PDMS layers.

Highlighting the medical application of EOG, a wearable system was created for the detection of eye vergence during eye therapies (Fig. 8.10G) [120]. An array of electrodes used to detect eye position with precision was printed onto a soft membrane with silver nanoparticles as conductive ink. The raw signals initially showed only 34% accuracy, but after adding reference electrodes on the same implant area (around the nose bridge), the accuracy rose to 91%. The low-profile electronics were compatible with a VR system for mobile eye therapy. A software was created to train and correct eye vergence with both setups acting together to instruct the user on how to perform correcting exercises.

8.2.3.4 EEG

EEG, which involves the brain's electric impulses, can be recorded with scalp electrodes. Different periodic waves can be identified and tested, and by comparing them, doctors can diagnose conditions such as seizures, strokes, inflammation, tumors, or abnormal sleeping conditions. In one case, a kirigami-patterned electrode was coated with EGaIn [121]. The electrode withstood 820% strain and increased only 33% in resistance. The alloy is liquid at room temperature, and it is self-healing as long as it is contained by the patch. The membrane is made of silicone and coated with a thin gold film, to which the liquid metal alloy is added. In another project, MXenes were the conducting agent inside a chitosan matrix [122]. While conductivity and crosslinking speed were improved, the MXenes became much more stable against oxidation due to the physical barrier the polymer and the solvent offered. The material showed a conductivity higher than the underlying skin even after 1000 cycles of 30% strain loading, and the electrodes recorded EEG signals. Transparent electrodes are possible with acrylate elastomers and conductive PEDOT:PSS electrodes (Fig. 8.11A) [123]. The electrode endured strains up to 1500% and can also be used with a blood flow oxygenation sensor, based on light absorbance. The electrode patches presented good skin adherence and fatigue resistance. The sensor was compared with a commercially available option, and a spectrogram (Fig. 8.11B) was made to separate the different signals originating in the brain. With it, the amplitude of the α waves' signals was captured, showing a similar performance to wet patches despite the absence of a wet electrode.

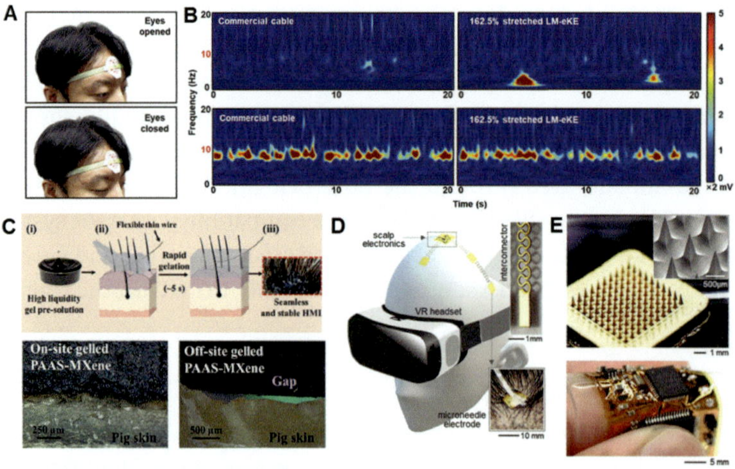

Figure 8.11 EEG measurement systems. (A) Electrode headband for EEG and (B) the spectrogram of the captured signal [121] (© Wiley, 2023). (C) Hydrogel electrode that can be applied to the skin to achieve conformal contact [122] (© American Chemical Society, 2022). (D) Applied electrode in the scalp with details of the local applied and serpentine pattern [125] (© Wiley, 2021). (E) Sensor microneedle array for application and composite foldable sensor [125] (© Wiley, 2021).

To achieve better conformal contact on the scalp despite hair, a hydrogel solution was formed that rapidly gels on the surface of the skin in 5 s (Fig. 8.11C) [124]. The low viscosity of the solution conforms to the skin before drying, whereas lamination of previously gelled hydrogel results in gaps in contact. With a similar purpose, by altering the proportion of polymer to water, it is also possible to create paintable electrodes. This work uses gelatin as the polymeric matrix to create a hydrogel whose consistency changes from solid to liquid at relatively low temperatures, making it perfect for printing directly on the skin surface. Glycerol was added for consistency control and stability, and salt was used to increase conductivity. The reference electrode can be added to the matrix and fixed in place without glue or other gels being needed. Projects involved in virtual reality and haptics are also interested in EEGs, as they can be used to interpret motor data to control real structures. In this work, a scalp implant was created (Fig. 8.11D) with microneedles (Fig. 8.11E) for sensing the electric potential and relaying it to an imbued wireless system [125]. The received data can then be processed in a CNN to interpret the reaction of the patient to specific pictures. After training the neural network, the system had a reported 93.22% accuracy, showing the precision of wearable EEG devices.

8.3 Biohybrid Electronics

While soft electronics have been extensively studied in the scope of wearable and implantable applications, there is current and emerging interest to integrate electronics with living biological systems to create biohybrid electronics with applications in disease models and bionic organs [126–128]. Biohybrid electronics, which integrate biological materials with soft electronics, offer advanced capabilities as embedded electronics offer monitoring or stimulation of biological materials. Such systems are often fabricated jointly, merging techniques from biofabrication of tissues and microfabrication of electronics. A prominent fabrication method is to combine the printing of electronics with bioprinting [128]. As illustrated in Fig. 8.12A, a biohybrid 3D printing method was created to print a tissue-sensor platform [15]. An electroconductive elastomeric ink was used to create a strain gauge, while a cardiac cell bioink was used to create heart tissue (Fig. 8.12B). The strain sensor was connected to a wireless data acquisition system to offer wireless and embedded monitoring of contractile force from the heart tissue (Fig. 8.12C). The integrated soft electronics and tissue were placed in an incubator for maturation. Similar to how soft electronics offer unobtrusive monitoring of human health with wearable applications, soft electronics can offer unobtrusive monitoring of disease and tissue models.

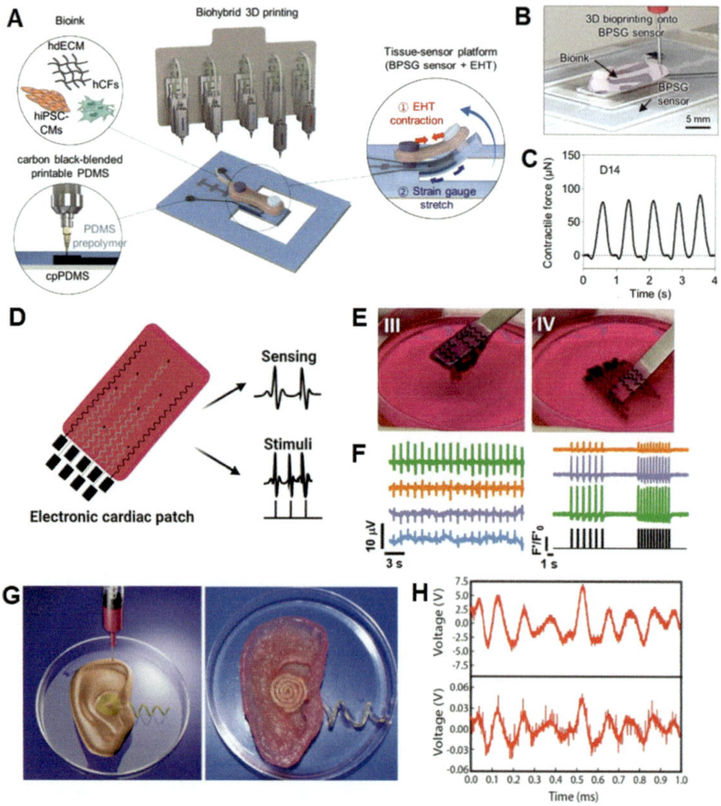

Figure 8.12 Biohybrid electronics. (A) A fully printed biohybrid device for sensing of cardiac tissues [15] (©Wiley, 2023). (B–C) The (B) tissue and sensor device could (C) monitor contractile forces in real-time [15] (© Wiley, 2023). (D) Printed electronics within a fabricated tissue for sensing and stimulation [129] (© Wiley, 2021). (E) The soft electronics and tissue device imaged being placed into and removed from culture media [129] (© Wiley, 2021). (F) Data showing the simultaneous stimulation and recording of the tissue [129] (© Wiley, 2021). (G) Bionic organ with printed electronics and bioprinted tissue [130] (© American Chemical Society, 2013). (H) Auditory signals recorded from the embedded electronics [130] (© American Chemical Society, 2013).

While the prior device showed the ability for embedded sensing of tissues, the process of printing tissues and electronics can also enable embedded stimulation of tissues [129]. In Fig. 8.12D shows an illustration of a cardiac patch printed with a PDMS-graphite composite for electronics and hydrogel-based cardiac cell bioink for heart tissue. Patterning soft strain sensors and stimulation electronics within tissue enabled control of the sensing and stimulation areas of the tissue. The integrated system is pictured in Fig. 8.12E being removed and placed into a culture medium. Figure 8.12F shows the recording of extracellular potentials from four locations in the tissue, along with the resulting changes in calcium after stimulation from the embedded electronics. The integration of soft electronics with tissue models will offer new avenues for advanced control and monitoring of tissues. In addition to tissue models, bionic organs have been developed with sensing functionality embedded within the organ. For example, Fig. 8.12G shows a bionic organ, fabricated with printed tissues and printed electronics [130]. An inductive coil was printed within the tissue to serve as an antenna for the ear and enable auditory sensing (Fig. 8.12H).

In addition to disease and tissue models, biohybrid robotics have also been developed [131, 132]. Biohybrid soft robots use living materials to mechanically actuate. For example, a swimming robot was developed using skeletal muscle, which could self-stimulate and induce contractions to generate movement [133]. Akin to the devices described above for tissue models, the robotic device was 3D printed with a cell-laden hydrogel and patterned to optimize force generation from the living tissue. Biohybrid robotics have also been created where embedded electronics generate controlled stimulation. One device could be wirelessly stimulated to contract cardiac tissue and enable movement [132].

8.4 Conclusion and Future Outlook

This chapter reviews recent developments of wearable electronics, with an emphasis on electrochemical sensing and electrophysiological monitoring. Wearable electronics that comfortably integrate with the skin and promote comfortability and reusability continue to be developed at a rapid pace as there is demand for continuous and at-home health monitoring. Key requirements for developing these electronics have been discussed in this chapter.

A variety of use cases of wearable electrochemical sensors were reviewed with applications across sweat, saliva, and wound monitoring. Key parameters of electrochemical biosensors were outlined, including lifetime, response time, reusability, and multi-sensor capability. Future directions for electrochemical sweat sensors encompass several key areas of development. These include the exploration of innovative materials for sensor construction, enhancement of biomarker detection capabilities, refinement of sweat collection techniques, and advancements in self-powering mechanisms. The latter is particularly crucial for continuous monitoring applications. Notably, self-powered systems, such as BFCs and energy harvesting technologies, represent a promising means to power wearable sensors. These advancements aim to improve the longevity and practicality of wearable applications, potentially revolutionizing personal health monitoring. Integration of NFC technologies and smartphone applications can enhance practicality and user-friendliness. Additionally, the development of multiplex sensors capable of simultaneously detecting multiple biomarkers could provide a more comprehensive understanding of wound healing processes.

Regarding biopotential sensing, the field still offers multiple opportunities for technological development with the use of different combinations of materials for the electrodes. One material that stands out is PEDOT:PSS, a polymeric blend that is conductive, highly elastic, and can substitute for traditional metal electrodes for a more comfortable experience. Liquid metals are a viable alternative for self-healing and highly elastic sensors, but they need to be contained within the polymeric matrix to avoid leakages. Regarding circuit encapsulation and adhesives, there are many options that can be used, ranging from single-use biodegradable polymers to elastomers for indefinite use. Continued development of materials and packaging schemes that promote switchable adhesion and reusability are necessary to better translate wearable electronics from research settings to commercial uses. Additionally, continued miniaturization and designs of wireless, wearable circuits are necessary to improve long-term viability and durability of wearable monitoring systems.

As techniques and capabilities of soft electronics and biofabrication continue to develop, the field of biohybrid electronics will continue to grow. The field possesses excellent potential within robotics and tissue or disease models. Embedding miniaturized soft electronics within tissue will create advanced monitoring and stimulating methods that are currently unachievable with bulky electronic systems. To this end, there is a major need to continue developing biocompatible electronics and materials that can endure long-term studies within culture medium and physiological conditions. Toxicity, material leakage, and swelling currently limit the choice of materials. In addition, merging the printing of electronics with bioprinting has been performed, although more work is needed to continue patterning within three-dimensional spaces.

Collectively, the use of stretchable and flexible materials and designs has been established as an excellent means of integrating with biological systems, whether laminated onto skin, implanted in organs, or embedded into laboratory tissue models. A variety of challenges still exist, and there remains steadfast growth and advancements to realize the full potential of soft electronics for interfacing with biological systems.

References

1. Gu, J. W., J. H. Lee, and S. K. Kang, 3D electronic sensors for bio-interfaced electronics and soft robotics. *Advanced Sensor Research*, 2023. 2(11): p. 2300013.
2. Herbert, R., et al., Soft material-enabled, flexible hybrid electronics for medicine, healthcare, and human-machine interfaces. *Materials*, 2018. 11(2): p. 187.
3. Chen, J., et al., Recent progress in essential functions of soft electronic skin. *Advanced Functional Materials*, 2021. 31(42): p. 2104686.
4. Lim, H. R., et al., Advanced soft materials, sensor integrations, and applications of wearable flexible hybrid electronics in healthcare, energy, and environment. *Advanced Materials*, 2020. 32(15): p. 1901924.
5. Wang, P., et al., The evolution of flexible electronics: from nature, beyond nature, and to nature. *Advanced Science*, 2020. 7(20): p. 2001116.

6. Lopes, P. A., et al., Reversible polymer-gel transition for ultra-stretchable chip-integrated circuits through self-soldering and self-coating and self-healing. *Nature Communications*, 2021. 12(1): p. 4666.
7. Kim, Y. S., et al., All-in-one, wireless, stretchable hybrid electronics for smart, connected, and ambulatory physiological monitoring. *Advanced Science*, 2019. 6(17): p. 1900939.
8. Heo, J. S., et al., Recent progress of textile-based wearable electronics: a comprehensive review of materials, devices, and applications. *Small*, 2018. 14(3): p. 1703034.
9. Lee, J., et al., Recent advances in 1D stretchable electrodes and devices for textile and wearable electronics: materials, fabrications, and applications. *Advanced Materials*, 2020. 32(5): p. 1902532.
10. Wang, L., K. Jiang, and G. Shen, Wearable, implantable, and interventional medical devices based on smart electronic skins. *Advanced Materials Technologies*, 2021. 6(6): p. 2100107.
11. Ling, Y., et al., Disruptive, soft, wearable sensors. *Advanced Materials*, 2020. 32(18): p. 1904664.
12. Gao, F., et al., Wearable and flexible electrochemical sensors for sweat analysis: A review. *Microsystems & Nanoengineering*, 2023. 9(1): p. 1–21.
13. Yao, Y., et al., Flexible and stretchable organic electrochemical transistors for physiological sensing devices. *Advanced Materials*, 2023. 35(35): p. 2209906.
14. Wu, H., et al., Materials, devices, and systems of on-skin electrodes for electrophysiological monitoring and human–machine interfaces. *Advanced Science*, 2021. 8(2): p. 2001938.
15. Yong, U., et al., Biohybrid 3D printing of a tissue-sensor platform for wireless, real-time, and continuous monitoring of drug-induced cardiotoxicity. *Advanced Materials*, 2023. 35(11): p. 2208983.
16. Shim, H. J., et al., Functionalized elastomers for intrinsically soft and biointegrated electronics. *Advanced Healthcare Materials*, 2021. 10(17): p. 2002105.
17. Lee, Y., et al., Soft electronics enabled ergonomic human-computer interaction for swallowing training. *Scientific Reports*, 2017. 7(1): p. 46697.
18. Yuan, Z., et al., Flexible and stretchable strategies for electronic skins: Materials, structure, and integration. *ACS Applied Electronic Materials*, 2021. 4(1): p. 1–26.

19. Li, G., et al., Highly conducting and stretchable double-network hydrogel for soft bioelectronics. *Advanced Materials*, 2022. 34(15): p. 2200261.

20. Pan, S., et al., Mechanically interlocked hydrogel–elastomer hybrids for on-skin electronics. *Advanced Functional Materials*, 2020. 30(29): p. 1909540.

21. Kim, H., et al., AR-enabled persistent human–machine interfaces via a scalable soft electrode array. *Advanced Science*, 2024. 11(7): p. 2305871.

22. Shi, Y., et al., Freestanding serpentine silicon strips with ultrahigh stretchability over 300% for wearable electronics. *Advanced Materials*, 2024. 36(24): p. 2313603.

23. Jeong, J.-W., et al., Materials and optimized designs for human-machine interfaces via epidermal electronics. *Advanced Materials*, 2013. 25(47).

24. Rodeheaver, N., et al., Breathable, wireless, thin-film wearable biopatch using noise-reduction mechanisms. *ACS Applied Electronic Materials*, 2022. 4(1): p. 503–512.

25. Zhang, Y. and T. H. Tao, Skin-friendly electronics for acquiring human physiological signatures. *Advanced Materials*, 2019. 31(49): p. 1905767.

26. Jinkins, K. R., et al., Thermally switchable, crystallizable oil and silicone composite adhesives for skin-interfaced wearable devices. *Science Advances*, 2022. 8(23): p. eabo0537.

27. Matthews, J., et al., Cloud-integrated smart nanomembrane wearables for remote wireless continuous health monitoring of postpartum women. *Advanced Science*, 2024. 11(13): p. 2307609.

28. Park, H., W. Park, and C. H. Lee, Electrochemically active materials and wearable biosensors for the in situ analysis of body fluids for human healthcare. *NPG Asia Materials*, 2021. 13(1): p. 23.

29. Lee, J., et al., Advances in electrochemical sensors for detecting analytes in biofluids. *Advanced Sensor Research*, 2023. 2(8): p. 2200088.

30. Wu, J., et al., Device integration of electrochemical biosensors. *Nature Reviews Bioengineering*, 2023. 1(5): p. 346–360.

31. Broza, Y. Y., et al., Disease detection with molecular biomarkers: From chemistry of body fluids to nature-inspired chemical sensors. *Chemical Reviews*, 2019. 119(22): p. 11761–11817.

32. Brasier, N. and J. Eckstein, Sweat as a source of next-generation digital biomarkers. *Digital Biomarkers*, 2020. 3(3): p. 155–165.

33. Sim, D., et al., Biomarkers and detection platforms for human health and performance monitoring: A review. *Advanced Science*, 2022. 9(7): p. 2104426.

34. Wu, Z., et al., Interstitial fluid-based wearable biosensors for minimally invasive healthcare and biomedical applications. *Communications Materials*, 2024. 5(1): p. 33.

35. Ferrari, A. G.-M., S. J. Rowley-Neale, and C. E. Banks, Screen-printed electrodes: Transitioning the laboratory in-to-the field. *Talanta Open*, 2021. 3: p. 100032.

36. Ribet, F., et al., Minimally invasive and volume-metered extraction of interstitial fluid: Bloodless point-of-care sampling for bioanalyte detection. 2020.

37. Morales, M. A. and J. M. Halpern, Guide to selecting a biorecognition element for biosensors. *Bioconjugate Chemistry*, 2018. 29(10): p. 3231–3239.

38. Salim, A. and S. Lim, Recent advances in noninvasive flexible and wearable wireless biosensors. *Biosensors and Bioelectronics*, 2019. 141: p. 111422.

39. Legner, C., et al., Sweat sensing in the smart wearables era: Towards integrative, multifunctional and body-compliant perspiration analysis. *Sensors and Actuators A: Physical*, 2019. 296: p. 200–221.

40. McCaul, M., T. Glennon, and D. Diamond, Challenges and opportunities in wearable technology for biochemical analysis in sweat. *Current Opinion in Electrochemistry*, 2017. 3(1): p. 46–50.

41. Bariya, M., H. Y. Y. Nyein, and A. Javey, Wearable sweat sensors. *Nature Electronics*, 2018. 1(3): p. 160–171.

42. Yang, Y. and W. Gao, Wearable and flexible electronics for continuous molecular monitoring. *Chemical Society Reviews*, 2019. 48(6): p. 1465–1491.

43. Mannoor, M. S., et al., Graphene-based wireless bacteria detection on tooth enamel. *Nature Communications*, 2012. 3(1): p. 763.

44. Tan, C. L. and Z. A. Knight, Regulation of body temperature by the nervous system. *Neuron*, 2018. 98(1): p. 31–48.

45. Hendricks, A. J., et al., Sweat mechanisms and dysfunctions in atopic dermatitis. *Journal of Dermatological Science*, 2018. 89(2): p. 105–111.

46. Dervisevic, M., et al., Skin in the diagnostics game: Wearable biosensor nano-and microsystems for medical diagnostics. *Nano Today*, 2020. 30: p. 100828.

47. Jadoon, S., et al., Recent developments in sweat analysis and its applications. *International Journal of Analytical Chemistry*, 2015. 2015(1): p. 164974.

48. Kim, J., et al., Simultaneous monitoring of sweat and interstitial fluid using a single wearable biosensor platform. *Advanced Science*, 2018. 5(10): p. 1800880.

49. Choi, J., et al., Skin-interfaced systems for sweat collection and analytics. *Science Advances*, 2018. 4(2): p. eaar3921.

50. Sonner, Z., et al., The microfluidics of the eccrine sweat gland, including biomarker partitioning, transport, and biosensing implications. *Biomicrofluidics*, 2015. 9(3).

51. Li, M., et al., A highly integrated sensing paper for wearable electrochemical sweat analysis. *Biosensors and Bioelectronics*, 2021. 174: p. 112828.

52. Bandodkar, A. J., I. Jeerapan, and J. Wang, Wearable chemical sensors: present challenges and future prospects. *ACS Sensors*, 2016. 1(5): p. 464–482.

53. Wilke, K., et al., A short history of sweat gland biology. *International Journal of Cosmetic Science*, 2007. 29(3): p. 169–179.

54. Emaminejad, S., et al., Autonomous sweat extraction and analysis applied to cystic fibrosis and glucose monitoring using a fully integrated wearable platform. *Proceedings of the National Academy of Sciences*, 2017. 114(18): p. 4625–4630.

55. Kim, J., et al., Noninvasive alcohol monitoring using a wearable tattoo-based iontophoretic-biosensing system. *ACS Sensors*, 2016. 1(8): p. 1011–1019.

56. Campbell, A. S., J. Kim, and J. Wang, Wearable electrochemical alcohol biosensors. *Current Opinion in Electrochemistry*, 2018. 10: p. 126–135.

57. Buono, M. J., Sweat ethanol concentrations are highly correlated with co-existing blood values in humans. *Experimental Physiology*, 1999. 84(2): p. 401–404.

58. Kim, J., A. S. Campbell, and J. Wang, Wearable non-invasive epidermal glucose sensors: A review. *Talanta*, 2018. 177: p. 163–170.

59. Lee, H., et al., Wearable/disposable sweat-based glucose monitoring device with multistage transdermal drug delivery module. *Science Advances*, 2017. 3(3): p. e1601314.

60. Nyein, H. Y. Y., et al., A wearable microfluidic sensing patch for dynamic sweat secretion analysis. *ACS Sensors*, 2018. 3(5): p. 944–952.

61. Choi, J., et al., Soft, skin-integrated multifunctional microfluidic systems for accurate colorimetric analysis of sweat biomarkers and temperature. *ACS Sensors*, 2019. 4(2): p. 379–388.

62. Lin, S., et al., Natural perspiration sampling and in situ electrochemical analysis with hydrogel micropatches for user-identifiable and wireless chemo/biosensing. *ACS Sensors*, 2019. 5(1): p. 93–102.

63. Nyein, H. Y. Y., et al., A wearable patch for continuous analysis of thermoregulatory sweat at rest. *Nature Communications*, 2021. 12(1): p. 1823.

64. Kwon, K., et al., An on-skin platform for wireless monitoring of flow rate, cumulative loss and temperature of sweat in real time. *Nature Electronics*, 2021. 4(4): p. 302–312.

65. Huang, X., et al., Intelligent soft sweat sensors for the simultaneous healthcare monitoring and safety warning. *Advanced Healthcare Materials*, 2023. 12(15): p. 2202846.

66. Zhao, J., et al., A fully integrated and self-powered smartwatch for continuous sweat glucose monitoring. *ACS Sensors*, 2019. 4(7): p. 1925–1933.

67. Escalona-Villalpando, R. A., et al., Clean energy from human sweat using an enzymatic patch. *Journal of Power Sources*, 2019. 412: p. 496–504.

68. Lv, J., et al., Sweat-based wearable energy harvesting-storage hybrid textile devices. *Energy & Environmental Science*, 2018. 11(12): p. 3431–3442.

69. Yu, Y., et al., Biofuel-powered soft electronic skin with multiplexed and wireless sensing for human-machine interfaces. *Science Robotics*, 2020. 5(41): p. eaaz7946.

70. Liu, D., et al., A constant current triboelectric nanogenerator arising from electrostatic breakdown. *Science Advances*, 2019. 5(4): p. eaav6437.

71. Song, Y., et al., Wireless battery-free wearable sweat sensor powered by human motion. *Science Advances*, 2020. 6(40): p. eaay9842.

72. Yin, L., et al., A self-sustainable wearable multi-modular E-textile bioenergy microgrid system. *Nature Communications*, 2021. 12(1): p. 1542.

73. Kim, S. B., et al., Soft, skin-interfaced microfluidic systems with wireless, battery-free electronics for digital, real-time tracking of sweat loss and electrolyte composition. *Small*, 2018. 14(45): p. 1802876.

74. Liu, X. and P. B. Lillehoj, Embroidered electrochemical sensors for biomolecular detection. *Lab on a Chip*, 2016. 16(11): p. 2093–2098.

75. Siddiqui, A. R. and J. M. Bernstein, Chronic wound infection: Facts and controversies. *Clinics in Dermatology*, 2010. 28(5): p. 519–526.

76. Mota, F. A., et al., Biomarkers in the diagnosis of wounds infection: An analytical perspective. *TrAC Trends in Analytical Chemistry*, 2021. 143: p. 116405.

77. O'Callaghan, S., et al., 'Smart'wound dressings for advanced wound care: a review. *Journal of Wound Care*, 2020. 29(7): p. 394–406.

78. Brown, M. S., B. Ashley, and A. Koh, Wearable technology for chronic wound monitoring: Current dressings, advancements, and future prospects. *Frontiers in Bioengineering and Biotechnology*, 2018. 6: p. 47.

79. Tegl, G., et al., Biomarkers for infection: Enzymes, microbes, and metabolites. *Applied Microbiology and Biotechnology*, 2015. 99: p. 4595–4614.

80. Jiang, Y., et al., Wireless, closed-loop, smart bandage with integrated sensors and stimulators for advanced wound care and accelerated healing. *Nature Biotechnology*, 2023. 41(5): p. 652–662.

81. Farooqui, M. F. and A. Shamim, Low cost inkjet printed smart bandage for wireless monitoring of chronic wounds. *Scientific Reports*, 2016. 6(1): p. 28949.

82. Shi, Z., et al., Wearable battery-free smart bandage with peptide functionalized biosensors based on MXene for bacterial wound infection detection. *Sensors and Actuators B: Chemical*, 2023. 383: p. 133598.

83. Arcangeli, D., et al., Smart bandaid integrated with fully textile OECT for uric acid real-time monitoring in wound exudate. *ACS Sensors*, 2023. 8(4): p. 1593–1608.

84. Yang, P., et al., Orange-emissive carbon quantum dots: Toward application in wound pH monitoring based on colorimetric and fluorescent changing. *Small*, 2019. 15(44): p. 1902823.

85. Rahimi, R., et al., Highly stretchable potentiometric pH sensor fabricated via laser carbonization and machining of carbon–polyaniline composite. *ACS Applied Materials & Interfaces*, 2017. 9(10): p. 9015–9023.

86. Brown, M. S., et al., Adhesive-free, stretchable, and permeable multiplex wound care platform. *ACS Sensors*, 2022. 7(7): p. 1996–2005.

87. Tang, N., et al., Highly efficient self-healing multifunctional dressing with antibacterial activity for sutureless wound closure and infected wound monitoring. *Advanced Materials*, 2022. 34(3): p. 2106842.

88. Li, X., et al., Electrochemical sensors for analyte in saliva: Recent update. *Sensor Review*, 2024. 44(1): p. 35–67.

89. Swetha, P., U. Balijapalli, and S.-P. Feng, Wireless accessing of salivary biomarkers based wearable electrochemical sensors: A mini-review. *Electrochemistry Communications*, 2022. 140: p. 107314.

90. Karimi-Maleh, H., et al., State-of-art advances on removal, degradation and electrochemical monitoring of 4-aminophenol pollutants in real samples: A review. *Environmental Research*, 2023. 222: p. 115338.

91. Kim, J., et al., Wearable salivary uric acid mouthguard biosensor with integrated wireless electronics. *Biosensors and Bioelectronics*, 2015. 74: p. 1061–1068.

92. Kim, K. R., et al., All-in-one, wireless, multi-sensor integrated athlete health monitor for real-time continuous detection of dehydration and physiological stress. *Advanced Science*, 2024. 11(33): p. 2403238.

93. Lee, Y., et al., Wireless, intraoral hybrid electronics for real-time quantification of sodium intake toward hypertension management. *Proceedings of the National Academy of Sciences*, 2018. 115(21): p. 5377–5382.

94. Lim, H.-R., et al., Wireless, flexible, ion-selective electrode system for selective and repeatable detection of sodium. *Sensors*, 2020. 20(11): p. 3297.

95. Lim, H.-R., et al., Smart bioelectronic pacifier for real-time continuous monitoring of salivary electrolytes. *Biosensors and Bioelectronics*, 2022. 210: p. 114329.

96. Sunwoo, S.-H., et al., Soft bioelectronics for the management of cardiovascular diseases. *Nature Reviews Bioengineering*, 2024. 2(1): p. 8–24.

97. Hwang, W., et al., A Breathable and stretchable metastructure for a versatile hybrid electronic skin patch with long-term skin comfort. *Advanced Materials Technologies*, 2023. 8(1): p. 2200477.

98. Chang, B., et al., Conductive elastic composite electrode and its application in electrocardiogram monitoring clothing. *ACS Applied Electronic Materials*, 2023. 5(4): p. 2026–2036.

99. Pless, C. J., et al., Soft electronic block copolymer elastomer composites for multi-material printing of stretchable physiological sensors on textiles. *Advanced Electronic Materials*, 2023. 9(5): p. 2201173.

100. Zu, W., et al., A comparative study of silver microflakes in digitally printable liquid metal embedded elastomer inks for stretchable electronics. *Advanced Materials Technologies*, 2022. 7(12): p. 2200534.

101. Rodeheaver, N., et al., Strain-isolating materials and interfacial physics for soft wearable bioelectronics and wireless, motion artifact-controlled health monitoring. *Advanced Functional Materials*, 2021. 31(36): p. 2104070.

102. Tan, P., et al., Solution-processable, soft, self-adhesive, and conductive polymer composites for soft electronics. *Nature Communications*, 2022. 13(1): p. 358.

103. Jiang, Y., et al., Topological supramolecular network enabled high-conductivity, stretchable organic bioelectronics. *Science*, 2022. 375(6587): p. 1411–1417.

104. Wang, S., et al., Self-adhesive, stretchable, biocompatible, and conductive nonvolatile eutectogels as wearable conformal strain and pressure sensors and biopotential electrodes for precise health monitoring. *ACS Applied Materials & Interfaces*, 2021. 13(17): p. 20735–20745.

105. Ji, H., et al., Skin-integrated, biocompatible, and stretchable silicon microneedle electrode for long-term EMG monitoring in motion scenario. *npj Flexible Electronics*, 2023. 7(1): p. 46.

106. Su, H., H. Moeinnia, and W. S. Kim, 3D printed electromyography sensing systems. *Advanced Sensor Research*, 2023. 2(9): p. 2300003.

107. Vaněčková, E., et al., 3D printed polylactic acid/carbon black electrodes with nearly ideal electrochemical behaviour. *Journal of Electroanalytical Chemistry*, 2020. 857: p. 113745.

108. Helú, M. A. B. and L. Liu, Fused deposition modeling (FDM) based 3D printing of microelectrodes and multi-electrode probes. *Electrochimica Acta*, 2021. 365: p. 137279.

109. Chen, K.-W., M.-J. Tsai, and H.-S. Lee, Multi-nozzle pneumatic extrusion-based additive manufacturing system for printing sensing pads. *Inventions*, 2020. 5(3): p. 29.

110. Christ, J. F., et al., 3D printed highly elastic strain sensors of multiwalled carbon nanotube/thermoplastic polyurethane nanocomposites. *Materials & Design*, 2017. 131: p. 394–401.

111. Zhao, Y., et al., All 3D-printed soft high-density surface electromyography electrode arrays for accurate muscle activation mapping and decomposition. *Advanced Functional Materials*, 2024. 34(14): p. 2312480.

112. Lee, H., et al., Stretchable array electromyography sensor with graph neural network for static and dynamic gestures recognition system. *npj Flexible Electronics*, 2023. 7(1): p. 20.

113. Wang, C., et al., Stretchable, multifunctional epidermal sensor patch for surface electromyography and strain measurements. *Advanced Intelligent Systems*, 2021. 3(11): p. 2100031.

114. Creel, D. J., The electrooculogram. *Handbook of Clinical Neurology*, 2019. 160: p. 495–499.

115. Ban, S., et al., Soft wireless headband bioelectronics and electrooculography for persistent human–machine interfaces. *ACS Applied Electronic Materials*, 2023. 5(2): p. 877–886.

116. Won, Y., et al., Biocompatible, transparent, and high-areal-coverage kirigami PEDOT: PSS electrodes for electrooculography-derived human–machine interactions. *ACS Sensors*, 2021. 6(3): p. 967–975.

117. Wan, S., et al., Highly stretchable starch hydrogel wearable patch for electrooculographic signal detection and human–machine interaction. *Small Structures*, 2021. 2(12): p. 2100105.

118. Yu, X., et al., Self-adhesive, self-healing, and water-soluble polymeric dry electrodes for robust physiological signal monitoring. *ACS Applied Electronic Materials*, 2024. 6(7): p. 5201–5211.

119. Kim, M. P., Y.-R. Kim, and H. Ko, Anisotropic silver nanowire dielectric composites for self-healable triboelectric sensors with multi-directional tactile sensitivity. *Nano Energy*, 2022. 92: p. 106704.

120. Mishra, S., et al., Soft, wireless periocular wearable electronics for real-time detection of eye vergence in a virtual reality toward mobile eye therapies. *Science Advances*, 2020. 6(11): p. eaay1729.

121. Choi, H., et al., Highly stretchable and strain-insensitive liquid metal based elastic kirigami electrodes (LM-eKE). *Advanced Functional Materials*, 2023. 33(30): p. 2301388.

122. Luo, J., et al., MXene-enabled self-adaptive hydrogel interface for active electroencephalogram interactions. *ACS Nano*, 2022. 16(11): p. 19373–19384.

123. Araki, T., et al., Skin-Like Transparent sensor sheet for remote healthcare using electroencephalography and photoplethysmography. *Advanced Materials Technologies*, 2022. 7(11): p. 2200362.

124. Wang, C., et al., On-skin paintable biogel for long-term high-fidelity electroencephalogram recording. *Science Advances*, 2022. 8(20): p. eabo1396.

125. Mahmood, M., et al., Wireless soft scalp electronics and virtual reality system for motor imagery-based brain–machine interfaces. *Advanced Science*, 2021. 8(19): p. 2101129.

126. Lee, J., et al., Large-scale smart bioreactor with fully integrated wireless multivariate sensors and electronics for long-term in situ monitoring of stem cell culture. *Science Advances*, 2024. 10(7): p. eadk6714.

127. Leng, Y., et al., Advances in in vitro models of neuromuscular junction: Focusing on organ-on-a-chip, organoids, and biohybrid robotics. *Advanced Materials*, 2023. 35(41): p. 2211059.

128. Prinz, V. Y. and K. B. Fritzler, 3D printed biohybrid microsystems. *Advanced Materials Technologies*, 2023. 8(2): p. 2101633.

129. Asulin, M., et al., One-step 3D printing of heart patches with built-in electronics for performance regulation. *Advanced Science*, 2021. 8(9): p. 2004205.

130. Mannoor, M. S., et al., 3D printed bionic ears. *Nano Letters*, 2013. 13(6): p. 2634–2639.

131. Shin, M., et al., Human motor system-based biohybrid robot-on-a-chip for drug evaluation of neurodegenerative disease. *Advanced Science*, 2024. 11(4): p. 2305371.

132. Tetsuka, H., et al., Wirelessly powered 3D printed hierarchical biohybrid robots with multiscale mechanical properties. *Advanced Functional Materials*, 2022. 32(31): p. 2202674.

133. Guix, M., et al., Biohybrid soft robots with self-stimulating skeletons. *Science Robotics*, 2021. 6(53): p. eabe7577.

Index

acute kidney injury (AKI) 274
adhesion 7, 46, 284, 334–335, 337
adipose tissue-derived stem cell (ASC) 209, 211–212, 214–221, 223–225, 232
AFM *see* atomic force microscopy
AKI *see* acute kidney injury
albumin 267–269
alcoholic liver disease (ALD) 270
ALD *see* alcoholic liver disease
alginate 250, 277, 284
Alizarin Red 220–222
ambient mass spectrometry (AMS) 85
AMS *see* ambient mass spectrometry
analysis
  automated image 256
  comprehensive exosome 141
  flow cytometric 219, 300
  integrative 116
  molecular 141
  on-skin 340
  real-time 80, 120
analyte 22, 27, 34–35, 48, 51, 133, 153–154, 164, 166, 337–339, 346
annealing 91, 93
antibody 21, 23, 113, 141, 145, 164, 166–167, 169, 171, 173–174, 176, 197, 200
  acetyl-lysine-specific 114
  homogeneous 11
  monoclonal 149
  polyclonal rabbit primary 222
  printed 170–171
application 2–3, 21–22, 24–25, 27–29, 33–34, 39–41, 43–45, 47–49, 83, 85–86, 139–141, 145–149, 154, 229–231, 233, 252–253, 255–257, 262, 265–266, 280–281
  athletic 346
  bioanalytical 3
  biological 198
  biomedical implant 46
  biomolecular sequencing 198
  cell biology 150, 152
  clinical 118, 253
  continuous monitoring 338, 360
  critical 338
  diagnostic 4, 27
  electronic 45
  fat regeneration 222
  field 139
  high-resolution HSI 161
  high-throughput 47, 140, 148
  implantable 357
  industrial 14
  medical 250, 355
  microscopic HSI 161
  nanoengineering 198
  point-of-care 48, 231
  wearable 358, 360
array 16, 77, 117, 132–133, 138, 140–141, 145, 161, 187, 339, 342
  biosensor 141
  microelectrode 251
  microfluidic cell culture 16
  sensing pixel 80
  surface-mounted 90
artificial trachea 254–255

ASC *see* adipose tissue-derived stem cell
assay 11, 21, 81, 154, 233
  endothelium permeability 272
  endpoint 81, 83
  label-free EV microarray 173
  microarray-based 160
  molecular manipulation 148
  nano-plasma exosome 141
  single-molecule 143
  surface modification 154
atomic force microscopy (AFM) 23–24, 45, 52, 183, 302
automation 5, 13–14, 33, 36, 41, 51–52, 231
autosampler 81–82

bacteria 151, 153, 187, 197
beads 80, 83, 90, 92
bile salt export pump (BSEP) 269
binding 21, 40, 81, 140, 148, 153, 167, 169–170, 173, 175
  biomolecular 24
  molecular 20, 135, 140
  single-molecule 146
  target-molecule 160
binding events 26–27, 133, 146–148
binding kinetics 21, 32, 132, 147, 149, 151
biofluid 337, 339
bioimaging 246, 248
bioink 227–228, 230, 246, 249–250, 255, 267, 269, 272–273, 280–281, 284–287, 290–292, 302–303
  cell-scaffold-based 280–281
  cell-scaffold-free 280–281
  low viscosity 290–291
  next-generation 302
  thermo-responsive 273
biological process 24, 31, 37, 115–116, 140, 164, 182, 246–247, 251, 299
biomarker 14–15, 37–39, 47, 96, 120, 159, 161, 337, 340, 343, 345–346
biomaterial 43–44, 207, 227–228, 230, 266, 280, 286–287, 290, 303–304, 331
  dynamic 229
  hydroxyapatite 220
  natural 277
  synthetic 277
biomolecule 15, 18–19, 23, 25, 27, 86–87, 132–133, 140, 160, 165, 167, 173, 182–200
bioprinting 227–231, 249–250, 267, 273, 276–277, 280–281, 284–288, 290–292, 294, 303–305, 357, 361
  core-shell 272
  droplet-based 291
  extrusion 269
  extrusion-based 292
  inkjet 277, 291
  laser-assisted 227
  layered 267
  multi-cell 305
  organoid 303
  organoids 287, 290, 295, 302–304
  scaffold-free 281
  stereolithography-based 287, 290, 292
biosensing 4, 31, 40, 197, 199
  real-time 24
  single-molecule 198
  ultrasensitive 197
biosensor 25, 27, 33, 48, 159, 164–165, 177, 197, 339, 342, 345
  capacitance-based 32
  electrical 4, 27, 32–33, 39
  electrochemical 168, 338, 360
  FET 28, 47
  high-sensitivity 26
  high-throughput 145

optical 159–160, 176
optical resonant 176
PC 160–161, 164–168, 170
sensitive 27
wearable 337
BMP *see* bone morphogenetic protein
bone morphogenetic protein (BMP) 232
bovine serum albumin (BSA) 173–174
brain 40, 259–260, 262–263, 276, 355–356
human 259–260, 262
breathability 336–337, 345, 348, 354
BSA *see* bovine serum albumin
BSEP *see* bile salt export pump

CAD *see* computer-aided design
CAFM *see* conductive atomic force microscopy
cancer 2, 18, 31, 39, 41, 48, 112, 149, 153
cancer stem cell (CSC) 78, 80
cardiomyocyte 249–251, 302
cardiomyopathy 103–104, 251
cartilage 43–44, 109, 207, 211, 291
CCD *see* charge-coupled device
cell 15–16, 28–29, 43–44, 73–75, 80–81, 83, 94–98, 100–102, 106–112, 119, 207–208, 211–212, 217–220, 246–247, 253–256, 271–272, 276–277, 279–281, 286–293, 297–298
amacrine 276
auto-fluorescent 256
bacterial 152–153
basal 254, 298
biofuel 342
bipolar 276, 298
bone-related 264
Bowman's capsule 272
cancer 33, 48, 255–256
cardiac 249
ciliated 254
cultivated 208
daughter 73
dead 217–218, 225, 269
dendritic 254
endothelial 249–250, 254, 261, 267, 273, 284–285, 305
endothelium 273
epithelial 255, 258, 273
fibroblast 305
ganglion 276, 279
germ 95, 210
healthy 256
immune 249, 254
inflammatory 226, 252
kidney-derived 272
kupffer 267
light-sensitive 276
living 32, 112, 149–150, 227, 286, 299
lymphoid 249
mast 254
mature 301–302
mesenchymal 270–271
miPS-LLCcm 80
miPS-P19c 78, 80
miPS-P19cm 78
muscle 249, 289, 305
nerve 211
neural 96–97, 289
neuronal 210, 276
osteogenic 232
pancreatic 210
parental 168–169
pluripotent 110
progenitor 109, 211, 230, 254–255, 260, 268, 276, 288–289
retinal 276
retinal ganglion 276, 278–279
secretory 254
seeded 208, 285
somatic 94, 101–102, 110, 112

stellate 267–268
stem 31, 40, 43, 70–120, 208, 210–211, 228, 230–232, 248, 252, 254–255, 259, 264
tumor 33
vascular 261
cell behavior 41, 43, 160, 231
cell growth 277, 280, 284, 289, 293
cell sheet 208–210, 213–219, 221, 232–233
charge-coupled device (CCD) 147, 161
charge transport 182–183, 191
CHD *see* congenital heart defect
clinical diagnostics 1–2, 33–34, 38–39
computer-aided design (CAD) 227, 250, 286–287
conductive atomic force microscopy (CAFM) 183, 185
congenital heart defect (CHD) 103–104
conventional SPR (cSPR) 131, 133–134, 137, 139–140, 143, 154
CSC *see* cancer stem cell
cSPR *see* conventional SPR
culture 40–41, 100, 209, 212, 217, 220, 225, 259, 268, 278, 301
cytokine 71–72, 141, 249, 301, 340

detection capability 133, 142, 146, 149, 153
detector 29, 84, 161
device 7–8, 11, 48–49, 330, 332, 336–337, 342–343, 345–346, 348–349, 352–353, 359
adhesive-free 345
adhesive-free electronic 344
assay 139
battery-free 341
benchtop 10
charge-coupled 147
colorimetric 341
compact HSI 231
conventional 330
cSPR 136
cytometer-connected 81
electronic 27, 194, 332
energy-harvesting 342
high-throughput 37
microelectronic 48, 194
microfluidic 39, 47, 117, 139, 145, 340
micro-organ 267
mouthguard 346
multimodal 342
optical biosensor 141
plasmonic 160
point-of-care 27
printed biohybrid 358
reusable 343
robotic 359
semiconductor logic 194
smartwatch-based 342
soft electronic 344
soft material 345
specimen-processing 81
ultrathin 349
wearable 329–332, 334, 336–337, 340–343, 348–350, 353
wearable ECG 335–336
wearable EEG 357
wearable electronic 336
diabetes 233, 270, 337
diagnosis 21, 27, 38, 40, 169, 177, 256, 265, 295, 305, 337
disease 31, 33, 39, 140, 168, 177, 247, 305
direct ink writing (DIW) 292, 351
disease 37, 39, 41, 100–101, 103, 119, 246–247, 251–253, 255, 275–277, 279, 339–340, 358–359
Alzheimer's 15, 263
brain 263
cardiovascular 39, 252

congenital heart 251
Fabry 274
genetic 101, 271
genetic heart 251
infectious 41, 255
ischemic heart 252
kidney 233, 273–275
multifactorial 253
noninvasive 293
ocular 279
Parkinson's 337
pulmonary 255
retinal 276, 278–279
disease state 38, 90, 119, 169, 270
DIW *see* direct ink writing
DNA 17, 20, 23, 26–27, 34, 86, 92, 164, 168, 186–195, 197–200
DNA conductance 190–191, 196–197
DNA methylation 18, 110, 112, 116
DNA tetrahedral framework (DTF) 145
double-stranded DNA (dsDNA) 187, 189–193, 195, 197
drug 32, 36, 40–41, 45, 100, 105–107, 109, 112–113, 148–149, 253, 267, 270–271, 274
anticancer 33
growth-promoting 266
high-throughput 86
immunosuppression 275
drug development 1–2, 5, 15–16, 21, 34, 38, 47, 105–106, 252
drug discovery 1–3, 5, 14, 16, 18, 20, 34–35, 52–53, 84–85, 103, 105, 107, 109, 119–120, 131
drug prediction 252–253, 266
drug screening 40, 72, 85, 140, 228, 253, 262–263, 267, 270, 275, 278–279
dsDNA *see* double-stranded DNA
DTF *see* DNA tetrahedral framework
E-cadherin 107–108
ECG *see* electrocardiography
ECM *see* extracellular matrix
ECM-PEG hydrogel 222, 224–227
E. coli 187, 197–198
EEG *see* electroencephalography
elastomer 332, 334, 345, 348–349, 360
electrical detection 4, 26–27, 34, 39–40, 44–46, 48–52, 182–200
electrocardiography (ECG) 330, 337, 348, 351
electroencephalography (EEG) 330, 355–357
electrolyte 135, 330, 339–340, 345
electromyography (EMG) 330, 351–353
electron-multiplex charge-coupled device (EMCCD) 161–162
electrooculography (EOG) 330, 353–355
embryo 94, 96, 111
embryonic stem cell (ESC) 98, 210, 253
EMCCD *see* electron-multiplex charge-coupled device
EMG *see* electromyography
EMT *see* epithelial-mesenchymal transition
enzyme 27, 92, 106, 112–113, 197, 267, 337, 339
EOG *see* electrooculography
epithelial-mesenchymal transition (EMT) 106–107
ESC *see* embryonic stem cell
EV 165, 168–176
exosome 141, 143–144, 159, 161, 164–168, 172
exosome microarray 160–162, 165–169
extracellular matrix (ECM) 207–208, 219–220, 223–226, 249, 264–265, 273, 281, 284

fabrication 44, 161, 164–165, 224, 227, 290
cell sheet 212–213, 233
FBS *see* fetal bovine serum
FDM *see* fusion deposition modeling
FET *see* field effect transistor
fetal bovine serum (FBS) 211, 214, 220, 222
field effect transistor (FET) 26–27, 48
flow cytometry 3, 70, 72, 75–76, 80–83, 96, 247, 297, 300
fluorescence 10–12, 91, 132, 199, 247, 293, 300
formyl peptide receptor (FPR) 81
FPR *see* formyl peptide receptor
fusion deposition modeling (FDM) 351

gene 37–38, 73, 75, 77–78, 80, 95, 99, 101–104, 110, 112, 119, 274–275, 280, 300–301, 303
gene expression 17, 37, 73, 75–78, 98, 101, 109–110, 188, 247, 268, 297
gene therapy 280
genomics 1–2, 14, 34, 36–37, 47, 69, 83, 90, 115
GFP *see* green fluorescent protein
green fluorescent protein (GFP) 99, 256
growth factors 227–228, 232, 254, 259, 276, 286–287, 291
guanine 190, 198

HBV *see* hepatitis B virus
HCS *see* high content screening
heart 30–31, 77, 94, 100, 103–104, 207, 249–250, 252, 304, 330, 348
hepatitis B virus (HBV) 270–271
hepatocytes 267–270
high content screening (HCS) 5, 8
highest occupied molecular orbital (HOMO) 190, 193
high-throughput detection 132–134, 138, 140, 143–144, 149, 153–154
high-throughput imaging (HTI) 3, 6, 29–30
high-throughput screening (HTS) 3, 5, 10, 13–14, 17–18, 36, 39, 41, 46–47, 49, 76–78, 83–85, 90, 96, 101, 106–107
high-throughput sequencing 17, 36–37, 90
high-throughput technology 14, 32, 34, 46–47, 49, 71, 75–76, 115, 117
hiPSC *see* human-induced pluripotent stem cell
HOMO *see* highest occupied molecular orbital
HSI *see* hyperspectral imaging
HTI *see* high-throughput imaging
HTS *see* high-throughput screening
HTS platform 32–33, 35–36, 41, 98
human-induced pluripotent stem cell (hiPSC) 248–250, 253, 258, 260, 264, 268, 272, 275, 277–279, 289
human umbilical vein endothelial cell (HUVEC) 255, 267–268, 272, 289
HUVEC *see* human umbilical vein endothelial cell
hydrogel 208–209, 222–228, 250, 268–269, 273, 276, 280–281, 284–285, 292, 302, 332, 334, 340, 343, 351
hyperspectral imaging (HSI) 2, 4, 29–31, 39–40, 44–46, 48–52, 159–163, 165–167, 176–177, 207–208, 229–231, 246–248, 251–252, 256–257, 265, 269–270, 278, 293, 295–296, 302–305

IAV *see* influenza A virus
induced pluripotent stem cell (iPSC) 98–102, 252–253, 262–263
infection 258, 263, 270, 280, 343, 345
influenza A virus (IAV) 152, 258
integration 20, 22, 32–33, 36, 40, 43–45, 50–51, 76, 78, 83–84, 115–117, 153–154, 329–332, 343, 345
interactions 3, 6, 21, 23, 27, 30, 37–38, 40, 77, 199–200, 207
  biomaterial 16
  biomolecular 18, 20–21, 26, 32, 39, 48, 77, 132, 143, 154
  intermolecular 194
  model interregional 260
  molecular 1–2, 4, 14, 20, 23–24, 40, 45, 47, 131, 133, 140–141, 143
  π-π stacking 189–190
iPSC *see* induced pluripotent stem cell

kidney 30–31, 40, 94, 271, 274–275, 297, 304

label-free detection 24, 27, 32, 46–49, 77, 131–132
label-free image 163, 166, 170–175, 177
laser 6–8, 11, 80, 87, 292, 300
laser scanning confocal microscope (LSCM) 70, 72, 298–299
ligand 81, 83, 142, 148, 151, 286
light microscopy 73, 247, 297
light source 29, 136–137, 151, 154, 161–162, 177
lipid 168, 227, 251, 267, 339
liver 31, 40, 94, 207, 210–211, 228, 267–271, 301, 304
localized surface plasmon resonance (LSPR) 132, 138–139, 143
LSCM *see* laser scanning confocal microscope
LSPR *see* localized surface plasmon resonance
lungs 254–256

machine learning 2, 33, 47, 49, 51–52, 71, 118–120, 231
machine learning algorithm 50, 75–76, 89, 118–120
macrophage 169, 173–176, 254
MALDI *see* matrix-assisted laser desorption ionization
marker 1–2, 70, 73, 75–76, 95–100, 107, 247, 262, 269, 297–298
mass spectrometry 4, 19–20, 22, 35–38, 84–85
material 3–4, 19, 29, 31, 44–46, 145–146, 177, 181, 280, 292–293, 329–330, 332–334, 348–349, 353–355, 360–361
  adhesive 348
  bio-reactive 143
  bone-grafting 231
  chemical 52
  conductive 87, 182, 349
  conventional 332–333
  elastomeric 332
  genetic 257
  living 359
  next-generation 46
  soft 330, 333, 345, 348
  spinal fusion 232
  vegetable-based 354
matrix-assisted laser desorption ionization (MALDI) 19, 85–86
MC *see* methylcellulose
mechanical properties 23, 43, 45, 227–228, 285–286, 288, 303, 332–333

medicine 77, 115, 229, 266
mesenchymal stem cell (MSC) 210, 219, 254–255
metabolism 36, 96, 140, 264
metabolite 36, 96–97, 340, 343, 345
metal 135–137, 332–333, 351
methylcellulose (MC) 209, 213, 215–217, 222, 233
microarray 15, 17, 25, 32, 77, 79, 166, 171, 173, 175, 177
  biomaterial 16
  fabricated 143
  fluorescent dye 78
  molecular 143
  polymer 15
microfluidics 26, 32, 48, 71, 106, 132–133, 138–139, 141, 143, 154, 347
microwell 149, 151, 153, 163–164, 166–167, 169–171, 173
miniaturization 5, 21, 27, 46–47, 348, 360
miPSC *see* mouse-induced pluripotent stem cell
model 41, 101, 104–105, 248, 250–251, 254, 258, 268, 270, 273, 279
  alveolar 254
  animal 40, 210, 250, 252, 259, 266, 275, 279
  atomistic 188
  bioprinted 273, 297
  bioprinted triculture 268
  bone defect 265
  bone osteoporotic 265
  CAD 287
  cardiac 105
  cardiomyopathic 251
  cell-based 36
  cell culture 246, 263
  computer-aided design 227
  diagnostic 76
  disease 100, 252, 263, 265, 268, 331, 357, 361
  hopping 192
  immunodeficient mouse 279
  lung tumor 256
  mouse 271, 275
  neurological disorder 96
  organoid 248, 251, 253, 255, 258–259, 270–271, 273–275, 292, 296–298, 304
  pharmacologic 105
  preclinical 253
  predictive 70
  proximal tubule 273
  rat 344
  theoretical 188
  tissue 329, 331, 358–359
  zebrafish 103–105, 107, 109
modeling 100, 117, 259, 272
  computational 70
  disease 40–41, 99–101, 253, 257, 262–263, 265–266, 270–271, 274, 305
  fusion deposition 351
  scientific 100
monitoring 18, 24, 27, 32–33, 168–169, 183–184, 230–231, 330–331, 336, 338, 345–347, 349, 357, 360–361
  continuous 176, 330, 338–339, 342
  ECG 349
  electrophysiological 329, 359
  environmental 24, 27, 31, 33, 49
  health 342
  medical analyte 339
  personal health 360
  postpartum 336
  quantitative 99, 116
  scaffold 230
  therapy response 38
  unobtrusive 329, 358
  wound 344–345, 360

mouse-induced pluripotent stem cell (miPSC) 78, 80
MSC *see* mesenchymal stem cell
muscle 100, 109, 210–211, 330

NAFLD *see* nonalcoholic fatty liver disease
nanomaterial 27, 45, 47, 51–52, 338
nanoparticle 139, 146–147, 160, 177, 349, 355
near-field communication (NFC) 342–343, 360
nephron 271–273, 275
neural network 261, 352–353, 357
neuron 96, 100, 110–111, 260–261, 276
next-generation sequencing (NGS) 89–91, 96
NFC *see* near-field communication
NGS *see* next-generation sequencing
NGS platform 90, 93–94
nonalcoholic fatty liver disease (NAFLD) 270
non-small cell lung cancer (NSCLC) 145, 257
NSCLC *see* non-small cell lung cancer
nucleic acid 92, 159, 197–198
nucleus 73–74, 77, 108, 225

OCNN *see* osteogenic convolutional neural network
OCT *see* optical coherence tomography
optical coherence tomography (OCT) 250–251
optical resonance 159–160, 164, 177
organ 40, 43, 246, 248, 253, 270, 275, 302, 305, 357–359, 361
organoid 5, 40–41, 43–44, 246–251, 253–262, 264–271, 274–305
  airway 254, 258, 297–298
  alveolar 254
  biliary 271
  bioprinted 267–270, 273–274, 286, 295–296, 304–305
  bone 264–266
  brain 259, 261–263
  cardiac 249–253, 302
  cerebral 261, 263, 298–299
  cortical 260
  epithelial 259, 297
  heart 251–252
  hepatic 269
  kidney 271–275
  liver 267–271, 300
  lung 254–256, 258
  neural 260
  pancreatic cancer 41
  retinal 247, 276–280, 297–298, 300
  ventral forebrain 260
osteogenesis 70, 212, 220–222
osteogenic convolutional neural network (OCNN) 70–72

patient 101–103, 227–228, 232, 257, 263, 271, 274, 277–279, 339–340, 344, 349–350, 354, 357
  cardiomyopathic 251
  hyperuricemia 346–347
  myocarditis 252
  postpartum 337
PBS *see* phosphate-buffered saline
PC *see* photonic crystal
PCA *see* principal component analysis
PCR *see* polymerase chain reaction
phosphate-buffered saline (PBS) 164, 170, 211, 213, 216, 222–223

photonic crystal (PC) 13, 87, 89, 160, 164–166, 169, 171
PKD *see* polycystic kidney disease
PLC *see* primary liver cancer
pluripotent stem cell (PSC) 78, 98–99, 101–102, 248–249, 252, 277, 279
polycystic kidney disease (PKD) 275
polymer 222, 281, 285, 288–292, 355, 357
polymerase chain reaction (PCR) 70, 72, 77, 92, 99, 198, 300
primary liver cancer (PLC) 270
principal component analysis (PCA) 96, 99
probes 21, 34–35, 83, 91, 93, 141, 185
proliferation 11, 208, 210, 217–218, 224–225, 227, 230, 281, 286, 288–289, 292
protein 6, 18–19, 21, 23, 26–27, 35, 37–39, 86–87, 98–99, 141–142, 164–166, 168, 267, 269, 337
  extracellular 246
  membrane 143, 147–149, 151, 166, 168, 176
  non-collagenous 265
PSC *see* pluripotent stem cell

QCM *see* quartz crystal microbalance
quartz crystal microbalance (QCM) 24, 33

rapid evaporative ionization mass spectrometry (REIMS) 19, 22
refractive index (RI) 18–19, 135–137, 149–150, 246
REIMS *see* rapid evaporative ionization mass spectrometry
renal tubular cell (RTC) 7
resonance wavelength 161, 163–166, 170–171, 173–175
retina 87–88, 276–277, 279–280
retinitis pigmentosa (RP) 276, 279
RI *see* refractive index
RNA 142, 186, 198–200
robotics 13–14, 52, 106, 277, 359, 361
RP *see* retinitis pigmentosa
RTC *see* renal tubular cell

saliva 164, 330–331, 339, 345–347, 360
sample 7–8, 13–15, 22–23, 32–33, 48, 81, 83–85, 96, 133–134, 139, 141, 160, 182–183, 223, 297–299, 339
  biochemical 2
  biological 48–49, 160, 164, 291
  chemical 2, 47
  clinical 48, 145, 168, 177
  clinical medical 339
  fluid 83
  fluorescent 299
  heterogeneous 11
  osteogenic 221
  postmortem 259
scaffold 43–44, 86, 208, 229–230, 250, 264, 277, 287
scanning electron microscopy (SEM) 73, 167, 169, 171, 247, 251, 255, 265, 273, 297, 302
scanning tunneling microscope (STM) 182–183, 190
SCLC *see* small cell lung cancer
SCNT *see* somatic cell nuclear transfer
secondary ion mass spectrometry (SIMS) 45–46
SEM *see* scanning electron microscopy
sensing 41, 44, 132, 135, 138, 331, 337, 339, 341–345, 347–348, 353, 357–360

sensitivity 19, 21, 24, 27, 32–34, 46–49, 85–87, 96, 98, 132–133, 141, 143, 145–146, 148, 154
  analytic 36
  antigen-antibody binding 149
sensor 24, 26–27, 39–40, 48, 78, 80, 164, 166, 338–339, 344–345, 349–350, 353, 356
  acoustic 43
  acoustic wave 24–25
  blood flow oxygenation 356
  chemical 48
  DNA 27
  elastic 360
  electrical 48
  electrical impedance 46
  electrochemical 47, 329–331, 337, 339, 343, 345, 360
  electrochemical sweat 360
  impedance-based electrical 28
  ion-selective sodium 346
  label-free 32
  laminating 330
  metasurface 161
  microfluidic sweat 341
  motion 353
  multiplex 360
  optical 161
  optical-based 47
  piezoelectric 24
  saliva 346
  surface plasmon resonance 161, 168
  wearable 330, 339, 343, 360
sequence 92–93, 113, 186–187, 190–193
sequencing 39, 73, 89–92, 119, 197, 199
signal 8, 50, 137, 139, 146, 276, 330, 339, 349, 353, 355–356
  ECG 349–350
  EMG 333–334
  EOG 353–354
SIMS *see* secondary ion mass spectrometry
single-molecule junction (SMJ) 184–186, 188, 197–200
single-stranded DNA (ssDNA) 187, 190
skin 109, 291, 304, 329–337, 340, 343–345, 347–349, 351–356, 359, 361
small cell lung cancer (SCLC) 257
SMJ *see* single-molecule junction
somatic cell nuclear transfer (SCNT) 94
SPP *see* surface plasmon polariton
SPR *see* surface plasmon resonance
SPRi *see* surface plasmon resonance imaging
SPRM *see* surface plasmon resonance microscopy
ssDNA *see* single-stranded DNA
stimulation 44, 329–331, 344, 357–359
STM *see* scanning tunneling microscope
substrate 113, 134, 160, 183, 190, 196, 290, 292, 335–337, 345, 348–349
surface plasmon polariton (SPP) 136–137
surface plasmon resonance (SPR) 3, 20–21, 32, 35, 131–154, 161
surface plasmon resonance imaging (SPRi) 132–136, 139, 143–146, 148–150, 154
surface plasmon resonance microscopy (SPRM) 132–137, 146–148, 151, 153–154
sweat 330–331, 336–337, 339–342, 347, 360
swelling 214–215, 361
system 11–14, 19, 47–48, 50–52, 80–82, 93, 98, 100, 151, 339, 342, 347, 357
  automated 7, 32, 41, 46

biological 18, 34, 116, 120, 181–182, 329–361
biomolecular 182
cardiovascular 103
cellular 116
chemical 14, 19
electronic 332, 361
gesture recognition 353
HSI 29, 48, 160–162, 164, 171
human organ 303
immunological 253
label-free 32
lab-on-chip 27
mammalian 105, 107
microfluidic 32, 36, 49, 340
microscopic 92
multi-enzyme 339
multi-nozzle 351
nervous 98, 210, 351
organoid 41
organ-on-a-chip 227
robotic automation 117
stretchable electronic 346
vascular flow 303
wearable 342, 355
wearable monitoring 360

technique 2, 18–19, 37, 45, 106, 132–133, 136, 141, 153–154, 182–185, 233, 247, 251, 291–292, 297
3D-bioprinting 290
biological 297
bioprinting 227, 229, 246, 290, 295, 302–304
label-free 2, 18–19, 31–33, 36, 38, 46, 51
microfluidic 303
organoid 260
passive 340
technology 2, 4, 12, 14, 18, 21, 24–26, 32, 38, 75–76, 159–160, 291–292, 303–304
3D-bioprinting 293, 304
automated 76
bioprinting 228, 268, 272, 276, 286, 290, 293–294, 303
label-free 43, 48
omics 115–116
organoid 41
robotics 13
sensor 44, 46
wearable 338
therapy 18, 21, 36, 98, 109, 118, 149, 248, 259, 270, 279
tissue 40, 43–44, 109, 207–208, 210–211, 228–233, 255–256, 260, 284–287, 289–290, 292–293, 303–305, 329–333, 357–359, 361
3D-printed 227
adipose 209, 211, 225, 232
aorta 252
bioengineered 43
biological 31, 39, 109, 246, 332
bioprinted 287, 358
bone 233, 264
cardiac 228, 249, 358–359
dermal 339
diseased 39, 48
engineered 208, 228–231, 249
fabricated 358
heart 250, 357, 359
liver 30, 44, 267–269
mimetic 303
muscle 228
myocardium 252
paraffin-embedded 78
proximal 273
pulmonary 255
sensitive 329
tumor 40, 52, 257
transplantation 228, 253, 271, 275

uric acid 271, 330, 337, 341, 343, 345

vascularization 227, 230, 261, 302

wearable electronics 329, 331–337, 339–348, 351, 353, 355, 359–360
WGBS *see* whole-genome bisulfite sequencing
whole-genome bisulfite sequencing (WGBS) 112
wound 330–331, 343–345

Young's modulus 354–355

Z-DNA 188–189
zebrafish 101, 103–105, 107, 109
Zika virus 263

For Product Safety Concerns and Information please contact our EU representative GPSR@taylorandfrancis.com Taylor & Francis Verlag GmbH, Kaufingerstraße 24, 80331 München, Germany